Smith's New Arithmetic.

ARITHMETIC

ON THE

PRODUCTIVE SYSTEM;

ACCOMPANIED BY A

KEY

AND

CUBICAL BLOCKS.

BY ROSWELL C. SMITH,

AUTHOR OF PRACTICAL AND MENTAL ARITHMETIC, THE PRODUCTIVE GRAMMAR,
THE PRODUCTIVE GEOGRAPHY, &c

STEREOTYPE EDITION

NEW-YORK:
PUBLISHED BY DANIEL BURGESS & CO.,
(LATE CADY & BURGESS,)
1852.

Adopted by the Brothers of the Christian Schools in the United States.

"IT IS NOT EASY TO DEVISE A CURE FOR SUCH A STATE OF THINGS, (THE DECLINING TASTE FOR SCIENCE;) BUT THE MOST OBVIOUS REMEDY IS TO PROVIDE THE EDUCATED CLASSES WITH A SERIES OF WORKS ON POPULAR AND PRACTICAL SCIENCES, FREED FROM MATHEMATICAL SYMBOLS AND TECHNICAL TERMS, WRITTEN IN SIMPLE AND PERSPICUOUS LANGUAGE, AND ILLUSTRATED BY FACTS AND EXPERIMENTS WHICH ARE LEVEL TO THE CAPACITY OF ORDINARY MINDS." QUARTERLY REVIEW.

"THE FIRST THING TO BE REQUIRED IN A SYSTEM OF POPULAR INSTRUCTION, IS. THAT IT SHOULD BE INTELLIGIBLE; THAT CHILDREN AND YOUTH SHOULD UNDERSTAND WHAT THEY LEARN. UNDERSTAND WHAT THEY LEARN? IT MAY BE ASKED; WHAT ELSE CAN THEY DO? WE ANSWER, THAT THEY MAY COMMIT IT TO MEMORY, MAY RECITE IT, MAY EVEN MAKE A FAIR SHOW OF KNOWLEDGE AND YET KNOW NOTHING. WE HAVE NOT THE LEAST HESITATION IN SAYING, THAT TWO OR THREE YEARS, IN THE EDUCATION OF ALMOST EVERY INDIVIDUAL IN THIS COUNTRY, HAVE BEEN THROWN AWAY UPON STUDYING WHAT THEY DID NOT UNDERSTAND."

NORTH AMERICAN REVIEW

HARTFORD,
STEREOTYPED BY
RICHARD H. HOBBS.

ARITHMETIC.[1]

PART FIRST:

BEING A MENTAL[2] COURSE FOR EVERY CLASS OF LEARNERS

NUMERATION.[3]

☞ Recite by the Questions.

I. 1. NUMBER, which shows how many are meant, is represented[4] by letters, by words, and by characters[5] called figures, as:—

One	I	1	One hundred	C	100
Two	II	2	Two hundred	CC	200
Three	III	3	Three hundred	CCC	300
Four	IV*	4	Four hundred	CCCC	400
Five	V	5	Five hundred	D *or* Iɔ†	500
Six	VI	6	Six hundred	DC *or* IɔC	600
Seven	VII	7	Seven hundred	DCC *or* IɔCC	700
Eight	VIII	8	Eight hundred	DCCC *or* IɔCCC	800
Nine	IX*	9	Nine hundred	DCCCC *or* IɔCCCC†	900
Ten	X	10	One thousand	M *or* CIɔ†	1000
Eleven	XI	11	Two thousand	$\overline{II}$ *or* MM	2000
Twelve	XII	12	Three thousand	$\overline{III}$ *or* MMM	3000
Thirteen	XIII	13	Four thousand	$\overline{IV}$ *or* MMMM	4000
Fourteen	XIV	14	Five thousand	$\overline{V}$ *or* Iɔɔ	5000
Fifteen	XV	15	Six thousand	$\overline{VI}$ *or* IɔɔM	6000
Sixteen	XVI	16	Seven thousand	$\overline{VII}$ *or* IɔɔMM	7000
Twenty	XX	20	Eight thousand	$\overline{VIII}$ *or* IɔɔMMM	8000
Thirty	XXX	30	Nine thousand	$\overline{IX}$ *or* IɔɔMMMM	9000
Forty	XL*	40	Ten thousand	$\overline{X}$ *or* CCIɔɔ†	10000
Fifty	L	50	Twenty thousand	$\overline{XX}$ *or* CCIɔɔCCIɔɔ	20000
Sixty	LX	60	Fifty thousand	$\overline{L}$ *or* Iɔɔɔ	50000
Seventy	LXX	70	One hundred thousand	$\overline{C}$ *or* CCCIɔɔɔ	100000
Eighty	LXXX	80	Five hundred thousand	$\overline{D}$ *or* Iɔɔɔɔ	500000
Ninety	XC*	90	One million	$\overline{M}$ *or* CCCCIɔɔɔɔ	1000000

* Or, IIII for 4; VIIII for 9; XXXX for 40; LXXXX for 90; and CM for 900,

† Every ɔ annexed to Iɔ increases its value 10 *times*: as Iɔ, is 500, Iɔɔ is 5000: in like manner the prefixing of C and the annexing of ɔ to CIɔ increases it 10 *times* as CIɔ, 1000, CCIɔɔ, 10000; lastly a line over any number increases it 1000 *times*: as, D, 500, $\overline{D}$, 500000.

NOTE. L. stands for the Latin language; G for the Greek and F. for the French.

1 ARITHMETIC, [G. *arithmétike.*] Computing, calculating or reckoning by numbers.

2 MENTAL, [L. *mentis.*] Pertaining to the mind; intellectually.

3 NUMERATION, [L. *numeratio.*] Numbering; the method or act of numbering.

4 REPRESENTED, Exhibited; described; personated; to supply the place of.

5 CHARACTER, A mark; a stamp; a letter reputation; a personage.

QUESTIONS ON THE FOREGOING.

2. What does Number show? 1. How is it represented? 1. What letters and what figures stand for one, four, five and nine? What for ten? *Ans.* The figure 1 and 0 called naught or cipher.

3. What letters and what figures stand for eleven? For twelve? Fifteen? Twenty? Forty? Fifty? Ninety? What different numbers may be represented by the figures 1 and 5 written together? *Ans* Fifteen and fifty-one; as, 15:51.

4. What different numbers may be expressed by the figures 1 and 9? 2 and 5? What number is expressed by C? by D? by M? What by $\bar{C}$, $\bar{D}$, and $\bar{M}$, with a dash over each.

5. What figures stand for one hundred? For two hundred? One thousand? Ten thousand? One hundred thousand? One million?

ADDITION.[1]

QUESTIONS.

II. 1. Thomas has 3 dollars and Rufus 5 dollars. How many dollars have they both? Say 3 and 5 are 8. *A.* 8 dollars.

2. A farmer has 5 cows in his yard, and 6 in the pasture. How many cows has he in both places?

3. A man bought a hat for 5 dollars, and a pair of boots for 7 dollars. How many dollars did he pay for both?

4. A man lost 7 dollars, and then had 8 dollars left? How many dollars had he at first?

5. A man gave 8 dollars for a saddle, 5 dollars for a bridle, and 2 dollars for a whip. What did the whole cost him?

6. Suppose there are 8 oranges in a basket, 5 on a table, and 4 in my pockets. How many will they all make?

7. A grocer sold to one man 6 barrels of flour, to another 9 barrels, and still had 5 barrels left. How many barrels had he at first?

8. In a certain class are 6 large boys, 7 small ones, and 10 girls. How many scholars are there in the class?

9. A boy has 10 dollars, his father gave him 10 more, and he has 5 owing to him. How many will they all make?

10. Thomas read 19 pages of history in one day, 12 in another, and 9 in another. How many pages did he read in all?

11. Suppose you are 10 years old, and that your brother was 10 years old when you were born. What is his age now?

12. A man gave 10 dollars more for his horse than for his wagon, and the wagon cost him 30 dollars. What did they both cost him?

13. How many are 10 and 30 and 40? 60 and 10 and 20? 100 and 200 and 400? 600 and 300 and 100?

---

1 ADDITION, [L. *additio.*] Any thing added; adding; joining; uniting two or numbers in one sum.

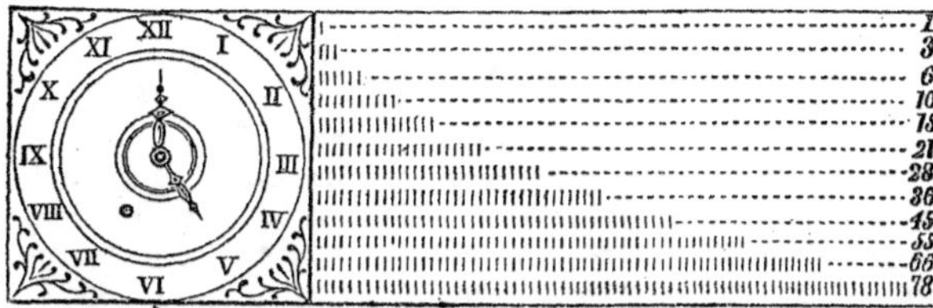

14. When it is 1 o'clock, a regular clock strikes once; when it is 2 o'clock, twice; when it is 3 o'clock, thrice. How many strokes will these make? *A.* 6 strokes: because 1 and 2 are 3, and 3 are 6.

15. Add together all the strokes that a clock strikes in 12 hours, as in the above picture of a clock. [Thus 1 and 2 are 3, and 3 are 6, and 4 are 10, and so on up to 12.]

In the following table the single figures are combined[1] by pairs in every possible manner; so that, if it be committed to memory, the learner can add with facility[2] any two numbers whatever.

ADDITION TABLE.

1 and 1 are **2**	2 and 9 or 3 and 8 or 4 and 7 or 5 and 6 are **11**
1 and 2 are **3**	3 and 9 or 4 and 8 or 5 and 7 or 6 and 6 are **12**
1 and 3 or 2 and 2 are **4**	4 and 9 or 5 and 8 or 6 and 7 are **13**
1 and 4 or 2 and 3 are **5**	5 and 9 or 6 and 8 or 7 and 7 are **14**
1 and 5 or 2 and 4 or 3 and 3 are **6**	6 and 9 or 7 and 8 are **15**
1 and 6 or 2 and 5 or 3 and 4 are **7**	7 and 9 or 8 and 8 are **16**
1 and 7 or 2 and 6 or 3 and 5 or 4 and 4 are **8**	8 and 9 are **17**
1 and 8 or 2 and 7 or 3 and 6 or 4 and 5 are **9**	9 and 9 are **18**
1 and 9 or 2 and 8 or 3 and 7 or 4 and 6 or 5 and 5 are . . . **10**	

16. Repeat the above table beginning at the top on the left. What are all the different pairs of single figures that together make 7? make 9? 10? 12? 13? 14? 15? 16? 17? 18?

17. How many are 8 and 4? 18 and 4? 28 and 4? 48 and 4? 7 and 6? 17 and 6? 26 and 7? 57 and 6? 86 and 7? 96 and 7?

18. How many are 9 and 5? 19 and 5? 55 and 9? 8 and 8? 38 and 8? 98 and 8? 5 and 7? 65 and 7? 9 and 8? 49 and 8? 99 and 8?

19. Add together audibly[3] 10 twos: 10 threes: 10 fours: 10 fives: 10 sixes: 10 sevens: 10 eights: 10 nines: 10 tens.

20. How many are 78 and 10 and 3? 78 and 13? 91 and 14 [10 and 4]? 105 and 15 [10 and 5]? 120 and 16? 136 and 17? 153 and 18? 171 and 19?

21. Add together audibly[3] all the different numbers under 20, beginning with the lowest.

---

1 Combined, United closely; associated; leagued; confederated.
2 Facility, [L. *facilitas.*] Easiness to be performed; readiness; affability.
3 Audibly, In a manner so as to be heard; in an audible manner.

SUBTRACTION.[1]

QUESTIONS.

III. 1. A boy having 15 cents lost 10; how many had he left? Say 10 from 15 leaves 5; because 10 and 5 are 15. *A.* 5 cents.

2. A man owing 12 dollars paid 7 dollars. How many dollars did he still owe? 7 from 12 leaves how many and why?

3. A grocer bought a barrel of molasses for 15 dollars, and sold it for 18 dollars. How much did he make on it?

4. Suppose your age to be 12 years, and your brother's 20 years What is the difference between your age and his?

5. A farmer bought a cow for 20 dollars and sold her for 15 dollars. What was his loss?

6. Thomas and William counted their nuts; the former had 30 and the latter 50. How many had one more than the other?

7. A certain cistern has one pipe by which 25 gallons run in every hour, and another pipe by which 19 gallons run out every hour. How many gallons will stay in every hour?

8. How many are left in taking 3 from 11? 3 from 21? 3 from 51? 4 from 13? 4 from 23? 4 from 53? 4 from 83?

9. How many does 5 from 12 leave? 5 from 32? 6 from 16? 6 from 36? 7 from 15? 9 from 17? 9 from 37? 9 from 107?

10. How many cents added to 100 cents will make 200 cents?

11. What is the difference between 300 cents and 700 cents?

12. How much smaller is 70 than 270? 500 than 1000?

13. How much larger is 590 than 90? 1275 than 275?

14. A gentleman paid 250 dollars for his carriage, and 50 dollars less for his horse. What was the price of his horse?

MULTIPLICATION.[2]

QUESTIONS.

IV. 1. A farmer gave 10 dollars for a calf, 10 dollars for a plough and 10 dollars for a load of hay. What did he pay for the whole?

Say 10 and 10 and 10 are 30; or as 10 is taken 3 times, rather say at once, 3 times 10 are 30, as in the Table. *A.* 30 dollars.

2. What will 5 hats cost at 4 dollars a-piece? *A.* 20 dollars.

3. A man bought 10 sheep for 3 dollars a-piece. What did he pay for the whole? How many are 3 times 10?

4. If a shoemaker can manufacture 4 pair of shoes in one day how many can he make at that rate in 8 days? In 10 days?

---

1 SUBTRACTION, [L. *subtractio.*] The act of taking a part from the rest
2 MULTIPLICATION. [*multiplicatio.*] The act of increasing in numbe

5. If 5 yards of cloth will make a suit of clothes, how many yards will it require to make 2 suits? 8 suits? 11 suits? Repeat the following multiplication table :—

MULTIPLICATION TABLE.

2	3	4	5	6	7	8	9	10
times	times	times	times	times	times	times	times	times
1 are 2	1 are 3	1 are 4	1 are 5	1 are 6	1 are 7	1 are 8	1 are 9	1 are 10
2 . 4	2 . 6	2 . 8	2 . 10	2 . 12	2 . 14	2 . 16	2 . 18	2 . 20
3 . 6	3 . 9	3 . 12	3 . 15	3 . 18	3 . 21	3 . 24	3 . 27	3 . 30
4 . 8	4 . 12	4 . 16	4 . 20	4 . 24	4 . 28	4 . 32	4 . 36	4 . 40
5 . 10	5 . 15	5 . 20	5 . 25	5 . 30	5 . 35	5 . 40	5 . 45	5 . 50
6 . 12	6 . 18	6 . 24	6 . 30	6 . 36	6 . 42	6 . 48	6 . 54	6 . 60
7 . 14	7 . 21	7 . 28	7 . 35	7 . 42	7 . 49	7 . 56	7 . 63	7 . 70
8 . 16	8 . 24	8 . 32	8 . 40	8 . 48	8 . 56	8 . 64	8 . 72	8 . 80
9 . 18	9 . 27	9 . 36	9 . 45	9 . 54	9 . 63	9 . 72	9 . 81	9 . 90
10 . 20	10 . 30	10 . 40	10 . 50	10 . 60	10 . 70	10 . 80	10 . 90	10 . 100
11 . 22	11 . 33	11 . 44	11 . 55	11 . 66	11 . 77	11 . 88	11 . 99	11 . 110
12 . 24	12 . 36	12 . 48	12 . 60	12 . 72	12 . 84	12 . 96	12 . 108	12 . 120

11				12			
times	times	times	times	times	times	times	times
1 are 11	4 are 44	7 are 77	10 are 110	1 are 12	4 are 48	7 are 84	10 are 120
2 . 22	5 . 55	8 . 88	11 . 121	2 . 24	5 . 60	8 . 96	11 . 132
3 . 33	6 . 66	9 . 99	12 . 132	3 . 36	6 . 72	9 . 108	12 . 144

6. It takes 4 pecks to make 1 bushel. How many pecks then are there in 3 bushels? In 5 bushels? In 7 bushels?

7. At 6 cents a quart, what will 6 quarts of cherries cost? What will 8 quarts cost? 9 quarts? 10 quarts? 11 quarts? 12 quarts?

8. When the board of a small family costs 7 dollars a week, what will the board for 7 weeks cost? For 8 weeks?

9. When flour is 8 dollars a barrel, how much must you pay for 8 barrels? For 9 barrels? For 11 barrels? For 12 barrels?

10. A man traveled by stage, at the rate of 7 miles an hour. How far did he go in 7 hours? In 9 hours? In 12 hours?

11. When hay is 9 dollars a ton, what will be the cost of 7 tons? 9 tons? 8 tons? 6 tons?

12. If 10 men can do a job of work in 10 days, how long will it take one man alone to do it, working at the same rate?

13. How many are 6 times 3? 9 times 5? 8 times 7? 7 times 6? 8 times 6? 12 times 5? 9 times 12?

14. If a cannon ball fly at the rate of 11 miles a minute, how far would it go in 7 minutes? In 9 minutes? In 12 minutes?

15. How many are 10 times 5? 10 times 10? 10 times 20? 10 times 30? 10 times 40? 10 times 90? 10 times 100?

16. If a regiment consists of 1000 men, of how many men would 2 regiments consist? 3 regiments? 5 regiments?

DIVISION.[1]

QUESTIONS.

V. 1. How many yards of cloth at 2 dollars a yard can you buy for 8 dollars, and why? *A.* 4 yards; because 4 times 2 are 8.

2. How many yards of cloth at 3 dollars a yard can you buy for 12 dollars? For 18 dollars? For 24 dollars? For 30 dollars?

3. At 5 dollars a hat, how many hats will 40 dollars buy? As many hats as 5 is contained times in 40. *A.* 8 hats.

4. A father having 24 books, gave 3 to each of his children. How many children must he have had?

5. A merchant bought a quantity of flour for 144 dollars, paying 12 dollars a barrel. How many barrels did he buy?

6. If one man alone can perform a piece of work in 100 days, how long would it take 10 such men to do the same?

7. If a man can travel 6 miles in an hour by stage, how long will it take him to perform a journey of 72 miles? Of 60 miles?

8. Suppose an orchard to have 132 trees in rows, with 12 trees in a row, of how many rows does the orchard consist?

9. How many times 8 in 96? 9 in 63? 8 in 56? 9 in 108? 11 in 132? 12 in 108? 12 in 132?

10. A man having 300 dollars, gave to his sons 100 dollars a-piece. How many sons had he? How many times 100 in 300?

11. How many times 100 in 500? 5 in 500? 200 in 1000? 4 in 800? 6 in 1200? 10 in 1000? 1000 in 10000?

12. If 1 barrel holds 5 bushels of rye; how many barrels will 21 bushels fill? *A.* 4 barrels and 1 bushel over.

13. How many times 5 in 42, and how many over? 6 in 39, and how many over? 7 in 46? 8 in 74? 9 in 98? 9 in 99?

14. Why not say 9 in 99 only 10 times and 9 over? *A.* Because 9 is contained in 99 all of 11 times.

15. How many times 6 in 41? 8 in 102? 9 in 112? 12 in 155?

FRACTIONS.[2]

QUESTIONS.

VI. 1. To divide 13 dollars equally among 2 persons; how would you proceed to find each man's exact part? Say 2 in 13, 6 times and 1 dollar over; the 1 over is considered as divided into 2 equal parts by writing the 2 under the 1, with a line between, making $\frac{1}{2}$ which is read 1-half. *A.* $6\frac{1}{2}$ dollars.

2. How is any number divided into 2, 3, 4, 5, &c. equal parts?

A. By writing under it as above the dividing numbers, 2, 3, 5, &c

1 DIVISION, [L. *divisio.*] The act of dividing any thing into parts; the state of being divided; that which divides or separates; partition; a part of an army or militia; dis-union; variance.

2 FRACTION. [L. *fractio.*] The act of breaking; the broken part of a number.

3. How many times 2 in 15? A. $7\frac{1}{2}$, read 7 and 1-half.
4. How many times 3 in 13? A. $4\frac{1}{3}$, read 4 and 1-third.
5. How many times 4 in 35? A. $8\frac{3}{4}$, read 8 and 3-fourths.
6. How many times 5 in 39? A. $7\frac{4}{5}$, read 7 and 4-fifths.
7. How many times 6 in 41? A. $6\frac{5}{6}$, read 6 and 5-sixths.

8. How many times 7 in 15? 8 in 30? 9 in 95? 10 in 109? 11 in [illegible]? 12 in 150? 12 in 155?

9. From what do *halves*, *thirds*, &c., take their names? A. From the figure below the line whether it be large or small.

10. What then, for example, is meant by 2, 3, 4, 5, 10, 20, 30, &c. with the figure 1 over each? A. 1-half [$\frac{1}{2}$]; 1-third [$\frac{1}{3}$]; 1-fourth [$\frac{1}{4}$]; 1-fifth [$\frac{1}{5}$]; 1-tenth [$\frac{1}{10}$]; 1-twentieth [$\frac{1}{20}$]; 1-thirtieth [$\frac{1}{30}$].

11. A man having 20 sheep sold $\frac{1}{2}$ of them. How many did he sell? A. 10 sheep.

12. How much is $\frac{1}{2}$ of 20? $\frac{1}{2}$ of 22? A. 10; 11.

13. How much is $\frac{1}{3}$ of 60? $\frac{2}{3}$ of 60? A. 20; 40.

14. How much is $\frac{1}{4}$ of 40? $\frac{3}{4}$ of 40? A. 10; 30.

15. A man deposited[1] in his cellar, in the fall, 60 bushels of apples, and all except $\frac{1}{5}$ of them rotted during the winter. How many bushels of sound apples had he in the spring?

16. How much is $\frac{1}{5}$ of 60? $\frac{2}{5}$ of 60? $\frac{4}{5}$ of 60?

17. How much is $\frac{1}{6}$ of 72? $\frac{2}{6}$ of 72? $\frac{4}{6}$ of 72?

18. How much is $\frac{1}{7}$ of 14? $\frac{3}{7}$ of 14? $\frac{6}{7}$ of 14?

19. A father gave his son 5 oranges, which was $\frac{1}{2}$ of all he had How many oranges had he? [2 times 5.] A. 10 oranges.

20. If $\frac{1}{2}$ of a number is 5, what is that number?

21. If $\frac{1}{3}$ of a number is 3, what is that number?

22. If $\frac{1}{4}$ of a number is 6, what is that number?

23. If $\frac{1}{5}$ of a number is 7, what is that number?

24. If $\frac{1}{6}$ of the length of a stick of timber is 5 feet, how long is the whole stick?

25. A merchant having 40 bushels of rye, sold $\frac{1}{2}$ of it. How many bushels had he left?

26. How many cents would you have left, if you had 24, and should lose $\frac{1}{3}$ of them? How many left if you lost $\frac{2}{3}$ of them?

27. Suppose you have 20 barrels of flour, and sell in one day $\frac{1}{4}$ of it; in another day $\frac{1}{5}$ of it. How many barrels will you sell in both days, and how many barrels will you have left?

28. When hay sells for 15 dollars a load, and you buy $\frac{1}{5}$ of a load how many more dollars will buy the whole load?

29. What is $\frac{1}{2}$ of 20? $\frac{2}{2}$ of 20? $\frac{1}{3}$ of 24? $\frac{3}{3}$ of 24? $\frac{4}{4}$ of 24? $\frac{5}{5}$ of 24?

30. If John owns $\frac{1}{3}$ of a vessel, and you the rest; how many thirds do you own? How many thirds make the whole?

31. Thomas owns $\frac{1}{5}$ of a factory, Rufus $\frac{2}{5}$, and Charles the rest What part does Charles own?

32. If $\frac{1}{5}$ of a mellon costs 2 cents, what will a whole one cost?

---

1 **Deposited**, [L. *depositum*.] Laid down; pledged; put.

33. When $\frac{1}{8}$ of a bushel of wheat costs 2 shillings; what would $\frac{5}{8}$ cost? $\frac{8}{8}$ or 1 bushel cost? $\frac{9}{8}$ cost? 2 bushels or $\frac{16}{8}$ cost?

34. When $\frac{1}{10}$ of a pew costs 10 dollars, how many tenths at that rate can be bought for 50 dollars? for 80 dollars? for 100 dollars?

35. What does $\frac{1}{2}$, $\frac{2}{3}$, $\frac{3}{4}$, $\frac{4}{5}$, &c. of any thing appear to mean? *A.* $\frac{1}{2}$ means 1 of its 2 equal parts; $\frac{2}{3}$ means 2 of its 3 equal parts; $\frac{3}{4}$, 3 of its 4 equal parts; $\frac{4}{5}$, 4 of its 5 equal parts, &c.

36. Which then is the greater fraction $\frac{1}{2}$ or $\frac{1}{3}$? $\frac{1}{4}$ or $\frac{1}{5}$? $\frac{1}{5}$ or $\frac{1}{6}$?

37. How many halves, or thirds, or fourths, or fifths, &c., make 1 whole? *A.* $\frac{2}{2}$ [2-halves,] or $\frac{3}{3}$ [3-thirds,] or $\frac{4}{4}$ [4-fourths,] or $\frac{5}{5}$ [5-fifths.]

38. How many halves then in 2 wholes? thirds in 4 wholes? fourths in 6 wholes? fifths in 8 wholes? sixths in 9 wholes?

39. How many whole ones in $\frac{4}{2}$ [4-halves]? in $\frac{12}{3}$ [12-thirds]? in $\frac{24}{4}$? in $\frac{40}{5}$? in $\frac{54}{6}$?

40. When the upper figure is the greater one, what does it indicate? *A.* That it is to be divided by the lower one.

41. How many wholes in $\frac{24}{4}$? $\frac{56}{8}$? $\frac{72}{9}$? $\frac{108}{12}$? $\frac{132}{11}$?

42. How many whole ones in $\frac{5}{2}$? [2 in 5]. *A.* $2\frac{1}{2}$.

43. How many halves in $2\frac{1}{2}$ wholes? [2 times 2 and $\frac{1}{2}$]. *A.* $\frac{5}{2}$.

44. How many wholes in $\frac{17}{3}$? [3 in 17]. *A.* $5\frac{2}{3}$.

45. How many thirds in $5\frac{2}{3}$ wholes? [3 times 5 and $\frac{2}{3}$]. *A.* $\frac{17}{3}$.

46. How many wholes in $\frac{40}{7}$? $\frac{50}{12}$? $\frac{90}{11}$? $\frac{100}{8}$? $\frac{155}{12}$? $\frac{130}{11}$?

47. How many fifths in 4? in $4\frac{1}{5}$? sevenths in $8\frac{3}{7}$? ninths in $10\frac{5}{9}$? enths in $8\frac{7}{10}$? twelfths in $12\frac{12}{12}$? twentieths in $2\frac{5}{20}$?

48. James has $\frac{5}{8}$ of a dollar; Rufus $\frac{6}{8}$, and Thomas $\frac{7}{8}$. How many dollars and how many eighths over have they all? *A.* $2\frac{2}{8}$ dollars.

49. Add together $\frac{1}{9}$, $\frac{2}{9}$, $\frac{3}{9}$, $\frac{4}{9}$, $\frac{5}{9}$, $\frac{6}{9}$, $\frac{7}{9}$, $\frac{8}{9}$, and $\frac{9}{9}$. *A.* 5.

50. Add together $\frac{2}{10}$, $\frac{3}{10}$, $\frac{4}{10}$, $\frac{5}{10}$, $\frac{6}{10}$, $\frac{7}{10}$, $\frac{8}{10}$, and $\frac{9}{10}$. *A.* $4\frac{4}{10}$.

51. A man having $\frac{7}{8}$ of a ship, sold $\frac{5}{8}$ of it. What part had he left?

52. Take $\frac{5}{9}$ from $\frac{8}{9}$; $\frac{9}{40}$ from $\frac{26}{40}$; $\frac{7}{50}$ from $\frac{49}{50}$.

53. If 1 gallon of molasses cost $\frac{1}{4}$ of a dollar, what will 2 gallons cost? 3 gallons cost? How many gallons will cost 1 dollar?

54. How much is 2 times $\frac{1}{2}$? [$\frac{2}{2}$ or 1]. 3 times $\frac{1}{2}$? [$\frac{3}{2}$ or $1\frac{1}{2}$] 4 times $\frac{2}{3}$? [$2\frac{2}{3}$]. 5 times $\frac{3}{4}$? 6 times $\frac{2}{5}$? 7 times $\frac{5}{6}$?

55. If 2 bushels of oats cost $\frac{6}{8}$ [6-eighths] of a dollar, how many eighths will buy 1 bushel? *A.* $\frac{3}{8}$. [3-eighths].

56. How much is $\frac{1}{2}$ of $\frac{6}{8}$? $\frac{1}{2}$ of $\frac{4}{8}$? $\frac{1}{3}$ of $\frac{9}{11}$? $\frac{1}{4}$ of $\frac{12}{19}$? $\frac{1}{5}$ of $\frac{20}{29}$? $\frac{1}{6}$ of $\frac{18}{30}$? $\frac{1}{7}$ of $\frac{28}{45}$? $\frac{1}{10}$ of $\frac{120}{145}$?

57. At 6 dollars a yard, what will 2 yards of cloth cost? What will $\frac{1}{2}$ a yard cost? What will $2\frac{1}{2}$ yards cost then? How much then is $2\frac{1}{2}$ times 6?

58. How much is $3\frac{1}{4}$ times 8, or 3 times 8 and $\frac{1}{4}$ of 8?

59. How much is $3\frac{3}{4}$ times 4, or 3 times 4 and $\frac{3}{4}$ of 4?

60. How much is $7\frac{2}{3}$ times 9, or 7 times 9 and $\frac{2}{3}$ of 9?

61. How much is $8\frac{1}{5}$ times 10? $8\frac{4}{5}$ times 10? $7\frac{5}{6}$ times 12?

62. When flour is 12 dollars a barrel, what will $3\frac{1}{2}$ barrels cost? $4\frac{2}{3}$ barrels cost? $8\frac{3}{4}$ barrels cost? $10\frac{5}{6}$ barrels cost?

VII. TABLES

OF MONEY, WEIGHTS AND MEASURES.

FEDERAL[1] MONEY,[2]

1. Is the currency[2] or coin[3] of the United States.[9]

10 mills[4] [sign m.]	make	1 cent[5]	. . .	ct.
10 cents	make	1 dime[6]	. .	d.
10 dimes	make	1 dollar[7]	. .	$.
10 dollars	make	1 eagle[8]	. .	E

ENGLISH, OR STERLING[10] MONEY,

2. Is used by Great Britain and her dependencies.[11]

1 farthing[12] [sign qr.]	makes	¼ of a penny	¼ d.
2 farthings	[13]make	½ of a penny	½ d.
3 farthings	make	¾ of a penny	¾ d.
4 farthings	make	1 penny	d.
12 pence[13]	make	1 shilling	s.
20 shillings[14]	make	1 pound[15]	£.

QUESTIONS.

3. What is Federal Money?[9] 1. Repeat the Table. How many mills in 5 cents? cents in 1 dollar [10 dimes]? in 5 dollars? eagles in 20 dollars? in 70 dollars?

4. What is English Money? 2. Repeat the Table. How many farthings in 10 pence? pence in 6 shillings? pounds[15] in 40 shillings?

---

1 FEDERAL, [L. *fœdus.*] Of or belonging to a league or contract; united by agreement. Federal Money is so called from the Federal or United States.

2 MONEY, [L. *moneta.*] The medium of exchange in buying and selling, and thence called the currency or circulating medium.

3 COIN, [F. *coin.*] A species of gold, silver, copper, or other metal, stamped to be used as money; our national coins.

4 MILL is so called from *mille*, L., for 1000; because 1000 mills make 1 dollar.

5 CENT is so called from *centum*, L., for 100; because 100 cents make 1 dollar.

6 DIME is so called from *dime*, F. *tenth*; because 10 dimes make 1 dollar.

7 DOLLAR, from the Danish *daler*, for *dollar*; a coin of the value of 100 cents.

8 EAGLE is so called as a mark of distinction, it being the most valuable of American coins.

9 Our national coins are the Eagle, Half-Eagle, Quarter-Eagle, made of gold; the Dollar, Half-Dollar, Quarter-Dollar, Dime and Half-Dime, made of silver; the Cent and Half-Cent, made of copper; the Mill has no coin to represent it, being merely nominal.

10 STERLING MONEY, so called from the *Easterlings* who first coined it.

11 DEPENDENCY. The state of a thing hanging from a supporter; the state of being subject to. Dependencies are countries or states subject to the power of others.

12 FARTHING, from the Saxon *feorthing*, *the fourth*, and its sign qr. from *quadrans* Latin for the *fourth part*.

13 PENCE or PENNY, from the Sweedish *penning*, and *d.* from the Latin *denarium*, for *pence*.

14 SHILLING, from *scild*, the Saxon for shield, because the shilling piece had originally the picture of a shield on it.

15 POUND is from the Latin *pondo*, *weight*, and the £ stands for *Libra*, the Latin for *pound*.

TROY WEIGHT,

5. For weighing gold, silver, liquors, bread, &c

24 grains [gr.]	make	1 pennyweight	dwt.
20 pennyweights	make	1 ounce	oz.
12 ounces	make	1 pound	lb.

AVOIRDUPOIS WEIGHT,[1]

6. For weighing hay, grain, groceries, and all coarse articles

16 drams [dr.]	make	1 ounce	oz.
16 ounces	make	1 pound	lb.
25 pounds[2]	make	1 quarter	qr.
4 quarters	make	1 hundred weight	cwt.
20 hundred weight	make	1 ton	T

APOTHECARIES' WEIGHT,

7. Is used for mixing, but not for selling medicines.

20 grains [gr.]	make	1 scruple	℈
3 scruples	make	1 dram	ʒ
8 drams	make	1 ounce	℥
12 ounces	make	1 pound	℔

QUESTIONS.

8. For what is Troy Weight used? 5. Repeat the Table.

9. How many pounds in 63 ounces? *A.* 5 lb. 3 oz. or $5\frac{3}{12}$ lb.

10. Why are 3 ounces the same as $\frac{3}{12}$ of a pound? *A.* The pound consists of 12 equal parts called ounces, therefore 1 ounce is $\frac{1}{12}$ of a pound; 2 ounces, $\frac{2}{12}$; 3 ounces, $\frac{3}{12}$, &c.

11. How many ounces are there in $5\frac{3}{12}$ pounds? in $10\frac{11}{12}$ pounds? pennyweights in $2\frac{8}{20}$ ounces?

12. What is the use of Avoirdupois weight? 6. Repeat the Table.

13. How many drams in 1 ounce 5 drams? quarters in 30 pounds? [$1\frac{5}{25}$] pounds in 2 quarters 10 pounds? quarters in $2\frac{2}{4}$ hundred weight? pounds in 1 hundred weight, or 4 quarters? in 20 hundred weight?

14. For what is Apothecaries' Weight used? 7. Repeat the Table.

15. How many drams in $3\frac{5}{8}$ ounces? grains in 2 scruples? pounds in 155 ounces?

16. How many pounds in 27 ounces? [$2\frac{3}{12}$]Ounces in 100 drams? Drams in 300 scruples? Scruples in 29 grains?

---

1 175 Troy ounces are equal to 192 ounces Avoirdupois; 1 lb. Troy, to 5760 grains, and 1 lb. Avoirdupois to 7000 grains. The pound and ounce in Apothecaries' weight are the same as in Troy weight; the only difference is in their divisions and sub-divisions.

2 Formerly 28 pounds were reckoned for a quarter, making 112 pounds to the hundred, but the practice has become nearly obsolete

CLOTH MEASURE,

17. Is used for measuring goods sold by the yard, ell, &c.

$2\frac{1}{4}$ inches [in.]	make 1 nail	na
4 nails	make 1 quarter	qr.
4 quarters	make 1 yard	yd.
3 quarters	make 1 Flemish ell	Fl. e.
5 quarters	make 1 English ell	E. e.
6 quarters	make 1 French ell	Fr. e.

DRY MEASURE,

18. For dry goods, as grain, fruit, roots, coal, &c.

2 pints [pt.]	make 1 quart	qt.
4 quarts	make 1 gallon	gal.
8 quarts	make 1 peck	pk.
4 pecks	make 1 bushel.[1]	bu.
36 bushels[1]	make 1 chaldron	ch.
8 bushels	make 1 quarter	qr.
5 quarters	make 1 load	load.

WINE MEASURE,

19. For spirits, mead, vinegar, cider, oil, honey, &c.

4 gills [gi.]	make 1 pint	pt.
2 pints	make 1 quart	qt.
4 quarts	make 1 gallon	gal.
$31\frac{1}{2}$ gallons	make 1 barrel	bl.
42 gallons	make 1 tierce	tier.
63 gallons	make 1 hogshead	hhd.
84 gallons	make 1 puncheon	pun.
2 hogsheads	make 1 pipe or butt	p. or b.
2 pipes	make 1 tun	T.

QUESTIONS.

20. For what is Cloth Measure used? 17. Repeat the Table.

21. How many nails in 1 quarter? in 1 yard? in 1 Flemish ell? in 1 English ell? in 1 French ell?

22. For what is Dry measure used? 18. Repeat the Table.

23. How many pints in 7 quarts? in $10\frac{1}{2}$ quarts? in 2 gallons? in 1 peck? bushels in 38 pecks? in 3 quarters?

24. What is the use of Wine Measure? 19. Repeat the Table.

25. How many pipes in 4 tuns? Hogsheads in 4 tuns? Gallons in 8 quarts? in 49 quarts? Gills in 20 pints? in $10\frac{3}{4}$ pints?

1 The statute bushel is said to contain 2150.4 cubic inches; but the number is supposed to vary in the different states. In Connecticut, it is 2198 cubic inches. Congress has recently proposed to the States an uniform standard of weights and measures, but it has not yet been generally adopted.

ALE OR BEER MEASURE,[1]

26. Is used for measuring malt liquors.

2 pints [pt.] .... make 1 quart. ...... qt.
4 quarts .... make 1 gallon. ..... gal.
9 gallons .... make 1 firkin. ..... fir.
2 firkins .... make 1 kilderkin. ... kil.
2 kilderkins .... make 1 barrel. ..... bl.
36 gallons............ make 1 barrel. ..... bl.
54 gallons............ make 1 hogshead. ... hhd.
2 hogsheads .... make 1 butt, ... bt.

LONG MEASURE,

27. For measuring length without regard to breadth or depth.

1 barley corn ———; 3 barley corns ——————— 1 inch.

28. Of the foregoing lines the shorter one is exactly 1 barley corn in length; then 3 times the length of this line makes 3 barley corns, or 1 inch, which is the exact length of the longer line. In like manner, 12 times the length of 1 inch makes 1 foot; 3 times the length of 1 foot, 1 yard, &c.

3 barley corns [b. c.] ..... make 1 inch...... in.
12 inches .... make 1 foot. ..... ft.
3 feet .... make 1 yard...... yd.
$5\frac{1}{2}$ yards, or $16\frac{1}{2}$ feet ...... make 1 rod. ..... rd.
40 rods, or 220 yards ...... make 1 furlong.... fur.
8 furlongs, or 1760 yards .. make 1 mile. .... m.
3 miles .... make 1 league .... l.
$69\frac{1}{2}$ statute miles .... make 1 degree. ... deg.
60 geographical miles ..... make 1 degree. ... deg.
360 degrees make 1 circle, or the earth's circumference.

29. DISTANCES—DEPTHS—HEIGHTS.

4 inches ... make 1 hand, for measuring the height of horses.
6 points ... make 1 line, } used in measuring the length of pen-
12 lines.... make 1 inch, } dulums for clocks.
5 feet..... make 1 geometrical space, used for distances.
6 feet..... make 1 fathom, for measuring depths at sea.
3 miles.... make 1 league, for measuring distances at sea.

GUNTER'S CHAIN.

30. For measuring distances, and the length or breadth of land.

$7\frac{92}{100}$ inches .... make 1 link.
25 links .... make 1 pole.
100 links .... make 1 chain.
10 chains .... make 1 furlong.
8 furlongs .... make 1 mile.

---

1. The dry gallon contains $268\frac{4}{5}$ cubic inches; the wine gallon 231 cubic inches and the beer gallon 282 cubic inches. The same standards continued in use in Great Britain, as late as 1826, when the act of Parliament came into operation, by which the *Imperial* gallon of $277\frac{274}{1000}$ cubic inches was substituted for the dry, beer and wine gallons

31: The chain for measuring distances varies from 2 to 4 rods in length, reckoning 25 links to a rod.

QUESTIONS

32. Repeat the Table of Ale Measure. Its use? 26. Repeat the Table of Long Measure. Its use? 27. For what purpose do we use the league and fathom? the geometrical space, lines and hands?

33. How many barley corns in 20 inches? Inches in 7 feet inches? Feet in $10\frac{2}{3}$ yards? Yards in 2 rods?

34. How many leagues in 38 miles? Furlongs in $8\frac{5}{8}$ miles? Fathoms in 75 feet? Feet in 100 geometrical spaces?

LAND, OR SQUARE MEASURE,

35. For measuring superficies, that is, surfaces or things, whose length and breadth are considered without regard to depth; as, land, paving, flooring, plastering, roofing, slating, tiling, &c.

Fig. 1. Fig. 2. Fig. 3.

36. A SQUARE[1] has four equal sides, and four equal and square corners, commonly called Angles; consequently its length and breadth are equal; as, Fig. 1.

37. A PARALLELOGRAM[2] has only its opposite sides equal, and at least its opposite angles, but may have all its angles equal; consequently it has more length than breadth; as, figures 2 and 3.

Fig. 4. Small Squares.

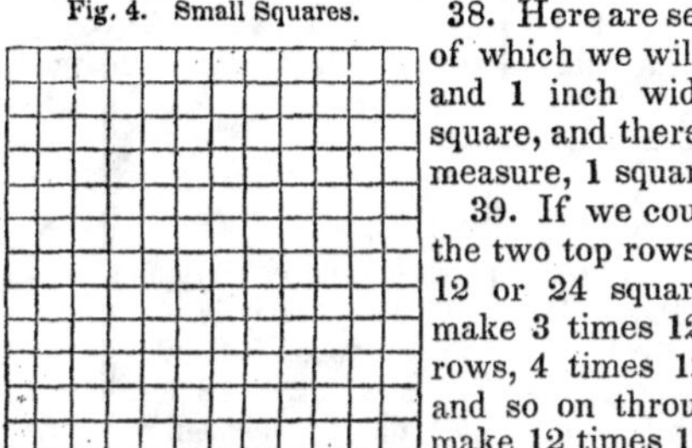

38. Here are several small squares, each of which we will suppose is 1 inch long and 1 inch wide, which make 1 inch square, and therefore is said to contain by measure, 1 square inch.

39. If we count the square inches in the two top rows, they will make 2 times 12 or 24 square inches; 3 rows will make 3 times 12 or 36 square inches; 4 rows, 4 times 12 or 48 square inches; and so on through the 12 rows, which make 12 times 12, or 144 square inches.

40. The same figure taken as a whole, is 1 foot square, for 12 inches are equal to 1 foot; then the whole contains 1 square foot therefore, 1 square foot is equal to 144 square inches.

---

1 SQUARE, [F. *quarre.*] A form like Figure 1, above; an area or the open surface with houses on four sides; a rule for measuring; a square body of troops; a squadron; a quarternion; equality; rule; conformity; accord.

2 PARALLELOGRAM is so called from *parallelos*, G. *equally distant*, and *gramma* G *a letter;* because its opposite sides are parallel; that is equally distant from each other in all their parts; of course parallel lines would not meet if continued ever so far. In *common use*, this word is applied to any quadrilateral, or four-sided figures of more length than breadth.

Fig. 5. A Parallelogram.

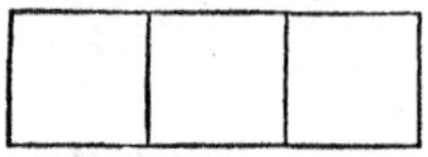

41. This figure we will suppose to be 3 feet long and 1 foot wide; then it will have 3 squares each 1 foot square, and will contain 3 square feet.

Fig. 6. Squares.

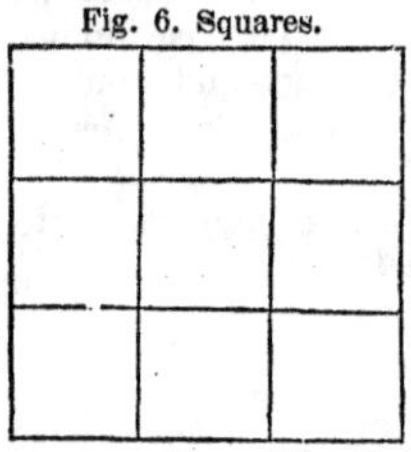

42. Three rows of 3 square feet in each row, will contain 3 times 3, or 9 square feet, but the whole figure will be only 3 feet square. Again, as 3 feet are equal to 1 yard; then 9 square feet are equal to 1 square yard.

43. Hence we see that 3 square feet is only one third as much as 3 feet square, making a difference of 6 square feet.

44. Again; suppose Figure 5 to contain 3 square miles of land, then 3 miles square of land; as, Figure 6, would be 3 times 3, or 9 square miles, making a difference of 6 square miles.

45 Hence a square foot, yard, &c. may be of any shape whatever, provided the foot contains exactly 144 squares, each 1 inch square, and the yard 9 squares, each 1 foot square, &c.

46. *From the above we learn that multiplying the length of any square or parallelogram by its breadth, gives its square measure, or, as it is sometimes called, its square contents, as in the following Table.*

144 square inches [sq. in.]	make 1 square foot	sq. ft.
9 square feet	make 1 square yard	sq. yd.
$30\frac{1}{4}$ sq. yd. or $272\frac{1}{4}$ sq. ft.	make 1 square rod or pole	sq. r.
40 square rods	make 1 rood	R.
4 roods	make 1 acre	A.
640 square acres	make 1 square mile	sq. m.

QUESTIONS.

47. Repeat the Table of Land or Square Measure. Its use? 35. What is a square? 36. Parallelogram? 37. What is meant by 1 square foot? 38. 39. 40. What by 1 square yard? 41. 42.

48. What is the difference between 3 square feet, and 3 feet square, and why? 42. 43.—between 3 square miles and 3 miles square? 44.

49. How many square feet in 12 square yards? square rods in $2\frac{10}{40}$ roods? acres in 38 roods?

50. What is said of the form of a square foot, square yard, &c. 45. How is the square content ascertained? 46.

51. How many square inches then, are contained in a small slate 4 inches long and 3 inches wide? [3 times 4]. In one 7 inches long and 5 inches wide? What is the form of such slates, and why? 37

52. How many square feet in a board 3 feet long and 1 foot wide? [3]. In one 7 feet long and 2 feet wide? In the floor of a room 10 feet long and 7 feet wide?

53. How many square yards in a piece of carpeting containing 54 square feet? Roods in a piece of land 10 rods long and 4 rods wide? [1]. In a piece 10 rods square?

SOLID OR CUBIC MEASURE,

54. For measuring solids, that is, things that have three dimensions, viz., length, breadth, and depth or thickness; as wood, timber, stones, masonry, &c.

55. A CUBE is a solid, whose length, breadth, and thickness are all equal. [See the Cubical block.]

56. Thus a small block 1 inch long, 1 inch wide, and 1 inch thick, is called a cube, and is said to contain 1 cubic or solid inch. Such a block has of course six equal sides

57. Now suppose a box to be 12 inches square in the inside; we can then place on the bottom 12 rows of 12 cubes, each containing 1 solid inch, making in all 144 cubic blocks.

58. If we lay another tier of 12 times 12, or 144 similar blocks upon the others, we shall have laid 2 times 144, or 288 blocks, and thus we might continue to do, till we had laid 12 tiers, which would make in all 12 times 144, or 1728 cubes; that is, so many solid inches, and all in the form of a perfect cube, for the whole pile would be 12 inches long, 12 inches wide, and 12 inches thick.

59. Then as 12 inches are equal to 1 foot, the above cube would be 1 foot long, 1 foot wide, and 1 foot thick, which makes 1 solid or cubic foot; hence it takes 1728 solid or cubic inches to make 1 solid or cubic foot.

60. *Hence it appears that a solid or cubic inch, foot, &c., arise from multiplying the length of a solid by its breadth, and that result by its thickness, as in the following Table*:

1728 cubic inches [c. in.]	make 1 cubic foot. . .	c. ft.
27 cubic feet	make 1 cubic yard. . .	c. yd.
50 cubic feet of round timber .	make 1 ton.	T.
40 cubic feet of hewn timber .	make 1 ton.	T.
42 cubic feet of shipping . . .	make 1 ton.	T.
16 cubic feet	make 1 cord foot. . .	c. ft.
8 cord feet, or 128 cubic feet	make 1 cord of wood.	C.

QUESTIONS.

61. How many dimensions has a cube? See 54. 55. How many sides has it? See 56. What is meant by one solid or cubic inch? See 56.

62. How many such blocks will exactly cover the space of 12 inches square? See 57. How many such blocks would 2 tiers require? How many would 12 tiers require? See 58.

63. What would be the proper name for the form of such a pile of blocks when taken as a whole, and why? 58. What is meant by 1 cubic foot, 59.

64. How is the solid content of a cube obtained? 60. How many solid feet in a cubic block 2 feet long, and 2 feet wide, and 2 feet thick? [8] In one 10 feet long, 10 feet thick, and 10 feet wide?

64 How many cord feet in 32 solid feet? in 64 solid feet? How many cubic feet in $\frac{1}{2}$ of a cord of wood? in 1 cord? Repeat the Table of Cubic Measure. Its use? 54.

TIME.

65. Which is reckoned by years, months, days, &c.

60 seconds [sec.]	make 1 minute.	m.
60 minutes	make 1 hour.	h.
24 hours	make 1 day.	d.
365 days	make 1 year.	Y.
7 days	make 1 week.	w.
4 weeks [in common reckoning,]	make 1 month.	mo.
52 weeks [in common reckoning,]	make 1 year.	Y.
30 days [in common reckoning,]	make 1 month.	mo.
12 months	make 1 year.	Y.
100 years	make 1 century.	C.

66. The number of days in each month are as follows:—

January 31 days,	May 31 days,	September 30 days,
February 28 days,	June 30 days,	October 31 days,
March 31 days,	July 31 days,	November 30 days,
April 30 days,	August 31 days,	December 31 days

67. The days in each month are often expressed thus:

Thirty days has September, April, June and November.
February has twenty-eight, and thirty-one the others rate,
Except in leapyear,[1] happening once in four,
When we give to February one day more.

68. A natural[3] day[3] is . . 24 hours.

A Lunar[4] month[5] . . . 4 weeks or 28 days.

A Solar[6] year[7] . . . 365 days, 5h. 48m. 48 sec. [nearly]

A Civil[8] year[9] . . . . . 12 calender[10] months[11] or 365 days.

A Julian[12] year[12] . . . . 13 lunar months, 1 d. 6 h. or 365¼ days

69. When any year can be divided by 4 without a remainder, it is leapyear,[1] except the centurial[2] years, which are explained below.

---

1 LEAPYEAR. So called, because the year *leaps* as it were from 365 to 366 days.

2 CENTURIAL, [L. *centuria, a century.*] Of, or belonging to a century.

3 A NATURAL DAY is the period in which the earth revolves (13) on its axis, (14) being once in every 24 hours.

4 LUNAR, [L. *luna, the moon.*] Pertaining to the moon.

5 A LUNAR MONTH, is the period of time required for the revolution (13) of the moon round the earth, being, strictly speaking, 27 days, 7 hours, 43 minutes, 5 seconds.

6 SOLAR, [L. *sol, the sun.*] Of, or belonging to the sun.

7 A SOLAR YEAR, or A YEAR, properly speaking, is the period which the earth revolves (13) round the *sun*, occurring once in every 365 days, 5 hours, 48 minutes and a trifle more than 48 seconds. This period of revolution is therefore called the SOLAR (6) or NATURAL YEAR.

8 CIVIL, [L. *civis.*] Relating to the community or government; polite.

9 A CIVIL YEAR, is the period of time established by law.

10 CALENDAR, [L. *calendarium.*] A register in which the months and days are set down in order; an almanac.

11 A CALENDAR MONTH, is a solar month, as it stands in almanacs.

12 A JULIAN YEAR, is so called from Julius Cæsar, Emperor of Rome, who from a desire to make the civil year correspond (15) with the solar, ordered it to consist of 13 months 1 day and 6 hours, or 365¼ days; and in common reckoning only 365 days Dropping the 6 hours, or ¼ of a day for 4 years, makes a loss of one day, which is

70. When there is a remainder, it denotes the number of years since the last leapyear· as, 1836, 1840, 1844, but in 1839, for instance, there will be 3 over; 1839, then, is the 3d year since the last leapyear.

71. How is Time reckoned? 65. Repeat the Table, 65. How many days [See 66] has January? February? March? April? June? August? December?

72. How are the days in each month expressed? 67. How may a leapyear be known? 69.

73. How many minutes in 93 seconds? in 1 hour and 37 minutes? days in $4\frac{3}{7}$ weeks? years in 60 weeks? months in 70 days? in 1 century?

CIRCULAR MOTION.

74. For calculating the motions of the planets,[1] and computing latitude[2] and longitude.[3]

60 seconds (″) . . . . . . make 1 minute. . . .
60 minutes . . . . make 1 degree. . . . °
30 degrees . . . . make 1 sign. . . . . s.
12 signs or 360 degrees, the whole circle of the Zodiac.[7]

75. Every circle, whether the greatest or least possible, is divided into 360 equal parts, called degrees, and each degree into 60 equal

1 PLANET, from *planao*, G. *to wander*, means a celestial (4) body, which revolves about the sun or other centre, or a body revolving about another planet as its centre.
2 LATITUDE, from *latus*, L. *side*, means the distance North or South of the equator; (5) breadth; room; space; extent of meaning; freedom from rules.
3 LONGITUDE, from *longus*, L. *long*, means in its application, the distance of any place on the globe from another place eastward or westward; the distance of any place from a given meridian.(6)
4 CELESTIAL, [L. *cœlestis*.] Heavenly; belonging to the upper regions, or that a part of which we see över our heads in a clear sky.
5 EQUATOR, L. is merely an imaginary (8) line or circle, which is supposed to pass quite round the earth from East to West.
6 Meridian, from *meridies*, L. *mid day*, is an imaginary circle, which is supposed to pass from North to South quite round the earth and through each pole.
7 ZODIAC, [L. *zodiacus*.] A girdle; a broad circle in the heavens, containing the 12 signs through which the sun passes in its annual (9) revolution.
8 IMAGINARY, Not real; not existing at all in fact.
9 ANNUAL, from *annus*, L. a *year*, means *yearly*; that returns every year.

---

made up by adding, every fourth year, one more day, making 366 days to the year, called Bissextile (16) or Leapyear. Bissextile or Leapyear then has 366 days, and happens once in 4 years, so that any year that can be divided by 4 without a remainder, is called Leapyear, as 1836, 1840, 1844, &c. with the following exception, viz: By this reckoning the civil year would have 11 minutes and about 12 seconds more than the true solar year; and every centurial year, that is, the year that completes a century would be a leapyear, therefore to make the two years correspond, it was ordered that 3 centurial years in succession should be reckoned common years, and the fourth one only, a leapyear.
13 REVOLVE, [L. *revolvo*.] To turn in a circle; to roll any thing; to consider.
14 AXIS, [L. *axis*.] Something passing through the centre of any thing on which it turns; an imaginary line running from North to South through the centre of the earth.
15 CORRESPOND, [L. *correspondeo*.] To answer to; communicate with.
16 BISSEXTILE, So called from the Latin *bis*, *twice*, and *sextilis* the *sixth*; because in that year the 6th of the kalends of March was repeated twice

parts, called minutes, and each minute into 60 equal parts, called seconds.

76. The circumference[1] of the earth is a great circle of 360 degrees. On this circle every minute is reckoned a mile, 60 of which are equal to 1 degree or about 69½ statute or common miles.

77. TABLE OF PARTICULARS.

12 things	make 1 dozen.
12 dozen	make 1 gross.
12 gross or 144 dozen .	make 1 great gross
20 things	make 1 score.
5 score	make 1 hundred.
24 sheets of paper . . .	make 1 quire.
20 quires	make 1 ream.
200 pounds	make 1 barrel of pork.
200 pounds	make 1 barrel of beef.
196 pounds	make 1 barrel of flour.
30 pounds	make 1 barrel of anchovies.
112 pounds	make 1 barrel of raisins.
256 pounds	make 1 barrel of soap.
200 pounds	make 1 barrel of shad or salmon.
7½ pounds	make 1 gallon of train oil.
11 pounds	make 1 gallon of molasses.
14 pounds	make 1 stone of iron or wood.
8 pounds	make 1 stone of meat.
28 pounds	make 1 tod.
56 pounds	make 1 firkin of butter.
94 pounds	make 1 firkin of soap.
112 pounds	make 1 quintal of fish
364 pounds	make 1 sack.
19½ cwt.	make 1 fother of lead.
30 gallons	make 1 barrel of fish.
32 gallons	make 1 barrel of cider.
32 gallons	make 1 barrel of herring, England.
42 gallons	make 1 bar. salmon, eels England.
7½ bushels	make 1 hhd. on shore.
8 bushels salt	make 1 hhd. at sea.

QUESTIONS.

78. How many single things in 5 dozen? In 1 gross? In 5 score? sheets of paper in 2½ quires? quires in 5¼ reams?

79. How many pounds in 2½ barrels of pork? In ⅗ of a barrel of shad? In 2⅗ barrels of shad?

1 CIRCUMFERENCE, from *circum*, L. *round*, and *fero*, L. to *bring*. The distance round the outside.

80. BOOKS.

A FOLIO[1] is when a sheet is folded in two leaves.
A QUARTO[2] or 4to., is when a sheet is folded in four leaves.
An OCTAVO[3] or 8vo., is when a sheet is folded in eight leaves
A DUODECIMO[4] or 12mo., is when a sheet is folded in 12 leaves
An 18mo.[5] is when a sheet is folded in 18 leaves.

ALIQUOT PARTS.

81. AN ALIQUOT PART is that number which is contained in another an exact number of times: thus 5 is an aliquot part of 15, for it is contained in 15 exactly 3 times; that is, 5 is $\frac{1}{3}$ of 15.

82. So 50 cents and 25 cents are both aliquot parts of a dollar; for 50 cents = (these 2 lines mean *equal to*) $\frac{1}{2}$ of 100 cents or 1 dollar, and 25 cents = $\frac{1}{4}$ of 1 dollar.

83. Again; every $6\frac{1}{4}$ cent piece is an aliquot part of 100 cents or 1 dollar, being exactly $\frac{1}{16}$ of a dollar, for $6\frac{1}{4}$ times 16 = 100.

84. Again; every $12\frac{1}{2}$ cent piece is an aliquot part of 1 dollar, being exactly $\frac{1}{8}$ of a dollar, for $12\frac{1}{2}$ times 8 = 100.

85. If one $6\frac{1}{4}$ cent piece is $\frac{1}{16}$, then two $6\frac{1}{4}$ cent pieces, which make $12\frac{1}{2}$ cents, are $\frac{2}{16}$, and $\frac{2}{16}$ are equal to $\frac{1}{8}$; for 8 times $\frac{1}{8}$ are $\frac{8}{8}$, which are equal to 1 whole; and 8 times $\frac{2}{16}$ are $\frac{16}{16}$, which are also equal to 1 whole.

---

86. TABLE OF ALIQUOT PARTS.

One	$6\frac{1}{4}$ cent piece, or $6\frac{1}{4}$ cents	$=\frac{1}{16}$ of a dollar.
Two	$6\frac{1}{4}$ cent pieces, or $12\frac{1}{2}$ cents	$=\frac{1}{8}$ of a dollar.
Three	$6\frac{1}{4}$ cent pieces, or $18\frac{3}{4}$ cents	$=\frac{3}{16}$ of a dollar.
Four	$6\frac{1}{4}$ cent pieces, or 25 cents	$=\frac{1}{4}$ of a dollar
Five	$6\frac{1}{4}$ cent pieces, or $31\frac{1}{4}$ cents	$=\frac{5}{16}$ of a dollar.
Six	$6\frac{1}{4}$ cent pieces, or $37\frac{1}{2}$ cents	$=\frac{3}{8}$ of a dollar.
Seven	$6\frac{1}{4}$ cent pieces, or $43\frac{3}{4}$ cents	$=\frac{7}{16}$ of a dollar.
Eight	$6\frac{1}{4}$ cent pieces, or 50 cents	$=\frac{1}{2}$ of a dollar.
Nine	$6\frac{1}{4}$ cent pieces, or $56\frac{1}{4}$ cents	$=\frac{9}{16}$ of a dollar.
Ten	$6\frac{1}{4}$ cent pieces, or $62\frac{1}{2}$ cents	$=\frac{5}{8}$ of a dollar.
Eleven . . .	$6\frac{1}{4}$ cent pieces, or $68\frac{3}{4}$ cents	$=\frac{11}{16}$ of a dollar.
Twelve . . .	$6\frac{1}{4}$ cent pieces, or 75 cents	$=\frac{3}{4}$ of a dollar.
Thirteen . .	$6\frac{1}{4}$ cent pieces, or $81\frac{1}{4}$ cents	$=\frac{13}{16}$ of a dollar.
Fourteen . .	$6\frac{1}{4}$ cent pieces, or $87\frac{1}{2}$ cents	$=\frac{7}{8}$ of a dollar.
Fifteen . . .	$6\frac{1}{4}$ cent pieces, or $93\frac{3}{4}$ cents	$=\frac{15}{16}$ of a dollar.
Sixteen . . .	$6\frac{1}{4}$ cent pieces, or 100 cents	= ONE DOLLAR.

---

1 FOLIO, from the Latin *folium* a *leaf*.
2 QUARTO, from the Latin *quartus*, the *fourth*.
3 OCTAVO, from the Latin *octavus*, the *eighth*.
4 DUODECIMO, Latin *duodecimus*, the *twelfth*.
5 An 18mo. from *octavus*, L. *eighth*, and *decimus*, L. *tenth*, both making *eighteen*.

One	$12\frac{1}{2}$ cent piece, or $12\frac{1}{2}$ cents=	$\frac{1}{8}$ of a dollar.
Two	$12\frac{1}{2}$ cent pieces, or 25 cents=	$\frac{1}{4}$ of a dollar.
Three . . .	$12\frac{1}{2}$ cent pieces, or $37\frac{1}{2}$ cents=	$\frac{3}{8}$ of a dollar
Four	$12\frac{1}{2}$ cent pieces, or 50 cents=	$\frac{1}{2}$ of a dollar.
Five	$12\frac{1}{2}$ cent pieces, or $62\frac{1}{2}$ cents=	$\frac{5}{8}$ of a dollar.
Six , ...	$12\frac{1}{2}$ cent pieces, or 75 cents=	$\frac{3}{4}$ of a dollar.
Seven . . .	$12\frac{1}{2}$ cent pieces, or $87\frac{1}{2}$ cents=	$\frac{7}{8}$ of a dollar.
Eight . . .	$12\frac{1}{2}$ cent pieces, or 100 cents=	ONE DOLLAR.

QUESTIONS.

87. What is meant by an aliquot part? 81. Why is 5 an aliquot part of 15? 81.

88. What aliquot part of a dollar is $6\frac{1}{4}$ cents? is 50 cents? is 25 cents? $12\frac{1}{2}$ cents? Why is each an aliquot part?

89. Why is $\frac{2}{16}$ equal to $\frac{1}{8}$? See 85.

90. Why is $\frac{4}{16}$ or $\frac{2}{8}$ equal to $\frac{1}{4}$? *A.* $\frac{4}{16}$ is equal to $\frac{1}{4}$, because it means 4 of 16 equal parts, which is $\frac{1}{4}$ of them: and $\frac{2}{8}$ is equal to $\frac{1}{4}$, because it means 2 of 8 equal parts, which is also $\frac{1}{4}$ of them.

91. How many cents in $\frac{2}{16}$ of a dollar? in $\frac{1}{8}$? in $\frac{3}{16}$? in $\frac{3}{8}$? in $\frac{5}{16}$ in $\frac{5}{8}$? in $\frac{15}{16}$? in $\frac{7}{8}$? in $\frac{3}{4}$? Repeat the Table.

92. What will 4 knives cost, at $6\frac{1}{4}$ cents apiece? At $12\frac{1}{2}$ cents apiece? At 25 cents? At 50 cents?

93. How many yards of cloth may be bought for 1 dollar, at 50 cents a yard? At 25 cents a yard? At $12\frac{1}{2}$ cents? At $6\frac{1}{4}$ cents?

94. How many yards then would 5 dollars buy at 50 cents a yard? At 25 cents? At $12\frac{1}{2}$ cents? At $6\frac{1}{4}$ cents?

95. Mary, having purchased 29 yards of ribbon for $6\frac{1}{4}$ cents a yard, gave the merchant a 2 dollar bill; how much change ought she to have received?

MISCELLANEOUS QUESTIONS.

VIII. 1. A man bought 12 bushels of wheat for 30 dollars, and sold 9 bushels for 25 dollars. How many bushels had he left, and what did they cost him?

2. Suppose a stage goes 120 miles in 12 hours, and a railroad car 3 times as fast, how far does each go in an hour? How far would they be apart in 2 hours after they had started?

3. How many times 12 in 27? $2\frac{3}{12}$ times 12 are how many? How many times 11 in 30? $2\frac{8}{11}$ times 11 are how many?

4. How many inches in 8 nails? yards in 10 rods? cents in 5 dollars? shillings in 2 pounds 2 shillings?

5. When a coat costs 2 pounds 2 shillings in London, how many guineas, at 21 shillings each, will pay for it?

6. If 1 quart of rum is enough to make one man act like a fool, how many at that rate may be made fools by 10 gallons?

7. What would a narrow strip of board only 4 inches wide and 3 feet or 36 inches long cost, at 2 cents a square foot? How many square rods make 1 acre? [4 times 40.]

8. When a piece of land is 40 rods long, how wide must it be to make 1 acre? *A.* 4 rods, [for 40 in 160 = 4 times.] How wide to make 2 acres? When a piece is 80 rods long, how wide must it be to make 1 acre?—to make 2 acres?

9. How many solid feet in a block whose three dimensions, viz 'ength, breadth and thickness, are each 2 inches? How many when each dimension is 3 inches? is 4 inches? is 5 inches?

10. How many cord feet in a small pile of wood, 4 feet long, 2 feet wide and 2 feet deep? How many in a pile 8 feet long, 2 feet wide and 2 feet high? In one 8 feet long, 4 feet wide and 4 feet high?

11. If a man earns 1 dollar a day, how much will he earn in the awful days for labor in 1 week? in $1\frac{2}{6}$ week? in $3\frac{2}{6}$ weeks?

12. At 1 dollar a day, how many weeks' labor may be hired for 13 dollars? [$2\frac{1}{6}$.] for 22 dollars? for 20 dollars? for 32 dollars?

13. What will 16 pounds of butter cost at $12\frac{1}{2}$ cents a pound? What will 24 pounds cost? 25 pounds cost?

14. What is the cost of 4 bushels of potatoes at 25 cents a bushel? of 20 bushels? 40 bushels? 41 bushels?

15. When 3 bushels of dried apples sell for 6 dollars, what are they a bushel? What would 2 bushels cost? 8 cost? 20 cost?

16. If 5 yards of broadcloth sell for 20 dollars, what will 8 yards sell for at the same rate? Find the price of 1 yard first. What would be the price of 9 yards? 11 yards? 12 yards? 20 yards?

17. If 8 pounds of sugar cost 1 dollar, what will 3 pounds cost? How many pounds may be bought for 4 dollars? for 5 dollars? for 10 dollars?

18. When 4 bushels of oats cost 2 dollars, how many bushels may be bought for 2 dollars 50 cents? for 3 dollars 50 cents? for 5 dollars? for 8 dollars 50 cents?

19. How many wholes are $\frac{1}{8}$, $\frac{5}{8}$, $\frac{7}{8}$ and $\frac{3}{8}$? $\frac{2}{10}$, $\frac{3}{10}$, $\frac{4}{10}$, $\frac{5}{10}$ and $\frac{6}{10}$? $\frac{1}{3}$, $\frac{2}{3}$ and $\frac{3}{3}$? $\frac{7}{8}$ and $\frac{2}{8}$? [$1\frac{1}{8}$] $\frac{1}{5}$, $\frac{2}{5}$, $\frac{3}{5}$, $\frac{4}{5}$, $\frac{5}{5}$ and $\frac{6}{5}$?

20. When a floor is 4 feet square, how many square feet does it contain? [4 times 4.]

21. What then is the square of 4? [16.] What the square of 5? [5 times 5.] What is the square of 6? of 7? of 8? of 10? of 20?

22. The 4, before it is multiplied by itself, is called the square root of 16;[1] what then is the square root of 25, and why?

A. 5, because 5 times 5 are 25

23. What is the square root of 4? of 9? 16? of 36? of 64? of 100? of 144? of 400? of 10000.

---

1 ROOT. That part of a plant which penetrates the ground and supports the plant· the first ancestors; the original cause of any thing. *To take root* means to become firmly fixed. The square root of any number is so called because it is the first number that is repeated or multiplied

24. In a square room which is calculated to accommodate 100 boys, how many must sit on a single bench?

25. Suppose 400 scholars should wish to form themselves into a solid phalanx,[1] or square body, how many must stand in each rank[2] and file?[3] *A.* 20.

26. What is the cubic or solid content of a regular cube 10 inches long, 10 inches wide, and 10 inches thick?

27. What is the 10 and 1000 each called? *A.* The 10 is called the cube root of the 1000, and the 1000 the cube of 10.

28. What then is the cube of 2, and why? *A.* 8, because 2 times 2 are 4, and 2 times 4 are 8.

29. What is the cube of 3? What is the cube of 4?

30. What is the cube root of 8, and why? *A* 2, because 2 times 2 are 4, and 2 times 4 are 8.

31. What is the cube root of 27? of 125? of 1000?

32. What is the length of each side of a cubical block which contains 1000 solid or cubic inches? What is the cube of 10?

33. 6 is $\frac{1}{2}$ of what number? 10 is $\frac{1}{3}$ of what number?

34. 12 is $\frac{1}{4}$ of what number? 11 is $\frac{1}{5}$ of what number?

35. What is that number $\frac{1}{6}$ of which is 9? $\frac{1}{7}$ of which is 8? $\frac{1}{9}$ of which is 10? $\frac{1}{10}$ of which is 12? $\frac{1}{12}$ of which is 11?

36. What number is that $\frac{2}{3}$ of which is 10? $\frac{1}{3}$ must be 5. *A.* 15.

37. What number is that $\frac{3}{4}$ of which is 24? Find $\frac{1}{4}$ first. *A.* 32.

38. What number is that $\frac{2}{50}$ of which is 16?—$\frac{3}{7}$ of which is 30? $\frac{7}{8}$ of which is 63? $\frac{5}{9}$ of which is 60? $\frac{6}{7}$ of which is 72?

39. If $\frac{2}{5}$ of a barrel of flour cost 4 dollars; what will $\frac{1}{5}$ of a barrel cost? What will the whole barrel cost [$\frac{5}{5}$]?

40. A man bought $\frac{3}{7}$ of a load of hay for 6 dollars, what was the whole load worth at that rate? Find the value of $\frac{1}{7}$ first.

41. Henry's age is 14 years, which is $\frac{7}{8}$ of his brother's age. How old is his brother? Find how much $\frac{1}{8}$ is first.

42. A man lost 15 dollars, which was $\frac{3}{11}$ of all the money he had; how much had he? How much had he left?

43. A man, who owed a certain sum of money paid 12 dollars, which was $\frac{3}{12}$ of the debt; how much remained unpaid?

44. A man, who lent a certain sum of money, could collect only 8 dollars, which was $\frac{4}{5}$ of it; how much did he lose?

45. If a man, having a quantity of flour on hand, sells 20 barrels, which is $\frac{2}{3}$ of it; how much will he have left?

46. Suppose a man sells $\frac{7}{8}$ of a barrel of flour, for 14 dollars, what will the remainder of the barrel bring at that rate?

---

1 PHALANX. A square battalion or body of soldiers, formed in ranks and files close and deep; any body of troops or men formed in close array, or any combination of people distinguished for their intrepidity and union.

2 RANK. A row or line; men standing side by side in a line; a line of things; degree class; order; degree of dignity.

3 FILE. A thread, string or line; a bundle of papers tied together with the title to each indorsed; a roll, list, or catalogue; a row of soldiers ranged one behind another from front to rear

47. A boy having a stick, broke it into two parts, one of which was 2 feet long, or $\frac{1}{5}$ of the length of both parts; what was the length of the stick before it was broken?

48. If $\frac{1}{5}$ and $\frac{3}{5}$ of a number are 16; what is tha number?

NOTE—Say $\frac{1}{5}$ and $\frac{3}{5}$ are $\frac{4}{5}$ which is 16; then $\frac{1}{5}$ is $\frac{1}{4}$ of 16, which is 4, and $\frac{5}{5}$ is 5 times 4, which is 20. *A.* 20.

49. If $\frac{2}{8}$ and $\frac{3}{8}$ of a number are 10; what is that number?

50 If $\frac{5}{9}$ and $\frac{2}{9}$ of a number are 70; what is that number?

51. There is a pole erected so that $\frac{1}{5}$ stands in the mud, $\frac{2}{5}$ in the water, and the rest which is 10 feet above water; what is the entire length of the pole?

52. There is a pole, $\frac{3}{7}$ above water, $\frac{2}{7}$ in the water; and 8 feet in the mud; what is the entire length of the pole?

53. Four men A, B, C, and D purchased a sloop together. A took $\frac{1}{8}$ of it, B. $\frac{2}{8}$, C. $\frac{4}{8}$ and D, the rest, which cost him 100 dollars. What was D's part? What did the other parts severally cost? What did the whole sloop cost?

54. The fifth part of an army was killed; $\frac{3}{5}$ of it taken prisoners, and 1000 fled; how many were there in the army? How many were killed? How many were taken prisoners?

55. In an orchard of fruit trees, $\frac{3}{6}$ of them bear apples, $\frac{2}{6}$ bear plums: 8 bear peaches, and 2 bear cherries: how many trees of each sort are there in the orchard? How many trees does the orchard contain? 8 trees and 2 trees are 10 trees which are $\frac{1}{6}$.

56. In a certain school $\frac{4}{7}$ of the pupils study Arithmetic, $\frac{2}{7}$ study Grammar and 10 only read and spell: what is the number of scholars in the school? What is the number in Arithmetic? What is the number in Grammar?

57. A man having 16 oranges would divide them so that his own son Samuel may have 4 more than his neighbor's son George; How many must he give to each?

NOTE—Give 4 to Samuel first then divide the rest equally between the two? *A.* George 6, and Samuel 10.

58. A gentleman bought a horse and carriage for 240 dollars, paying 40 dollars more for the horse than for the carriage; what did each cost?

59. A man and a boy were both hired for 20 dollars a month, the man receiving 4 dollars a month more than the boy; what would the wages of each amount to in a year?

60. A man, woman, and boy were hired a week for 21 dollars; the woman to receive 5 dollars more than the boy, and the man 5 dollars more than the woman; at that rate what would the wages of each amount to in one month.

3

ARITHMETIC.[1]

PART SECOND.

AS CONSISTING BOTH IN THEORY[2] AND PRACTICE[3]

QUANTITY AND NUMBER.

IX. 1. QUANTITY[4] is any thing that may be increased or diminished; as, a sum of money, a line, weight.

2. A QUANTITY is ascertained to be great or small, much or little only in comparison with a known quantity of the same kind, which is either greater or smaller.

3. For example, ten thousand hogsheads of water is a great quantity, compared with one gill of water, but quite a small quantity, compared with the water in the ocean.

4. A UNIT,[5] which represents[6] a single thing; as, 1 hat, 1 ounce, &c. is fixed upon as the criterion[7] or known quantity by which to measure all other quantities of that kind.

5. Thus 2 would express a quantity 2 times as great as 1, that is 2 units; 3, 3 times as great, or 3 units, and so on.

6. QUANTITIES then of every kind are properly expressed by NUMBERS; as 5 bushels of rye, 5 oranges, &c.

7. A CONCRETE[8] NUMBER has reference to some particular object or objects; as, 1 man, 2 dollars, 3 benches.

8. AN ABSTRACT[9] NUMBER has no reference to any object whatever; as 1, 2, 3.

IX. *Q.* What is Quantity? 1. How is a quantity ascertained to be great or small? 2. Give an example? 3. What is the criterion for estimating different quantities? 4. Illustrate it? 5. How are quantities expressed? 6. What is a Concrete Number? 7. An Abstract Number? 8.

---

1 ARITHMETIC, [G. *Arithmetike*.] Reckoning by numbers; calculating.
2 THEORY, [F. *theorie*. L. *theoria*.] Speculation; a system, plan, scheme; opposed to practice.
3 PRACTICE, [F. *pratique*.] Habit, use, dexterity, method.
4 QUANTITY, [*Quantitas*.] Any thing that may be increased or diminished; bigness; bulk, weight; measure.
5 UNIT, [L *unus*.] One; a word denoting a single thing.
6 REPRESENT. To show; to exhibit; to describe.
7 CRITERION, [G. *kriterion*.] A standard of judging; a distinguishing mark.
8 CONCRETE, [L. *concretus*.] United in one mass; a compound; a term involving both the thing and its quality; as, a *white fence*; 2 mellons; 1 cent.
9 ABSTRACT, [L. *abstractus*.] Separate· distinct; expressing only quality or number; as, *whiteness*; 1, 2, 3. &c

9. DENOMINATION[1] is a name given to units or things of the same sort or class; as, 4 dollars, 5 oranges, 10 pigeons.

10. A SIMPLE NUMBER is composed of units of the same value or denomination.

11. A COMPOUND[2] NUMBER is composed of two or more simple numbers of different denominations, but of the same genus,[3] kind, or general class.

12. Thus 2 pounds is a simple number, so is 5 shillings; but 2 pounds 5 shillings taken together, forms a compound number, for it has one denomination of pounds, another of shillings; but both are of the same kind, or general class, viz. money.

13. ARITHMETIC treats[5] of numbers: as a SCIENCE,[6] it explains their properties; and as an ART,[7] it teaches the method of computing[8] by them.

14. ARITHMETIC has five principal rules for its operation, viz. Numeration, Addition, Substraction, Multiplication, and Division, which are often called the *fundamental*[9] or *ground rules* of Arithmetic, because they are the foundation of all the other rules.[10]

Q. What is meant by denomination? 9. What is a Simple Number? 10. Compound Number? 11. Give an example of a compound number. See 12. What is Arithmetic? 13. When is it regarded as a Science? 13. When as an Art? 13. How many rules has it for its operations? 14. What general name have these rules, and why? 14.

---

1 DENOMINATION, [L. *denomino.*] The act of naming, a name; a vocal sound, a class, sort, or name of a species.

2 COMPOUND. Composed of two or more ingredients; united in one.

3 GENUS, [L. *genus.*] A general name for several species;[4] a class of greater extent than species; thus animal is a genus; embracing a great variety of species; as man, horse, beast, bird, reptiles, &c.

4 SPECIES. A kind, sort, class; a subdivision of a general sum called a *genus*, thus, things that resemble each other in several particulars form a *species*; when several species are compared together and we observe several particulars common to the whole, they form a *genus*; a species then is one class of a genus.

5 TREAT, [F. *traiter.*] To handle; to use; to discourse on; to entertain without expense; to negociate; to manage in the application of remedies.

6 SCIENCE, [L. *Scientia.*] Knowledge, a system comprising the theory and reasons without any practical application; and therefore stands opposed to Art.

7 ART, [L. *art.*] Human skill; a system of rules; skill; dexterity.

8 COMPUTING. Counting; numbering; reckoning, estimating.

9. FUNDAMENTAL, [L. *fundament, seat.*] Relating to the foundation or basis.

10 Addition alone may not inappropriately be styled the sole or fundamental rule of Arithmetic, for all the other rules are easily resolvable (11) into this single one.

10 Thus 4 in 20, 5 times, because 5 times 4 are 20. Division then involves the principle of Multiplication.

10 Again, 4 times 5 are 20, because 5 and 5 and 5 and 5, that is 5 added 4 times, makes 20. Hence Multiplication may be performed by Addition.

10 Subtraction too is virtually performed by Addition, for 5 from 20 leaves 15 only because 15 and 5 are 20.

11 RESOLVABLE, [L. *resolvo.*] That may be reduced to first principles.

NUMERATION.[4]

X. 1. There are three methods, as we have seen, by which numbers are represented; viz. by *words*, by *single letters*, and by characters usually termed *figures*[1] and sometimes *digits*.[2]

2. In the method by letters; which is called the Roman method because the Romans invented[5] it; are employed seven letters only, viz., I, V, X, L, C, D, M.

3. This method possesses some advantages over that by figures; but it has become nearly obsolete,[6] being confined principally to the numbering of chapters, hymns, &c.

4. The method by figures; which is called the Arabic method, because the Arabs invented it; is, taken as a whole, by far the shortest and best method ever devised.[7]

5. In this method are employed ten figures only; viz.

1 ,	2 ,	3 ,	4 ,	5 ,	6 ,	7 ,	8 ,	9 ,	0 ;
one	two	three	four	five	six	seven	eight.	nine	cipher

6. The first nine figures, which have each an absolute[8] value, are called *significant*[9] *figures*, to distinguish them from the *cipher;* which has no value in itself, being used merely to fill a vacant[10] place. The cipher is also called *naught* or *zero*.[3]

7. By variously combining[11] these ten characters, no number can be conceived of too great to be represented by them, as will appear by the sequel.[12]

X. *Q.* How are numbers represented? 1. Describe the Roman method? 2. Is it still used and to what extent? 3. Which is the best method? 4. Describe it? 4. What characters does it employ? 5. Of what use is the cipher? 6. What other names has it? 6. What names have the other characters and why? 6. Can a large number be represented by so few characters? 7.

---

1 Figures were introduced into Spain, by the Arabs, in the 8th century (13) and from Spain into England about the middle of the 11th century; most eight hundred years ago. On the continent their use had become quite extensive: they are now so common, that if you were to visit China, for instance, you would recognize (14) at once their numerals, (15) without understanding a word of their language.

2 DIGIT, [L. *digitus, a finger.*] The measure of a finger's breadth, or the fourth of an inch. Figures were so called trom counting the fingers in reckoning.

3 The character 0 is called a cipher, from the Arabic word *tsphara*, which signifies a *blank* or *void*. The uses of this character in numeration are so important, that its name *cipher*, has been extended to the whole art of Arithmetic, which has been called to *cipher*, meaning *to work with figures*.

4 NUMERATION, [L. *numeratio.*] Numbering; the act of numbering,

5 INVENTED, [L. *inventus.*] Found out; devised; contrived; forged.

6 OBSOLETE, [L. *obsoletus.*] Gone into disuse; disused, neglected.

7 DEVISED, [F. *deviser.*] Given by will; bequeathed; contrived; invented.

8 ABSOLUTE, [L. *absolutus.*] Complete; positive; unconditional; independent

9 SIGNIFICANT, [L. *significans.*] Having meaning; expressive; important.

10 VACANT, [F. *from L. vacans.*] Empty; not filled; exhausted of air; unoccupied

11 COMBINING, [F. *combiner.*] Uniting closely; joining; confederating in purpose

12 SEQUEL, [F. *sequelle.*] That which follows; consequence; event.

13 CENTURY, [L. *centuria.*] A period of one hundred years.

14 RECOGNIZE, [L. *recognitio.*] To know again; to revise.

15 Numerals, [L. *numeratio.*] Characters used for representing numbers

8. The UNIT, which occupies the lowest place in the scale of whole numbers, means a single thing; that is, *one;* as, 1 hat, 1 boy.

9. The TEN, which means 10 units, is the least number that is formed by the union[1] of two single characters, to wit: by annexing[2] the cipher to the figure 1, thus, 10; twenty, thus, 20; thirty, thus, 30, &c.

10. The HUNDRED, which is *ten* times 10 units, is formed by annexing[2] two ciphers to the figure 1, thus, 100; two hundred, thus, 200; three hundred, thus, 300, &c.

11. The THOUSAND, which means *ten* times 100 units, is formed by annexing three ciphers to the figure 1, thus, 1000; two thousand, thus, 2000; three thousand, thus, 3000, &c.

12. The TEN THOUSAND, which is *ten* times 1000 units, is written thus, 10000; one hundred thousand, thus, 100000, &c.

13. In these examples, every additional[3] cipher increases the value of the figure 1, ten times, by removing it one place further towards the left.

14. When a cipher or ciphers occur[4] on the extreme[5] left of other figures, they possess no value, as, 01; or 001, or 0001, each of which examples means simply 1.

15. *In general, the removal of any figure one place further towards the left, enhances*[6] *its value* TEN TIMES.

16. Thus in 1111, the first figure on the right means 1 unit; the next, on the left 10 times 1 unit, or 10; the next, 10 times 10 units or 100; the next, 10 times 100 units, or 1000; all making one thousand one hundred and eleven.

17. *Hence numbers increase from the right to the left in a* TENFOLD[7] *proportion,*[8] *as in the following Table.*

18. NUMERATION TABLE I.

10 units	make 1 ten.
10 tens	make 1 hundred.
10 hundreds	make 1 thousand.
10 thousands	make 1 ten thousand.
10 ten thousands	make 1 hundred thousand
10 hundred thousands	make 1 million.

Q. What is meant by the unit? 8. How is the ten formed? 9. How, the hundred? 10. How, the thousand? 11. How, the ten thousand, and so on? 12. What effect has the cipher in these examples? 13. What does the figure 1 with either one, or two, or three ciphers prefixed represent? 14. What is the effect of removing a figure to the left? 15. What does each figure 1 in 1111 mean? 16 What is the law of increase? 17.

1 UNION, [L. *unus, one.*] Forming into one; bond; affection; concord.
2 ANNEXING, [L. *annexus.*] Uniting at the end; placing after something
3 ADDITIONAL, [L. *additio.*] That which is added, or which increases.
4 OCCUR, [L. *occurro.*] Meet; come to the mind; appear; meet the eye.
5 EXTREME, [L. *extremus, the last.*] Outermost; fatherest; most pressing
6 ENHANCES, Raises; lifts; advances; increases; heightens; aggravates.
7 TENFOLD. [*ten* and *fold.*] Ten times more in degree or extent.
8 PROPORTION, [L. *proportio.*] Equal degree or equal rate; symmetry

19. Suppose a curious old miser to have laid up several bags of dollars containing the following sums, viz. 1 dollar, 10 dollars, 100 dollars, and 10000 dollars.

20. Then 1 bag of 1 dollar would represent 1 unit; 1 bag of 10 dollars, 1 ten; 1 bag of 100 dollars, 1 hundred; 1 bag of 1000 dollars, 1 thousand; and 1 bag of 10000 dollars, 1 ten thousand.

21. As the second bag and all the succeeding ones are each but a single collection, or but one thing, it may properly be called a unit, as well as the bag that contains but 1 dollar.

22. Hence, a series,[1] or a progressive[2] order[3] of units may be established in which each succeeding[4] one shall be ten times the value of a former one.

23. Simple units may then be denominated[5] the first order, tens, the second order; hundreds, the third order, and so on.

24. Thus in 4689, the 9 is 9 units of the first order; the 8, 8 units of the second order; the 6, 6 units of the third order; the 4, 4 units of the fourth order.

25. We see also that the value of figures depends on the places they occupy.

26. When 2 and 5, for instance, stand separately, they mean simply 2 units and 5 units; but placed together, they may mean either 25 units or 52 units.

27. The value of a figure standing alone, is called its *simple value;* when combined with other figures, its *local*[6] *value.*

28. To express two thousand three hundred and forty-five, we write them as follows, viz.

2 Thousands	3 Hundreds	4 Tens	5 Units

Write the 2 in the Thousands' place; the 3 in the Hundreds' place; the 4 in the Tens' place; and the 5 in the Units' place. This is called Notation.

29. To ascertain if we have correctly written the number, begin on the right and say; units, tens, hundreds, thousands; then begin on the left and read,

Q. Repeat the Table in which 10 units make 1 ten, &c.? 18. How many units are there in 2 tens? in 5 tens? tens in 50 units? in 100 units? in 89 units? [8 tens and 9 units.] tens in 95 units? in 105 units? [10 and 5 units.] tens in 165 units? hundreds, tens and units in 165 units? [1 hun. 6 tens and 8 units.] in 456 units? units in 4 hundreds 5 tens and 6 units? What is meant by a series of units? 22. Give an example? 20, 21. What constitute the several orders? 23. In 4689, for instance, point out the different orders? 24. On what does the value of a figure depend? 25. In expressing 2345, by figures, what places would each figure occupy? 28. How is it ascertained if it be correctly written? 29. What number will 1, 2, and 3 represent, taken together in the same order as they stand? *A.* One hundred and twenty-three. What number will 2, 3, 4, and 5 represent, taken in like manner?

1 SERIES. [L. *series.*] A regular succession of things; course; order
2 PROGRESSIVE. Going forward, advancing or increasing gradually.
3 ORDER, [L. *ordo.* F. *ordre.*] Method; a mandate; rule, rank, class.
4 SUCCEEDING. Following in order; following in the place of another
5 DENOMINATED, [L. *denominatus.*] Named, called; styled.
6 LOCAL. [L. *locus, a place.*] Of or belonging to a place.

giving to each figure the name of the place against which it stands; thus 2 thousand 3 hundred and 45; which we find corresponds with the given number. This is called Numeration.

30. Write in words on the slate, the following numbers:—

1 0	1 0 0	1 0 0 0
1 5	1 2 5	3 4 5 6
2 5	5 2 1	6 5 4 3
8 9	8 9 1	8 7 5 2

31. It is customary to separate large numbers by a comma, into parts or portions called periods of three figures each, beginning on the right.

32. The first period; as it contains units, tens of units, hundreds of units, is called the period of Units.

33. The next left hand period, for a similar reason, is called the period of Thousands, and so on as in the following.

34. NUMERATION TABLE II.

Period of BILLIONS.			Period of MILLIONS.			Period of THOUSANDS.			Period of UNITS.			
Hundred Billions.	Ten Billions.	BILLIONS.	Hundred Millions.	Ten Millions.	MILLIONS.	Hundred Thousands.	Ten Thousands.	THOUSANDS.	Hundreds.	Tens.	UNITS.	
1	0	0	0	0	0	0	0	0	0	0	0	*read* 100 billion
	2	0	0	0	0	0	0	0	0	0	0	*read* 20 billion.
		3	0	0	0	0	0	0	0	0	0	*read* 3 billion.
			4	0	0	0	0	0	0	0	0	*read* 400 million.
				5	0	0	0	0	0	0	0	*read* 50 million.
					6	0	0	0	0	0	0	*read* 6 million.
						7	0	0	0	0	0	*read* 700 thousand.
							8	0	0	0	0	*read* 80 thousand.
								9	0	0	0	*read* 9 thousand.
									9	0	0	*read* 9 hundred.
										9	0	*read* ninety.
1	2	3	4	5	6	7	8	9	9	9	9	*read* one hundred twenty

three billion, four hundred fifty-six million, seven hundred eighty nine thousand, nine hundred and ninety-nine.

Q. What are periods of figures? 31. What are the first, second, third, &c. periods called? 32. 33. Repeat the Numeration Table II; as, units, tens. hundreds, &c., as far as hundred billions? 34. What figures in the Table represent ninety? Nine hundred? Nine thousand? Eighty thousand? Seven hundred thousand? Six million? Fifty million? Four hundred million? Three [illegible] Twenty billion? One hundred billion?

35. Write in words on the slate the following numbers.

36.	9 0 , 0 0 0	44.	9 0 0 , 0 0 0 , 0 0 0
37.	7 8 , 6 4 3	45.	7 6 2 , 3 4 5 , 2 7 1
38.	2 0 0 , 0 0 0	46.	8 , 0 0 0 , 0 0 0 , 0 0 0
39.	5 4 3 , 7 8 2	47.	7 , 3 6 2 , 4 9 1 , 7 2 3
40.	3 , 0 0 0 , 0 0 0	48.	5 0 , 0 0 0 , 0 0 0 , 0 0 0
41.	2 , 4 6 5 , 7 8 9	49.	4 9 , 4 1 6 , 5 3 1 , 2 7 2
42.	4 0 , 0 0 0 , 0 0 0	50.	8 0 0 , 0 0 0 , 0 0 0 , 0 0 0
43.	2 3 , 4 5 6 , 8 9 2	51.	9 8 7 , 3 6 5 , 2 1 4 , 7 1 8

52. 8 , 0 0 8 , 0 0 8 , 0 0 8 *read* 8 billion, 8 million, &c.

53. 7 0 , 0 7 0 , 0 7 0 , 0 7 0 *read* 70 billion, 70 million, &c.

54. 8 0 0 , 8 0 0 , 8 0 0 , 8 0 0 *read* 800 bill'n, 800 mill'n, &c.

55. 6 , 0 0 0 , 6 0 0 , 0 0 0 *read* 6 bill'n, and 600 thousand.

56. 9 0 0 , 0 0 0 , 0 0 0 , 0 0 9 *read* 900 billion and 9.

57. By canceling[1] all the ciphers in the last example the 900 billion and 9 becomes only 99.

58. Hence be careful to fill all vacant plans with ciphers.*

RECAPITULATION.

59. NOTATION is the writing of numbers in figures ; NUMERATION, the reading of them expressed in figures.

RULE FOR WRITING NUMBERS.

60. *Begin on the left and write each figure according to its value in the Numeration Table, taking care to supply all vacant places with ciphers.*

Q. What caution is given in respect to vacant places? 58. Give an example of its importance? 57. What is Notation? 59. Numeration? 59. What is the rule for writing numbers ? 60.

* The practice of reckoning only three figures to a period, is derived from the French. The English reckon six figures for a period, which would carry the millions' place in the above Table, into the billions' place. One billion then, by the French mode, expresses a number one thousand times smaller than by the English method; which, as you may perceive, greatly diminishes the power of figures. But as the former is most convenient it generally has the preference.

ENGLISH METHOD OF NUMERATING.

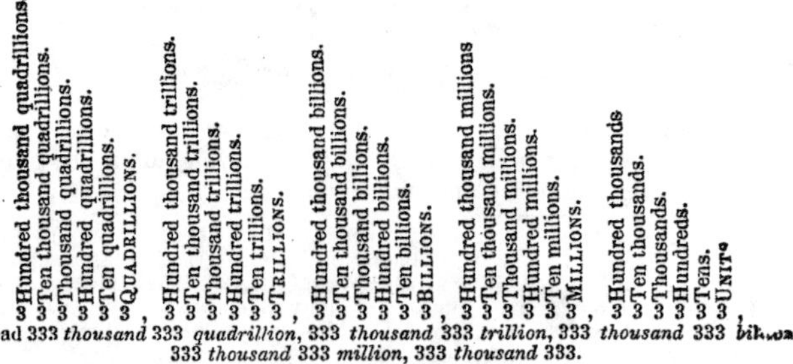

Read 333 *thousand* 333 *quadrillion*, 333 *thousand* 333 *trillion*, 333 *thousand* 333 *billion* 333 *thousand* 333 *million*, 333 *thousand* 333.

1 CANCELING, [F. *canceller.*] Crossing ; obliterating, annulling.

PROOF, OR RULE FOR READING NUMBERS.

61. *Begin on the right and numerate by saying units, tens, hundreds, &c.; then begin on the left and read, joining the name of its place to each figure; which, if it correspond with the given number, is correctly written.*

NUMERATION TABLE III.

Hundred Octillions.	Ten Octillions.	OCTILLIONS.[14]	Hundred Septillions.	Ten Septillions.	SEPTILLIONS.[14]	Hundred Sextillions.	Ten Sextillions.	SEXTILLIONS.[14]	Hundred Quintillions.	Ten Quintillions.	QUINTILLIONS.[14]	Hundred Quadrillions.	Ten Quadrillions.	QUADRILLIONS.[14]	Hundred Trillions.	Ten Trillions.	TRILLIONS.[14]	Hundred Billions.	Ten Billions.	BILLIONS.[14]	Hundred Millions.	Ten Millions.	MILLIONS.[13]	Hundred Thousands.	Ten Thousands.	THOUSANDS.[13]	Hundreds.[11]	Tens.[9]—[10]	UNITS.[8]
5	5	5,	5	5	5,	5	5	5,	5	5	5,	5	5	5,	5	5	5,	5	5	5,	5	5	5,	5	5	5,	5	5	5

62. Read 555 *octillion*, 555 *septillion*, 555 *sextillion*, 555 *quintillion*, 555 *quadrillion*, 555 *trillion*, 555 *billion*, 555 *million*, 555 *thousand*, 555.

63. Write in figures on the slate, the following numbers
64. Ninety-seven.
65. Four hundred and twenty-five.
66. Three thousand and five.
67. Forty-nine thousand five hundred and twenty.
68. Six hundred and fifty-two thousand five hundred.
69. Eight million nine hundred and forty thousand.
70. One hundred and one.
71. Five thousand and five.
72. Four thousand two hundred and eight.

Q. What for reading numbers? 61. Repeat the Numeration Table III. How are thirty 5s in succession read? How would thirty 3s be read?

---

1 PRIMITIVE. Original, not derived from any thing; primary.
2 PREFIXING. Uniting at the beginning; placing before.
3 NUMERALS. Of or belonging to number; consisting of numbers.
4 TERMINATION. Limiting; bounding; ending; end of a word.
5 MODIFICATIONS. Changing the forms; altering the appearance.
6 EUPHONY, [G. *eu, good*, and *phone, sound*.] An agreeable sound.
7 PREFIX. A letter, syllable, or word, put at the beginning of a word.
8 ONE, TWO, THREE, and up to TWELVE, are reckoned primitive (1) words.
9 THIRTEEN, FOURTEEN; THEE and TEN, FOUR and TEN, &c.
10 TWENTY, THIRTY, &c. are derived from TWO TENS, THREE TENS, &c
11 THE HUNDRED is from the Latin *hun* or *hundred*.
12 THE THOUSAND is derived from the Saxon *thousand*. This and the two preceding umerals (3) are usually considered as primitive in our language
13 THE MILLION is derived from the French *million*.
14 THE BILLION, TRILLION, QUADRILLION, &c. are formed by prefixing (2) the Latin numerals to the termination (4) *illion*, with such slight modifications (5) as euphony (6) requires. The Latin prefixes (7) are *bis, twice; tres, three; quartuor, for; quinque, five; sex, six; septem, seven; octo, eight; novem, nine; decem, ten; undecim, eleven, duodecim, twelve; tredecim, thirteen*, &c. These prefixes, with *illion*, make *Billion, Trillion, Quadrillion, Quintillion, Sextillion, Septillion, Octillion, Nonillion, Undecillion, Duodecillion, Tredecillion*

73. Three hundred thousand five hundred.
74. Six million one hundred thousand.
75. Four million four thousand and forty-nine.
76. Seventeen million one hundred and twenty-five.
77. One billion, one million, one thousand and one.
78. Five hundred and twenty-one billion, three hundred million, three hundred thousand and one.
79. Five trillion, five billion, five million, five thousand and five hundred and fifty-five.
80. Six quadrillion, six hundred million, four hundred and fifty-nine thousand and sixteen.
81. Two hundred and fifty quintillion, six quadrillion, two billion, three hundred and forty thousand.†

Figures on the slate are written thus,—

1 2 3 4 5 6 7 8 9 0

SIMPLE ADDITION.[1]

XI. 1. Add together 9 dollars, 7 dollars, 5 dollars, 8 dollars, 6 dollars, 4 dollars, and 3 dollars, thus;

(1.)	(2.)	(3.)	(4.)	(5.)	(6.)
9 dollars.	8 tons.	3 cents.	8 ounces.	5 mills.	8 hats.
7 dollars.	5 tons.	4 cents.	9 ounces.	8 mills.	9 hats.
5 dollars.	6 tons.	9 cents.	7 ounces.	3 mills.	7 hats.
8 dollars.	9 tons.	7 cents.	8 ounces.	9 mills.	8 hats.
6 dollars.	7 tons.	8 cents.	5 ounces.	6 mills.	6 hats.
4 dollars.	3 tons.	9 cents.	9 ounces.	7 mills.	7 hats
3 dollars.	4 tons.	5 cents.	6 ounces.	8 mills.	9 hats.
**A.* 42 dollars.	*	*	*	*	*

XI. *Q.* How much is 33 and 9? What is the 42 called? See 7.

---

† *Remarks to the Learner.* As very high numbers are somewhat difficult to apprehend; it may not be amiss to illustrate, by a few examples the value of the words *million, billion, trillion,* and *quadrillion,* according to the English notation.

Suppose that a person employed in telling money, reckons a hundred pieces in a minute, and continues to do so twelve hours each day, he will take nearly fourteen days to reckon a *million.* A thousand men would take more than 38 years to reckon a *billion.*

The inhabitants of the United States in 1820, were about 10 million. Now if we suppose all these persons had been constantly employed in counting money since the birth of Christ, they could not as yet have reckoned a *trillion.*

Though we admit the earth, from the creation, to have been as populous as it is at present (being about 800 million,) and the whole human race to have been counting money without intermission; they could scarcely, as yet, have reckoned the five hundredth part of a *quadrillion* of pieces.

1 ADDITION, [L. *additio.*] Any thing added; adding; joining; uniting two or more numbers in one sum.

* (1). *A.* 42 dollars. (2). *A.* 42 tons. (3.) *A* 45 cents. (4.) *A* 52 ounces. (5.) *A* 46 mills. (6.) *A.* 54 hats

7. The Answer in adding is called the *sum* or *amount.*

8. A drover bought 5 cows at one time, 8 at another, 9 at another. How many cows did he buy in all?

		PROOF. Having added upwards as before; which makes 22, draw a line under the 5 at top; then add downwards all the figures under the 5, thus, 8 and 9 are 17; Then if the 17 and the 5 at top make 22 as they do, the work is right.
	5 cows.	
	8 cows.	
	9 cows.	
A.	22 cows.	
	17 cows.	
P.	22 cows.	

In like manner perform and prove the following examples. 8

(9.)	(10.)	(11.)	(12.)	(13.)
8 horses.	3 oxen.	9 calves.	8 shillings.	7 pence
9 horses.	2 oxen.	8 calves.	9 shillings.	5 pence
7 horses.	5 oxen.	7 calves.	8 shillings.	9 pence
6 horses.	6 oxen.	6 calves.	9 shillings.	4 pence.
7 horses.	8 oxen.	5 calves.	3 shillings.	2 pence.
8 horses.	9 oxen.	4 calves,	2 shillings.	4 pence.
4 horses.	7 oxen.	3 calves.	5 shillings.	3 pence
5 horses.	4 oxen.	2 calves.	4 shillings.	2 pence.
6 horses.	9 oxen.	9 calves.	6 shillings.	9 pence.
5 horses.	8 oxen.	8 calves.	7 shillings.	6 pence

14. Since 1 ten and 1 unit, for instance, make neither 2 tens, nor 2 units, (although 1 and 1 are 2); but 1 ten, which is 10 units, added to 1 unit will make 11 units, thus keeping the 1 ten in its proper place therefore :—

15. *Write units under units, tens under tens, hundreds under hundreds, &c., then add each column separately.*

(16.)	(17.)	(18.)	(19.)
3 2 pints	3 1 2 4	2	1 9 2 2 2 2 2 2
3 5 pints	4 1 1	5 2 4 3	4 1 3 2 1 1
2 1 pints.	2 2 3	1 2 2	2 4 3 4
1 pint.	1 1	2 1 1	2 1

20. *Ciphers are passed over in adding, because they are used to fill vacant places. See* x. 6.

(21.)	(22.)	(23.)	(24.)
4 0 0	4 0 0 0	5 1 0 0 4 0	4 0 0 3 0 0 5 0 0 5
9 0 0	3 0 0	3 0 6 0 5 4	4 2 0 1 0 0
1 0 3	2 0 0 0	6 0 0	6 0 0 1 0 0 0 2 0 1
2 0 4	4 0 0 1	9 0 0 3 0 0	5 1 9 0 0 0 2 0 9 1

25. A man bought a farm for 4,000 dollars; he paid for his cows 405 dollars; for his horses 320 dollars; for his farming utensils 60

Q. What do 1 ten and 1 unit make added together? See 14. Why not 2 tens or 2 units? How then should units, tens, &c., be written and added? 15. How are ciphers to be regarded in adding? 20

dollars; and his expenses for securing his title were **3 dollars.** What was the amount of the whole? *A.* 4,788 dollars.

26. Add into one sum forty, two hundred, three hundred and nineteen, and nine hundred and forty. *A.* 1499.

27. What is the amount of one thousand, thirty-three thousand three hundred and sixty-one, five hundred thousand and ten, five million and five thousand? *A.* 5,539,371.

28. What is the sum of five, fifty, five hundred, five thousand, fifty thousand, five hundred thousand, five million, fifty million, five hundred million, and five billion? *A.* Ten 5s.

29. Add together 8 trillion, 800 billion, 80 billion, 8 billion, 800 million, 80 million, 8 million, 800 thousand, 80 thousand, 8 thousand, 8 hundred, 80 and 8. *A.* Thirteen 8s.

30. Find the sum of 3 trillion, 3 billion, 3 million, 3 thousand and 3; 20 billion, 20 million, 20,020; 200 million; 200 thousand and 200; 16 million, 16 thousand and 16. *A.* 3,023,239,239,239.

31. What is the amount of 4 million and six, 300 thousand two hundred, 90 thousand three hundred and one, 4 thousand two hundred and ten, 1 hundred and 70, eleven and 1? *A.* 4,394,899.

32. A cashier has in one drawer 5,305 dollars, in another 406 dollars, in another 7,312, in another 2,309, and in another 42. What amount has he in all the drawers?

```
     5 3 0 5
       4 0 6
     7 3 1 2
     2 3 0 9
         4 2
   ---------
A. 1 5 3 7 4
```

The first column makes 24 units or 2 tens and 4 units; write down only the 4 units and add in the 2 tens with the column of tens, which is called *carrying 1 for every ten.* The 2 to carry to 4 tens makes 6 and 1 are 7 tens, and none to carry. The next column makes 13 hundreds, or 1 thousand and 3 hundreds; carry the 1 thousand to the thousands; therefore;—

33. When the amount of any single column is 10 or more:—*Write down only the right hand figure, and carry the left hand figure or figures to the next column.*

(34.)	(35.)	(36.)	(37.)
3906	4207	4128	2634
4827	9353	201	209
3339	408	7095	3215
7408	7819	4320	81
9512	208	9090	9
A. 28992			

38. From the above it appears that we begin on the right and carry 1 for every 10; *because figures increase from the right to the left in a tenfold proportion.*

Q. When, in adding several columns, the right-hand column makes 24, for instance; what is to be done with it? 32. What is this process called? 32. What is the direction for carrying? 33. What is the reason for beginning and carrying in this manner? 38. How many do you carry when the sum is 59? 115?

39. Add together 28,992; 21,995; 24,834 and 5,148. *A.* 80,969.

40. A man bought a suit of clothes for 57 dollars, a pair of boots for 8 dollars, and a secretary for 28 dollars. What is the amount of the whole? *A.* 93 dollars.

41. In an orchard, 20 trees bear pears, 54 bear peaches, and 6 bear plums. How many trees are there in the orchard? *A.* 80 trees.

42. A man bought a barrel of flour for 10 dollars, a barrel of molasses for 29 dollars, and a barrel of pork for 19 dollars. What did the whole cost him? *A.* 58 dollars.

43. What is the sum of eighty-seven, two hundred and seventeen, eight thousand nine hundred and eighty-six, and nine? *A.* 9,299.

44. Find the sum of eight thousand and twenty, four hundred and seventy-nine, thirty thousand and sixty-five. *A.* 38,564.

45. General George Washington was born A. D.[1] 1732, and lived 67 years. In what year did he die? *A.* 1799.

RECAPITULATION.

46. ADDITION is the uniting of two or more numbers in one number, which is called their SUM OR AMOUNT.

47. SIMPLE ADDITION is the adding of numbers of the same denomination.

RULE.

48. *Write the numbers in columns, so that units may be added to units, tens to tens, hundreds to hundreds, &c.*

49. *Begin on the right and place underneath each column its whole amount, unless it be 10 or more; in which case set down the right-hand figure only, and carry the left-hand one to the next column*

50. *Do the same with each column to the last, under which write the whole sum.*

51. PROOF. Omit the top line, and find the sum of all the rest, adding downwards; if the numbers were before added upwards and *vice versa.*[2] Then if this amount added to the top line, corresponds with the first amount, the work is supposed to be right.

52. Or, more practically, add the figures downwards without omitting the top line, and if the two amounts agree they will probably be right.

53. The rule as well as the proof is based[3] on the well known axiom,[4] *that the whole is equal to the sum of all its parts.*

Q. What is Addition? 46. Simple Addition? 47. Rule? 48, 49, 50. What are the two methods of proof? 51. 52. What is the reason for both? 53.

---

1 A. D. The A. stands for *anno*, L. for *year;* and D. for *Domini*, L. *of our Lord.* Hence *Anno Domini* means, *in the year of our Lord;* and A. D. 1732 means so many years since Christ, or our Saviour came on earth.

2 VICE VERSA. That is, upwards if the figures were before added downwards Generally *vice versa*, from the Latin, means *the terms being exchanged;* thus, the generous should be rich and *vice versa*, that is, the rich should be generous.

3 BASED, [L. *basis, foundation.*] Founded; reduced in value.

4 AXIOM, [G. *axioma.*] A self-evident truth: that which is so plain, that no proof can make it any plainer

(54.)	(55.)	(56.)	(57.)	(58.)
35	313	1645	13213	456732
64	280	321	24512	121212
21	741	4610	52108	784503
18	240	5380	60389	908762
12	391	5210	78978	736542

59. Add into one sum 150; 1,965; 17,172; 229,200, and 3,007,751. *A.* 3,256,238.

60. A father gave to his oldest son 4,200 dollars, to his second 2,300 dollars, and to his youngest 1,560 dollars. What is the amount of these several sums? *A.* 8,060 dollars.

(61.)	(62.)
987654321	876542103459652
98765432	1300000000
9876543	3210123456789
987654	9999
98765	88885555
9876	444444444444444
987	32107520
98	786743512
9	815
1	39767

63. Find the sum of 1 billion, 97 million, 393 thousand, 686, and 1 quadrillion 324 trillion, 198 billion, 879 million, 148 thousand, and 53. *A.* 1,324,199,976,541,739.

64. A gentleman purchased a ship for 25,000 dollars, and sold it for 3,715 dollars more than it cost him. What did he get for it? *A.* 28,715 dollars.

65. A gentleman sold a tract of wild land for 13,000 dollars, which was 1,750 dollars less than it cost him. What did he give for it? *A.* 14,750 dollars.

66. A merchant had a store-house in which he had at one time 6,000 bushels of corn, 5,756 bushels of wheat, 1,375 bushels of rye, 8,750 bushels of oats, and had room enough left for 2,000 bushels more of corn. How many bushels would the store-house hold? *A.* 23,881 bushels.

67. Two men started from New York and traveled in opposite directions. The one was to go 37 miles a day, and the other 35 miles. How far would they be apart the first night? *A.* 72 miles.

68. How far were they apart the 2d night? *A.* 144 miles.

69. How far were they apart the 3d night? *A.* 216 miles.

70. How far were they apart the 4th night? *A.* 288 miles.

71. How far were they apart the 5th night? *A.* 360 miles.

72. How far were they apart the 6th night? *A.* 432 miles.

SIMPLE SUBTRACTION.[1]

XII. 1 A man owing 9 dollars paid 3 dollars. How many dollars did he still owe?

(1.)	(2.)	(3.)	(4.)
From 9 dollars	3 8	7 3 9	9 4 6 7
Take 3 dollars	1 5	4 3 2	7 1 3 7
* *A.* 6 dollars.	*	*	*

5. If we take 6 miles from 27 miles, we have 21 miles left; because 21 and 6 are 27: thus,—

	(6.)	(7.)	(8.)	(9.)
27 miles.	1 4 5	4 5 9 2	9 8 6 4	4 9 2 5 7
6 miles.	2 1	3 5 1	2 3 5 1	6 1 1 2
A. 21 miles.				
P. 27 miles.				

10. Observe that like Addition, *units must be placed under units tens under tens, hundreds under hundreds, &c.*

11. From 6235 dollars take 2111 dollars? *A.* 4124 dollars.

12. From 89659 bushels take 7216 bushels? *A.* 82443 bushels.

13. A merchant having 13,069 bags of coffee sold 9,020 bags; how many bags had he remaining on hand?

Minuend,[2]	1 3 0 6 9	Say 0 from 9 leaves 9; 2 from
Subtrahend,[3]	9 0 2 0	6 leaves 4; 0 from 0 leaves 0;
Remainder,	4 0 4 9	9 from 13 leaves 4.
Proof.	1 3 0 6 9	*A.* 4049 bags.

14. In like manner perform and prove the following examples,—

(15.)	(16.)	(17.)
18350000	79527890000	16656213708
9300000	3121800000	9604000501

18. Two brothers engaged in mercantile business; one gained 1,584 dollars and the other 920 dollars. How much did one gain more than the other? *A.* 628 dollars.

19. A father gave 36,540 dollars to his son, and 25,000 to his daughter. How much greater was his son's portion than the daughter's? *A.* 11,540 dollars.

20. Subtract 13 million from 27 million. *A.* 14,000,000.

XII. *Q.* How are units, tens, &c., to be placed in subtraction? 10. Why does 6 from 27 leave 21? 5. How do you subtract 0 from 9? 13. 0 from 0? 13. 9 from 13? 13. In subtracting 9,020 from 13,069 which number is the minuend? 13. Which is the subtrahend? 13. What is the answer called? 13

1 SUBTRACTION, [L. *subtractio.*] The taking a part from the rest
2 MINUEND, [L. *minuendus.*] To be diminished or lessened
3 SUBTRAHEND, from *subtrahendus* L. to be taken from.
* (1.) *A* 6 dollars. (2.) *A.* 23. (3.) 307 (4.) *A* 2330

21. Subtract 85 thousand from 96 thousand *A.* 11,000.

22. What is the difference between one million eight hundred thousand, and six hundred thousand? *A.* 1,200,000.

23. A man bought a chaise for 262 dollars, and a harness for 39 dollars. How much did one cost more than the other?

```
    2 6 2
      3 9
A.  2 2 3
P.  2 6 2
```

We can't take 9 from 2, but we can take 1 ten, which is 10 units, from 6 tens. The 10 units (borrowed) added to the 2 units make 12. Then 9 from 12 leaves 3. Now instead of calling the 6 a 5, we may add the 1 ten, which we borrowed, to the 3 tens, the next lower figure, since it can make no difference in the result, for 4 from 6 is the same as 3 from 5. Therefore say, 1 to carry to 3 is 4, which from 6 leaves 2, &c.

24. Hence, when the lower figure is greater than the one above it, *add* 10 *to the upper figure, and only* 10 *in any case, then take the difference, being particularly careful to carry* 1 *to the next lower figure.*

	(25.)	(26.)	(27.)	(28.)	(29.)
From	3 5 6	7 8 3	5 2 3 7	1 4 6 5 7	1 5 7 8 2
Take	4 9	1 4 7	1 0 1 8	3 9 0 1	1 9 0
A.	3 0 7				

30. If we take 4,508 from 67,297, how many will remain?

	(31.)	(32.)	(33.)
	6 7 2 9 7	3 5 7 8 6 3 4	4 5 6 3 2 8 5 8
	4 5 0 8	2 7 1 0 9 1 6	2 9 0 4 1 9 3 9
A.	6 2 7 8 9		

33. A gentleman purchased a farm for 4,000 dollars, and paid 5 dollars as "earnest money."[1] How much did he still owe?

```
    4 0 0 0
          5
A.  3 9 9 5
```

Say 5 from 10 leaves 5; 1 to carry from 10 leaves 9; 1 to carry again from 10 leaves 9; 1 to carry from 4 leaves 3.

34. Since in the last example, there are no tens nor hundreds in the upper line, we do in reality borrow the 10 from the 4 thousands.

35. For if we call the 4 thousands 10 less, making 3990, and subtract the 5 without carrying any, the result will be the same as before.

36. *Hence the* 10 *is always borrowed from the first significant figure on the left in the upper line.*

37. From 10,000 cents take 1 cent. *A.* 9,999 cents.

38. From 10,000 mills take 9,999 mills. *A.* 1 mill.

Q. In taking 39 from 262, how do you subtract the 9 from 2? 23. Do you borrow the 10 units from the upper figure 6, or the lower figure 3? 23. To which figure do you pay it again, or carry it? 23. How happens it that you borrow from one and pay to another? 23. How then do you proceed in all such cases? 24. In subtracting 5 from 4000, for instance, from what is the 10 borrowed, and why? 34. From which figure is the 10 always borrowed? 36.

[1] EARNEST. Seriousness; a reality; first fruits; *money given to bind a bargain.*

39. How much does 110,001 exceed 99,999? *A.* 10002.
40. How much does 1,234,567 lack of 2 million? *A.* 765,433.
41. How much greater is 1 million than 1,000? *A.* 999,000.
42. How much smaller is 5,000 than 500,000? *A.* 495,000.
43. Find how much must be added to fifteen thousand and five to make twenty-three thousand. *A.* 7,995.
44. The deluge happened A. M.[1] 1656, and our Saviour was born A. M. 4,004. How many years intervened between these tw events? *A.* 2348 years.
45. How many years from the birth of our Saviour, to the discovery of America by Columbus, which occurred A. M. 5,496? *A.* A. D.[2] 1492.
46. Find by the following list, the length of time each President of the United States held that office.
47. George Washington, from 1789 to 1797. *A.* 8 years.
48. John Adams, from 1797 to 1801. *A.* 4 years.
49. Thomas Jefferson, from 1801 to 1809. *A.* 8 years.
50. James Madison, from 1809 to 1817. *A.* 8 years
51. James Monroe, from 1817 to 1825. *A.* 8 years.
52. John Quincy Adams, from 1825 to 1829. *A.* 4 years.
53. Andrew Jackson, from 1829 to 1837. *A.* 8 years.
54. Martin Van Buren, from 1837 to 1841. *A.* 4 years.
55. William Henry Harrison, from March 4th, 1841. *A.*

RECAPITULATION.

56. SUBTRACTION is the taking of a less number from a greater.
57. SUBTRACTION then is exactly the reverse of *Addition.*
58. The MINUEND is the greater number, and the one from which the subtraction is to be made.
59. The SUBTRAHEND is the smaller number, and the one which is to be subtracted.
60. The REMAINDER is the difference between any two numbers.
61. SIMPLE SUBTRACTION is the subtracting of one number from another, when both are of the *same denomination.*

RULE.

62. *Place the less number under the greater, so that units may stand under units, tens under tens, &c.*
63. *Begin on the right and take each figure separately, from the figure over it.*
64. *If the upper figure be too small, add* 10 *to it, then subtract, and carry* 1 *to the next lower figure.*

Q. What is Subtraction? 56. Does it resemble Addition? 57. What is the Minuend? 58. Subtrahend? 59. Remainder? 60. Simple Subtraction? 61 Rule? 62, 63, 64. Order? 65. Proof? 66. Reason of the rule? 67.

1 A. M. The A. stands for *Anno*, L., meaning, *in the year*, and the M. for *Mu[illegible]* *the world;* hence A. M. 1656, means so many years since the creation of the wo[illegible]
2 A. D. *in the year of our Lord.* See xi. 45.

65. ORDER. WRITE DOWN: SUBTRACT: AND CARRY.

66. PROOF. Add together the remainder, and subtrahend, and if their sum equal the minuend, the work is right.

67. The rule and proof proceed both on the same general principle, viz.; "*That the sum of the parts is equal to the whole.*"

(68.)	(69.)	(70.)
4 5 6 7 8 9 3 7	4 2 3 5 4 0 0 7	6 5 0 0 1 0 3 9
1 1 7 0 9 0 0 2	5 3 4 2 1 0 9	2 7 3 4 2 0

71. From one million take 999. *A.* 999,001.

72. From one million take nine. *A.* 999,991.

73. What number added to four hundred and fifty-nine thousand will make one million? *A.* 541,000.

(74.)	(75.)
5 0 0 3 2 7 8 0 0 9 1 0 1 7 3 0	6 8 1 2 3 4 5 6 7 0 8 0 9 0 8 0
4 9 1 7 5 0 9 0 8 0 3 9 0 9 5 0	1 0 0 0 9 0 0 9 0 0 9 0 0 9 9 0

76. Subtract 37,408 from 197,000. *A.* 159,592.

77. Suppose a man is worth 100,000 dollars lacking 1000 dollars How much is he worth? *A.* 99,000.

78. Subtract 376,000 from 9,567,000. *A.* 9,191,000

79. A father divided his estate, amounting to 75,000 dollars, between his son and daughter, giving to the former 37,560 dollars. How many dollars did he give to the daughter? *A.* 37,440 dollars.

80. A merchant bought goods worth 125,000 dollars, on four months credit, but is offered a discount[1] of 825 dollars for cash down. What would be the cost of the goods for cash? *A.* 124,175 dollars.

81. Suppose you buy goods worth 35,765 dollars on six months credit, and are offered the same goods for 35,565 dollars ready money.[2] What is the discount offered you? *A.* 200 dollars.

MISCELLANEOUS EXAMPLES.

	(1.)	(2.)
XIII.	Add One hundred,	Add Six hundred and twenty-seven,
	Two thousand,	Two thousand and fifteen.
	Two million,	Thirty thousand and seventy,
	Four million,	Forty-two million and eighty,
	Ten thousand,	Fifty-one thousand and nine,
	Ten million,	One billion one million and one,
	Forty million,	Seventeen trillion and seventeen
	A. 56,012,100	*A.* 17,001,043,083,819

1 DISCOUNT, Deduction; allowance for prompt pay
2 READY MONEY. Cash down; money paid down

(3.)	(4.)
From Ten million,	From One billion and fifty,
Take Forty-five.	Take One million and one.
A. 9,999,955.	*A.* 999,000,049.

5. A farmer having 500 dollars, paid away 200 dollars for a chaise and 100 dollars for a harness. How many dollars had he left? *A.* 200 dollars.

6. A gentleman has an annual income of 4,000 dollars; his family expenses are 1,500 dollars, and he contributes 500 dollars to charitable objects. How much can he lay up at the year's end? *A.* 2,000 dollars.

7. A merchant owing 2,345 dollars, paid 1,400 dollars at one time and 549 dollars at another. How much more will it require to pay the debt? *A.* 396 dollars.

8. A father divided his property, which amounted to 35,000 dollars, among his three children. He gave to the oldest 13,456 dollars; to the youngest 10,000 dollars. What was the other child's part? *A.* 11,544 dollars.

9. A man bought a house, a horse and a chaise for 3,700 dollars. He paid 115 dollars for the horse and 210 dollars for the chaise. What did the house cost him? *A.* 3,375.

10. Suppose the distance from Boston to New York be 210 miles; thence to Philadelphia 90 miles, and thence to Baltimore 100 miles, how many miles are Philadelphia and Baltimore from Boston? *A.* Pa. 300 miles; Be. 400 miles.

11. The population of the United States in 1820, was 9,638,166; in 1830 it was 12,858,670; and in 1840, it is supposed by some that it will not vary much from 18,000,000. At these estimates what was the increase of population from the first period of time to the second; from the second period to the third; and from the first period to the third? *A.* 3,220,504; 5,141,330; 8,361,834.

SIMPLE MULTIPLICATION.

XIV. 1. Suppose a man earns 9 dollars a week, for 5 successive weeks; how many dollars will he earn in all?

9 dollars.
9 dollars.
9 dollars.
9 dollars.
9 dollars.
A. 45 dollars.

	(1.)	
Multiplicand.	9 dollars.	
Multiplier.	5 times.	
Product.	45 dollars.	*A.*

XIV. *Q.* By what two methods can you find how many 5 times 9 are? 1. Which is the shortest method? 1. Multiplying 9 by 5 making 45; what are the numbers severally called? 1.

1 MULTIPLICATION, [L. *Multiplicatio.*] The act of multiplying or of increasing any number a given number of times

2. Suppose a farmer has 3 pastures with 502 sheep in each pasture, how many sheep will they all make?

(2.)	(3.)	(4.)	(5.)
5 0 2 sheep.	9 4 0 2	5 2 0 3 0 0 2	6 0 1 0 2 0 1 0 1
3 times.	2	3	4
1 5 0 6 sheep.	*	*	*

8	1 2
1 2	8
9 6	9 6

(6.) When two numbers are to be multiplied together, if either be made the multiplier, the product will be the same; but it is more convenient to make the smaller one the multiplicand.

7. *Recollect to write units under units, tens under tens, &c.*

8. Suppose a steamboat receive 702 dollars for every trip she makes; when she has made 3 trips, how many dollars will she have received? *A.* 2,106 dollars.

9. If a man earns 3 dollars a day, how much will he earn in a year, which contains 313 working days *A.* 939 dollars.

10. If 8,011 men could build a bridge in 9 days, how long would it require 1 man alone to do it, working at the same rate? *A.* 72,099 days.

11. If a man, traveling 9 hours a day, performs a journey in 210 days, how many hours is he in doing it? *A.* 1,890.

12. An agriculturist[1] sold 8,101 trees "of the genuine Chinese Morus[2] Multicaulis"[3] species,[4] for 8 cents a tree. What sum did he get for the whole? *A.* 64,808 cents.

13. There are 365 days in one year; how many days then are there in 5 years?

```
   3 6 5
       5
A. 1 8 2 5
```

Say 5 times 5 are 25; set down the 5 and carry 2, as in Addition. Next say 5 times 6 are 30 and 2 to carry are 32; write down the 2 and carry 3 as before, and so on.

14. *Hence observe the same rule for carrying as in Addition.*

15. How many are 2 times 9,313,654? *A.* 18,627,308.

16. How many are 3 times 8,252,180? *A.* 24,756,540.

17. How many are 4 times 3,008,309? *A.* 12,033,236.

18 How many are 5 times 8,090,876? *A.* 40,454,380.

Q. Which number is generally made the multiplier and why? 6. Give an example, See 6. How are the numbers to be written? 7. In multiplying 365 by 5 how do you commence? 13. How many do you carry for 25? See 13. What is done with the 2 "to carry," in multiplying 6 by 5? 13. What is the general direction? 14.

---

1 Agriculturist, from *ager*, L. *a field*, and *cultura*, L. *culture*. One who tills the ground.

2 Morus, [L. *morus*.] The mulberry-tree.

3 Multicaulis, from *multi*, L. *many*, and *caulis*, L. the *stalk or stem*. Hence *multicaulis* means *having many stalks or stems*.

4 Species, L. Form; outward shape; sort; kind.

* (3.) *A.* 18,804. (4.) *A.* 15,609,006. (5.) *A.* 2,404,080.404

19. How many are 6 times 7,130,028? *A.* 42,780,168.
20. How many are 7 times 6,087,695? *A.* 42,613,865
21. How many are 8 times 4,795,732? *A.* 38,365,856

RECAPITULATION.

22. MULTIPLICATION is the repeating of one number as many times as there are units in another number.

23. MULTIPLICATION is a compendious method of performing Addition, when the numbers to be added are equal.

24. The MULTIPLICAND is the number to be repeated.

25. The MULTIPLIER is the number by which we multiply.

26. The PRODUCT is the result obtained by multiplying.

27. The FACTORS[1] of any number are such numbers as will, by multiplying them together, produce it: thus, the Factors of 72 are 6 and 12; for 6 times 12 are 72;

28. In Multiplication, then, both the *Multiplier* and *Multiplicand* are properly called *Factors*.

29. SIMPLE MULTIPLICATION is the multiplying of one simple number by another.

XV. When the multiplier does not exceed 12.

RULE.

1. *Write the smaller number under the greater, so that units may stand under units, tens under tens, &c.*

2. *Begin at the right and multiply each figure of the multiplicand by the multiplier separately, carrying as in Addition.*

3. ORDER. WRITE DOWN: MULTIPLY: AND CARRY.

4. What is the product of 40,936 multiplied by 9?

```
4 0 9 3 6
        9
-----------
3 6 8 4 2 4
```

5. Multiply 52,031 by 10. *A.* 520,310.
6. Multiply 67,098 by 11. *A.* 738,078
7. Multiply 60,359 by 12. *A.* 724,308.

8. A grocer bought 121 firkins of butter for 8 dollars a firkin What did the whole cost him? *A.* 968 dollars.

9. There are 10 dollars in every eagle; how many dollars then are there in 36,089 eagles? *A.* 360,890 dollars.

10. The earth is 25,000 miles in circumference; how many furlongs will they make, reckoning 8 furlongs to a mile? *A.* 200,000.

11. A father distributed all his property equally among 11 sons. One son's part was, 23,042 dollars. What was the value of the father's estate? *A.* 253,462 dollars.

Q. What is Multiplication? 22. What rule does it abridge, and when? 23. What is the Multiplicand? 24. Multiplier? 25. Product? 26. Factors? 27. What are the factors of 72? Of 40? Of 48? Of 96? Which terms may properly be considered factors? 28. What is Simple Multiplication? 29.

XV. *Q.* When the multiplier does not exceed 12, how do you proceed? 1, 2. What is the order of proceeding? 3.

---

1 FACTOR, [L. *factor.*] An agent. Factors are so called because they act as *agents* in forming the *product.*

12. There are 12 months in a year. How many months has i been since the creation up to A. D. 1840, making in all A. M. 5,844 years? *A.* 70,128 months.

XVI. When the multiplier exceeds 12.

1. Suppose a drover has bought 469 cows for 36 dollars apiece, what must he have paid for the whole?

```
  4 6 9
    3 6
  -------
  2 8 1 4
1 4 0 7
---------
1 6 8 8 4
```

We multiply by the 6 as before, also by the 3; but observe that the 7 is placed in the tens' place, because the 3 by being 3 tens, is ten times greater than if it were 3 units. Lastly, add 2,814 and 1,407 together as they stand, thus, 4 is 4; 7 and 1 are 8, &c.; for it takes both the product of 6 units and 3 tens or 30, to make the total product of 36. *A.* 16,884 dollars.

2. *Observe that the first figure in each product is placed directly under the figure by which you are multiplying.*

3. What is the cost of 675 barrels of pork at 23 dollars a barrel? *A.* 15,525 dollars.

4. What will 37 horses cost at 125 dollars apiece? *A.* 4,625.

5. In one year there are 365 days; how many days then are there in 4,167 years?

```
      4 1 6 7
        3 6 5
    ---------
    2 0 8 3 5
  2 5 0 0 2
1 2 5 0 1
-------------
A. 1 5 2 0 9 5 5
```

6. Multiply 5,426 by 423. *A.* 2,295,198.
7. Multiply 9,132 by 239. *A.* 2,182,548.
8. Multiply 6,799 by 425. *A.* 2,889,575.
9. Multiply 8,706 by 359. *A.* 3,125,454.
10. Multiply 9,508 by 698. *A.* 6,636,584.
11. Multiply 8,903 by 452. *A.* 4,024,156.

12. Suppose a certain army of 84,050 men are to receive their year's pay, being 153 dollars apiece; how much will they all receive? *A.* 12,859,650 dollars.

13. If 3,675 men will do a piece of work in 327 days, how many men will be required to do it in one day? *A.* 1,201,725 men.

GENERAL RULE.

14. *Multiply by each significant figure of the multiplier separately, placing the first figure in each product directly under the figure by which you are multiplying.*

15. *Then add together these partial[1] products in the same order as they stand; the amount will be the total product required.*

XVI. *Q.* In multiplying 469 by 36, how do you proceed with the 6? 1. How with the 3? 1. Why is the first figure in the second product placed in the tens' place? 1. What is to be done with these partial products? 1. What is the direction for placing each product, when there are several figures in the multiplier? 2. What is the General Rule? 14, 15.

---

1 PARTIAL, [L. *pars.*] Biassed to one party; affecting a part only; not total.

16. Multiply 8 5 6 7 8 9
by 1 3 6 0 5

```
           8 5 6 7 8 9
        by   1 3 6 0 5
        -------------
           4 2 8 3 9 4 5
       5 1 4 0 7 3 4
     2 5 7 0 3 6 7
     8 5 6 7 8 9
A. 1 1 6 5 6 6 1 4 3 4 5
```

17. Suppose 3,756,701 to be a multiplicand, and 34,005 a multiplier; what will be the product? *A.* 127,746,617,505.

18. When the multiplier is 6,035 and the multiplicand 732,006; what will be the product? *A.* 4,417,656,210.

19. There are 63 gallons in one hogshead; how many gallons then must there be in 4,086 hogsheads? *A.* 257,418 gallons.

20. There are 1,728 solid inches in 1 solid foot; how many solid inches are there in 3,056 solid feet? *A.* 5,280,768 solid inches.

21. In 7 pieces of cloth, each containing 25 yards, how many yards? *A.* 175 yards.

22. What will be the cost of 7 pieces of cloth, each piece containing 25 yards, at 12 dollars per yard? *A.* 2,100 dollars.

23. Suppose that a certain field has 475 hills of potatoes, and that each hill will average 11 potatoes; how many potatoes at that rate would a field 275 times as large contain? *A.* 1,436,875 potatoes.

24. Suppose a father can hoe in one day 3 times as many hills of corn as his son, who hoed 520 hills; how many hills could the father hoe in 212 days? *A.* 330,720 hills.

25. If 25 men dig a ditch in 375 days, how long will it take one man alone to dig 45 such ditches? *A.* 421,875 days.

26. There are 6 days in the week for labor and study; suppose you get 4 lessons a day for 1 week, how many at that rate could you get in one year of 52 weeks? *A.* 1,248 lessons.

27. Suppose there are 50 baskets in a row, each containing 157 ears of corn, and that each ear has 8 rows, and each row 20 kernels; how many kernels will the baskets contain in all? *A.* 1,256,000.

XVII. To multiply by a composite[1] number.

1. A COMPOSITE NUMBER *is one that is produced by multiplying any two smaller numbers together.*

2. Thus 6 times 7 are 42. Here 42 is a composite number, and the 6 and 7 its factors, or as they are sometimes called, its *component*[2] *parts.*

3. In the example, 5 times 1 are 5, the 5 is not a composite number, because there are no two smaller numbers that, multiplied together, will produce it.

4. A COMPOSITE NUMBER may have two, three, or more factors for 3 times 4 are 12, and 8 times 12 are 96. Here 96 is compose of the factors 3, 4 and 8.

XVII. *Q.* What is a composite number? 1. Give an example. 2. Why are 1, 2, 3, 5, &c. not composite numbers? 3. Does a composite number ever have more than one factor? 4. Give an example. 4.

---

1 COMPOSITE, [L. *compositus.*] Composed; formed from several others.
2 COMPONENT. *Literally*, setting or placing together; forming; composing

5. The finding what two or more numbers multiplied together will make a given number, is called *resolving*[1] *it into factors.*

6. What will 24 carriages cost at 375 dollars apiece.

```
 3 7 5
     4
-------
1 5 0 0
     6
-------
9 0 0 0
```

The first product is the cost of 4 carriages; then 6 times that product must be the cost of 24 carriages, for 4 times 6 are 24. *A.* 9000 dollars.

7. In the last example, any other factors would answer the same purpose; as, 2 and 12, or 3 and 8, or 2, 3, and 4.

NOTE. When composite numbers do not exceed 144, their factors are easily found by the Multiplication Table.

RULE.

8. *Resolve the composite number into two or more factors, and multiply by them successively.*[2]

9. Multiply 670,032 by 18. *A.* 12,060,576.
10. Multiply 724,081 by 27. *A.* 19,550,187.
11. Multiply 634,081 by 35. *A.* 22,192,835.
12. Multiply 754,038 by 96. *A.* 72,387,648.
13. Multiply 603,407 by 108. *A.* 65,167,956.
14. Multiply 708,936 by 144. *A.* 102,086,784.
15. Multiply 608 by 24, using three factors *A.* 14,592.
16. Multiply 37 by 120, using four factors. *A.* 4,440.

XVIII. To multiply by 10, 100, 1000, &c.

RULE.

1. *Annex to the multiplicand all the ciphers in the multiplier.* See x. 15.

2. Multiply 68,345 by 10. Annex 1 cipher. *A.* 68,3450.
3. Multiply 3,45 by 100. Annex 2 ciphers. *A.* 34,500
4. Multiply 678 by 1000. Annex 3 ciphers. *A.* 678,000.
5. Suppose 53,467 to be a multiplicand, and 10,000 a multiplier, what will be the product? *A.* 534,670,000.
6. Multiply 63,456 by 10. *A.*
7. Multiply 38,065 by 100. *A.*
8. Multiply 65,721 by 1000. *A.*
9. Multiply 37,561 by 10,000. *A.*
10. Multiply 91,509 by 100,000. *A.*
11. Find the sum of the last 5 answers. *A.* 9,596,672,060.
12. There are 10 mills in 1 cent, 10 cents in 1 dime, 10 dimes in 1 dollar, and 10 dollars in 1 eagle; how many mills then are there in 678,345 eagles? *A.* 6,783,450,000 mills.

Q. What is the finding of factors called? 5. How do you multiply by the factors of 24? 6. What is the Rule? 8.

XVIII. *Q.* How can you multiply by 10, or 100, or 1000 easily? Why? See x. 15.

---

1 RESOLVING. Separating into component parts; analyzing; determining
2 SUCCESSIVELY. One after another regularly.

XIX. When there are ciphers on the right of either or both of the factors.

1. All such terms are resolvable[1] into two factors, one of which will always be either 10, or 100, or 1000, &c.

2. Thus, the factors of 890 are 89 and 10; of 3500, they are 35 and 100.

3. Instead, however, of proceeding as in XVII., we can abbreviate[2] the process by XVIII. as follows.

RULE.

4. *Multiply only the significant figures together, and annex to the product all the ciphers on the right of both factors.*

(5.)

```
    3 6 3 0
      2 3 0 0 0
    ---------
    1 0 8 9
    7 2 6
  -------------
  8 3 4 9 0 0 0 0
```

6. There are 100 years in a century, how many years then in 32 centuries? *A.* 3,200 years.

7. There are 365 days in 1 year, how many days then in 3,200 years? *A.* 1,168,000.

8. There are 24 hours in 1 day; how many hours then in 2,168,000 days? *A.* 52,032,000 hours.

9. There are 60 minutes in 1 hour; how many minutes then in 52,032,000 hours? *A.* 3,121,920,000 minutes.

10. There are 60 seconds in 1 minute; how many seconds then in 3,121,920,000 minutes? *A.* 187,315,200,000 seconds.

11. Suppose a canon ball flies 1 mile in 8 seconds, how long would it be, at that rate, in flying round the earth, it being 25,000 miles? *A.* 200,000 seconds.

MISCELLANEOUS EXAMPLES.

XX. 1. What is the whole number of inhabitants in the world, there being, according to Hassel, in each grand division as follows

Europe, one hundred and eighty millions;
Asia, three hundred and eighty millions;
Africa, ninety-nine millions;
America, twenty-one millions;
Australasia, &c. two millions? *A.* 682,000,000.

2. If one man in a factory earns 375 dollars a year, how many dollars will 345 men earn at that rate in the same time? *A.* 129,375.

3. A father deceased, left an estate of 40,000 dollars to his 3 sons and his widow. He directed, 9,750 dollars to be paid to each son. What was the widow's part? *A.* 10,750 dollars.

XIX. *Q.* When any number has one or more ciphers on the right; what may always be one of its factors? 1. What are the factors of 890? 2. Of 3500? 2. What is the direction for multiplying by such numbers? 4.

---

1 RESOLVABLE. That may be resolved or reduced to first principles.
2 ABBREVIATE. To shorten; to make shorter by contracting the parts; to reduce to a smaller compass.

4. A gentleman paid for the building of his house by contract, 7,650 dollars; for the site 4,500 dollars; he was worth 50,000 dollars when he began to build; how much had he left? *A.* 37,850 dollars.

5. A farmer bought 769 sheep for 3 dollars apiece. After keeping them 3 years they doubled their number. What was their value at the price he paid for the original stock? *A.* 4,614 dollars.

6. A merchant bought a bale of goods consisting of 40 pieces, and each piece of 23 yards, at 7 dollars a yard. What did the bale cost him? *A.* 6,440 dollars.

7. A bookseller shipped for New Orleans 4 boxes of books, each box containing 400 books. Now suppose each book had 200 pages, and each page 45 lines, and each line 43 letters; how many letters then were there in the 4 boxes? *A.* 619,200,000 letters.

8. A speculator took out to the west 20,000 dollars with which he purchased 1,250 acres of land for 2 dollars an acre; 1,400 acres for 3 dollars an acre, and 3,000 acres for 4 dollars an acre. How much money had he left on his return? *A.* 1,300 dollars

---

SIMPLE DIVISION.[1]

XXI. 1. A man having 28 dollars, laid it all out in broadcloth, at 4 dollars a yard; how many yards could he buy?

2 8	
4	I.
2 4	
4	II.
2 0	
4	III.
1 6	
4	IV.
1 2	
4	V.
8	
4	VI.
4	
4	VII.
0	

2. For every 4 dollars he bought 1 yard; then he could buy as many yards as there are *fours* in 28.

3. Subtracting on the left 4 from 28 leaves 24: 4 from the remainder 24, leaves 20: and continuing to do so we find at last that nothing remains.

4. It appears by counting that 4 is subtracted 7 times; then there are 7 *fours* in 28, that is, 4 is contained in 28, just 7 times: consequently he could buy 7 yards; which is the answer.

5. The process however by subtraction would be, in most cases exceedingly tedious, compared with that by division.

(6.)

BY DIVISION.

Divisor.[2] 4) 2 8 *Dividend*[3].
7 *Quotient*[4].

(7.)

Say 4 in 28, 7 times; because 4 times 7 are 28: which is the PROOF.
A. 7 yards.

(8.)

PROOF.

7 *Quotient.*
4 *Divisor.*
28 *Dividend.*

---

1 DIVISION, [L. *divisio.*] The act of dividing or separating into parts; a part or distinct portion; a part of an army or fleet; disunion; discord.

2 DIVISOR, [L. *divisor.*] A dividing number, or that number which shows how many parts a given number is to be divided into.

3 DIVIDEND, [L. *dividendus.*] To be divided; a share of profits in banks, &c

4 QUOTIENT, [L. *quoties.*] The number showing how many parts any thing is divided into

9. In like manner perform and prove the following examples.

(10.)	(11.)	(12.)	(13.)	(14.)
6) 4 2	8) 5 6	9) 1 0 8	1 0) 1 1 0	1 1) 1 3 2
A. 7				

15. Divide 369 dollars equally among 3 men.

3) 3 6 9
A. 1 2 3

Say 3 in 3, 1 time; 3 in 6, 2 times; 3 in 9, 3 times writing each under that figure that contains it: for the 369 really means 300; 60; and 9: then 3 in 300, 100 times; 3 in 60; 20 times; 3 in 9, simply 3 times.

16. Hence, to preserve the different values of figures in the quotient, *each figure must be written under the figure that contains it.*

(17.)	(18.)	(19.)	(20.)
2) 1 8 0 6	4) 3 6 8 0 4	7) 4 9 0 7	8) 7 2 8 0 8
A. 9 0 3			

21. A farmer sold a quantity of wheat for 27,030 dollars, receiving at the rate of 3 dollars per bushel. How much wheat did he sell? *A.* 9010 bushels.

22. How long would it take 3 men, working at the same rate, to do what 1 man is 2,709 days about? *A.* 903 days.

23. How many times is 6 contained in 480,609? Thus,—

6) 4 8 0 6 0 9
A. 8 0 1 0 1 $\frac{3}{6}$
6
P. 4 8 0 6 0 9

The 3 *over* is properly called the Remainder,[1] under which is written the Divisor 6 See VI. 2. *Proof.* Say 6 times 1 are 6, &c., and 3 remaining are 9; 6 times 0 are 0, &c.

24. *Recollect to add in the remainder in proving the operation.*

(25.)	(26.)	(27.)
3) 2 7 0 3 7	4) 1 6 0 0 9	5) 3 0 0 5 0 0 5 4

28. How many piles of 4 dollars in each pile, will 21 dollars make? *A.* 5 piles and 1 dollar over or remaining.

29. If the dividend is dollars, the remainder will of course be dollars; if sheep, the remainder will be sheep, *it being always like the dividend.*

30. Suppose a divisor to be 3, the dividend to be 654, what will be the quotient?

XXI. *Q.* What are the two methods, for finding how many times one number is contained in another? 1, 6. Which is the shorter method? See 5 and 6. How many subtractions are necessary to find how many times 4 is contained in 28? See 4. How is the same result obtained by division? 6, 7. What are the 4; 28 and 7 severally called? 6. Which terms are multiplied together in the proof? 8. Which term in division is reproduced by the proof? See 6 and 8. How is each figure of the quotient to be written in dividing, and why? 16. What denomination is the Remainder? 29. What is to be done with it in proving? 23. When in dividing there are left 2 tens, for instance, and the next figure is 4, how do you proceed? 30. Why? 30.

1 REMAINDER, [L. *remaneo.*] That which is left after the separation into equal parts; relics; remains; the sum left after subtraction or any deduction.

```
3 ) 6 5 4
A.  2 1 8
        3
P.  6 5 4
```

Say 3 in 6, 2 times; 3 in 5, 1 time, and 2 tens over, which are joined to the 4, thus, 24; then 3 in 24, 8 times; for the 2 tens being 20 units make with the 4 units 24 units.

31. *Recollect then not to add what is over to the next figure, but* TO PREFIX[1] it.

32. How many times 4 in 96,877? A. 24,219$\frac{1}{4}$.
33. How many times 6 in 39,247? A. 6,541$\frac{1}{6}$.
34. How many times 7 in 79,189? A. 11,312$\frac{5}{7}$.
35. How many times 4 in 60,300?

```
4 ) 6 0 3 0 0
A.  1 5 0 7 5
```

Say 4 in 6, 1 time and 2 over; 4 in 20, 5 times; 4 in 3, 0 times and 3 over; 4 in 30, 7 times and 2 over; 4 in 20, 5 times.

36. How many times 8 in 90,246? A. 11,280$\frac{6}{8}$.
37. How many times 9 in 85,056? A. 9,450$\frac{6}{9}$.
38. How many times 8 in 33,001? A. 4,125$\frac{1}{8}$.

(39.)	(40.)	(41.)
1 0) 4 9 5 0 5	1 1) 8 9 0 4 7	1 2) 3 9 4 2 1

42. When 1 bushel of wheat will buy 12 bushels of potatoes; how many bushels of wheat will 3,705,602 bushels of potatoes buy? A. 308,800$\frac{2}{12}$ bushels.

RECAPITULATION.

43. DIVISION is finding how many times one number is contained in another.

44. DIVISION is exactly the reverse[2] of multiplication, and a concise method of performing many subtractions.

45. The DIVIDEND is the number to be divided.

46. The DIVISOR is the number to divide by.

47. The QUOTIENT shows the number of times the divisor is contained in the dividend.

48. The REMAINDER is always less than the divisor, and of the same denomination with the dividend.

49. Division may be performed by subtracting the divisor from the dividend as many times as there are units in the quotient.

50. Since the *divisor* and *quotient* multiplied together will reproduce[3] the *dividend*, therefore,—

51. The *divisor* and *quotient* correspond to the *factors* of Multiplication, and the *dividend* to the *product.*

Q. What direction is given in such cases? 31. What is Division? 43. What is said of Division in respect to Multiplication and Subtraction? 44. What is the Dividend? 45. Divisor? 46. Quotient? 47, Remainder? 48. How may Division be performed? 49.

1 PREFIX, [L. *prefigo.*] To put or fix before or at the beginning.
2 REVERSE, [L. *reversus.*] Change; vicissitude; a turn of affairs; misfortune; a contrary; opposite.
3 REPRODUCE, [*re* and *produco.*] To produce again; to renew.

52. HENCE THE PROOF. Multiply the divisor and quotient together, and add in the remainder; if this result be equal to the dividend the work is right.

53. SIMPLE DIVISION is the dividing of one simple number by another.

XXII. SHORT DIVISION is when the Divisor does not exceed 12.

RULE.

1. *Having written the divisor on the left, find how many times it is contained in the first figure, or figures of the dividend, and write it underneath, in the place of the quotient.*

2. *When any figure is too small, or when there is a remainder, prefix it to the next figure, then divide as before, supplying the vacant places of the quotient with ciphers.*

3. *Under the last remainder write the devisor, with a line between and annex it to the quotient.*

(4.)	(5.)	(6.)
2) 3 7 8 5 6 7 0	3) 3 7 8 5 6 7 0	4) 8 0 3 7 0 1 9
1 8 9 2 8 3 5		

7. A man purchased sheep at 4 dollars a head, and laid out 1,020 dollars. How many sheep did he buy? *A.* 255 sheep.

8. An Englishman, returning from a tour, found his whole expenses 1000 guineas, at an average expense of 3 guineas a day; how many days must he have spent in traveling? *A.* 333$\frac{1}{3}$ days.

9. Suppose a railroad car goes 7 miles while a footman is going 1 mile; how far will the footman go while the car is going 37,856 miles? *A.* 5,408 miles.

10. A father gave to his youngest son 8,506 dollars, which was 8 times the value of what the oldest had. What was the portion of the oldest son? *A.* 1,063$\frac{2}{8}$ dollars.

11. Suppose an apprentice is 9 times as long in performing the same piece of work as a journeyman; how long will it take the latter to accomplish what the former does in 1,460 days? *A.* 162$\frac{2}{9}$ days.

12. Suppose a man gives away 1 dollar as often as he makes 10 dollars; when he has accumulated[1] 608,350 dollars, how much will he have given away? *A.* 60,835 dollars.

13. Suppose the capital of a bank amounts to 5,000,000, dollars and that it is equally owned by 11 men; what is each one's interest in the bank? *A.* 454,545$\frac{5}{11}$ dollars.

Q. With what other terms of a different rule do the divisor and quotient correspond? 51. With what does the dividend correspond? 51. How do we ascertain this? 52. How many times 11 in 140? Which number is the dividend? Which the divisor? What is the quotient? What, the remainder? What is Simple Division? 53.

XXII. *Q.* What is Short Division? What is the Rule? 1, 2. How is the remainder to be expressed? 3. What is the Proof!

---

1 ACCUMULATED, [L. *accumulo.*] Collected into a heap or great quantity.

14. The circumference[1] of the earth is about 25,000 miles; what is $\frac{1}{2}$ of that distance? *A.* 12,500 miles.

15. What is $\frac{1}{3}$ of 25,000 miles? *A.* 8,333$\frac{1}{3}$ miles.

16. What is $\frac{1}{4}$ of 25,000 bushels? *A.* 6,250 bushels.

17. What is $\frac{1}{5}$ of 25,000 dollars? *A.* 5,000 dollars.

18. What is $\frac{1}{6}$ of 25,000 ounces? *A.* 4,166$\frac{4}{6}$ ounces.

19. What is $\frac{1}{7}$ of 25,000 pounds? *A.* 3,571$\frac{3}{7}$ pounds.

20. What is $\frac{1}{8}$ of 25,000 hours? *A.* 3,125 hours.

21. What is $\frac{1}{9}$ of 25,000 days? *A.* 2,777$\frac{7}{9}$ days.

22. What is $\frac{1}{10}$ of 25,000 drams? *A.* 2,500 drams.

23. What is $\frac{1}{11}$ of 25,000 inches? *A.* 2,272$\frac{8}{11}$ inches.

24. How much is $\frac{500}{7}$? Divide 500 by 7. *A.* 71$\frac{3}{7}$.

25. How much is $\frac{417}{8}$? *A.* 52$\frac{1}{8}$.

26. How much is $\frac{817}{9}$? *A.* 90$\frac{7}{9}$.

27. What is the value of $\frac{48759}{12}$? *A.* 4,063$\frac{3}{12}$.

28. What is the value of $\frac{39005}{11}$? *A.* 3,545$\frac{10}{11}$.

29. The salary of the President of the United States is 25,000 dollars a year; what is that a month, allowing 12 months to the year? *A.* 2,083$\frac{4}{12}$ dollars.

30. The President's salary then is about 2,083 dollars a month; what is it a week, allowing 4 weeks to a month? *A.* 520$\frac{3}{4}$ dollars.

31. Allow the President's salary to be just 520 dollars a week, what is it a day, there being 7 days in a week? *A.* 74$\frac{2}{7}$ dollars.

32. Division has hitherto been performed partly in the mind, and partly on the slate, and is called Short Division, to distinguish it from Long Division, in which the divisor is so large that we are obliged to write down the whole operation.

XXIII. Long Division is when the divisor exceeds 12.

1. *Recollect you are to proceed as in Short Division, excepting the entire work is to be written out.*

2. Divide 15 dollars equally among 13 poor persons.

```
                  Div.
Divisor, 1 3 ) 1 5 ( 1 Quo.
               1 3
               ---
    Remainder,   2
```

Proceeding as before, say 13 in 15, 1 time and 2 over, write the quotient figure 1, on the right of 15 to make room for 13 under the 15; then 13 from 15 leaves 2, as at first.

Proof. 13 times 1 are 13, and 2 are 15, the *dividend*. *A.* 1$\frac{2}{13}$ dollars.

3. Perform and prove the following examples in like manner.

(4.)	(5.)	(6.)	(7.)	(8.)
1 3) 1 6 (	1 4) 1 9 (	1 5) 2 2 (	1 6) 2 5 (	1 7) 3 3 (

XXIII. *Q.* What is Long Division? How does it differ from Short Division? XXII. 32. What is the first direction? 1. Where is the quotient written? 2 What is the proof that 13 is contained in 15, 1 time and 2 over?

[1] Circumference, [L. *circumferentia.*] The line that bounds a circle; distance round orbit; circle; any thing circular or orbicular.

9. *Recollect to place the divisor on the left; the quotient on the right; and having multiplied the divisor and quotient together, to place their product under the dividend, then subtract one from the other.*

10. How many times 27 in 40? *A.* $1\frac{13}{27}$

11. How many times 56 in 81? *A.* $1\frac{25}{56}$

12. How many times 75 in 96? *A.* $1\frac{21}{75}$.

13. At 39 dollars a month, how many months' labor can be procured for 75 dollars? *A.* $1\frac{36}{39}$ months

14. What is the quotient of 39 divided by 18?

```
18) 39 (2
    36
    --
     3
    ==
```

Here 18 is contained in 39; 2 times, for 3 times 18 are more than 39, and only 1 time 18 would leave a remainder larger than the divisor. See v. 14. *A.* $2\frac{3}{18}$ acres.

15. Divide 41 by 19. *A.* 2 and $\frac{3}{19}$ remainder.

16 Divide 70 by 16. *A.* 4 and $\frac{6}{16}$ remainder.

17. Divide 91 by 14. *A.* 6 and $\frac{7}{14}$ remainder.

18. At 13 dollars a barrel, how many barrels of flour can be bought for 99 dollars? *A.* $7\frac{8}{13}$ barrels.

19. At 25 dollars an acre, how much land would 103 dollars buy?

```
25) 103 (4
    100
    ---
      3
    ===
```

Here 25 is in 103, 4 times, for 5 times 25 are more than 103; and only 3 times 25 would leave a remainder larger than the divisor. *A.* $4\frac{3}{25}$ acres.

20. Hence if the divisor and the figure you put in the quotient make a product greater than the dividend; *rub out the work and place a smaller figure in the quotient.*

21. But if the remainder is as large, or larger than the divisor; *rub out the work and place a larger figure in the quotient.*

22. Divide 108 by 25. Remainder $\frac{8}{25}$.

23. Divide 125 by 30. Remainder $\frac{5}{30}$.

24. Divide 165 by 45. Remainder $\frac{30}{45}$.

25. Divide 250 by 67. Remainder $\frac{49}{67}$

26. Divide 510 by 95. Remainder $\frac{35}{95}$.

27. Divide 908 by 99. Remainder $\frac{17}{99}$.

28. Suppose 1,036 pounds of beef are to be shared equally among 125 soldiers, what is each man's portion?

```
125) 1036 (8
     1000
     ----
       36
     ====
```

We must always take figures enough at first to contain the divisor once at least even if it require a hundred, *A.* $8\frac{36}{125}$ pounds.

29. Divide 1,039 by 126. Remainder 31.

30. Divide 1,208 by 135. Remainder 128.

31. Divide 2,085 by 250. Remainder 85.

32. Divide 3,780 by 395. Remainder 225.

Q. How can you find what figure to place in the quotient? 20, 21.

33. Divide 6,901 by 941. Remainder 314.

34. How many building lots, at 950 dollars apiece, may be bought for 2,895 dollars? *A.* $3\frac{45}{950}$ lots.

35. Suppose a regiment contains 1624 men, how many companies of 203 men each would be required to make such a regiment? *A.* 8 companies.

36. Divide 10,835 by 2,083. Remainder 420.
37. Divide 26,008 by 5,041 Remainder 803.
38. Divide 59,246 by 7,085. Remainder 2,566.
39. Divide 56,738 by 20,301. Remainder 16,136.
40. Divide 98,304 by 10,605. Remainder 2,859.
41. Divide 90,090 by 20,303. Remainder 8,878.

42. A father divided an estate of 813,824 dollars equally among his sons, giving to each 203,456 dollars. How many sons had he? *A.* 4 sons.

43. When flour is 13 dollars a barrel, how many barrels will 329 dollars purchase?

```
13) 3 2 9 (2 5
    2 6
    ---
      6 9
      6 5
      ---
        4
```

Say 2 times 13 are 26, writing the 26 under the 32. The 6 over is by Short Division to be *prefixed* to 9; or which is the same thing, *bring down* the 9, and *annex*[1] it to the 6; then 13 in 69; 5 times, for 5 times 13 are 65, and 4 remainder. *A.* $25\frac{4}{13}$ barrels.

44. *Therefore, take at first only figures enough to contain the divisor, and having divided them, annex to the remainder the next figure of the dividend; after which divide as before, and so on till the figures of the dividend are all brought down.*

45. Divide 278 by 13. *A.* Quotient 21. Remainder 5.
46. Divide 985 by 25. *A.* Quotient 39. Remainder 10.
47. Divide 988 by 46. *A.* Quotient 21. Remainder 22.
48. What is the quotient of 3,369 divided by 25?

```
25) 3 3 6 9 (1 3 4
    2 5
    ---
      8 6
      7 5
      -----
      1 1 9
      1 0 0
      -----
        1 9
```

After having brought down the 6 and divided, bring down the 9 and annex it to the 11; which divide as before. *A.* $134\frac{19}{25}$.

49. Divide 3,386 by 25. *A.* Rem. 11.
50. Divide 6,798 by 45. *A.* Rem. 3.
51. Divide 8,241 by 35. *A.* Rem. 16.

52. When hay is 25 dollars a load, how much can be bought for 52,186 dollars?

Q. How many left-hand figures do you take first? 44. When there is a remainder in dividing, how do you proceed? 44.

---

1 ANNEX, [L. *annecto.*] To unite at the end; to subjoin; to affix, to connect with

```
25)52186(2087
   50
   ---
    218
    200
    ---
     186
     175
     ---
      11
```

Since 25 is not contained in 21, put a cipher in the quotient and bring down the 8, then divide as before. *A.* 2,087$\frac{11}{25}$ loads.

53. When then the remainder with one figure annexed is too small; *mark its place in the quotient by a cipher and bring down another figure, then divide as before*

54. Divide 209,491 by 102. Quotient 2,053$\frac{85}{102}$.
55. Divide 280,483 by 112. Quotient 2,504$\frac{35}{112}$.
56. Divide 401,123 by 125. Quotient 3,208$\frac{123}{125}$.

57. If the remainder, with two figures annexed, is still too small, *annex another figure of the dividend as before, and thus continue annexing till you obtain a number large enough to contain the divisor.*

```
        (58.)                          (59.)
165)495825(3005 A.          209)6270836(30004 A
    495                          627
    ------                       -------
       825                           836
       825                           836
```

60. Divide 115,611,740 by 2,312. *A.* 5,0005$\frac{180}{2312}$.
61. Divide 345,235,530 by 4,315. *A.* 8,0008$\frac{1010}{4315}$.

62. A gentleman expended 1,253,763 dollars for land, paying for each acre 125 dollars; how many acres did he buy?

Dividend.
Divisor, 125)1253763(10030$\frac{13}{125}$ acres; Quotient.

63. What is the quotient of 95,658 divided by 245? *A.* 390$\frac{108}{245}$.

64. When the divisor is 103 and the dividend 42,024, what is the quotient? *A.* 408.

65. When the quotient is 408 and the dividend 42,024, what is the divisor? *A.* 103.

66. When the divisor is 103 and the quotient 408, what is the dividend? (103 times 408). *A.* 42,024.

67. When the quotient is 408, the divisor 103, and the remainder 98, what is the dividend? *A.* 42,122.

68. When the dividend is 42,122, and the divisor 103, what is the quotient and remainder? *A.* 408$\frac{98}{103}$.

69. The salary of the President of the United States is 25,000 dollars per annum; what is that a day, allowing 365 days to the year? *A.* 68$\frac{180}{365}$ dollars.

GENERAL RULE.

70. *Begin on the left-hand of the dividend, and take the fewest figures that will contain the divisor, and write the number of times it is contained in them on the right of the dividend, for the first quotient figure.*

71. *Multiply the divisor by this quotient figure, and place their pro-*

duct under the figures of the dividend used, subtract it therefrom and annex to the remainder the next figure of the dividend, which divide as before, and so on.

72. *But if the remainder thus increased is still too small, write a cipher in the quotient and annex another figure, and so on till it does become large enough.*

73. *When the product of the divisor and quotient figure is too large, the latter must be diminished; but when the remainder is as large or larger than the divisor, the quotient figure must be increased.*

74. ORDER FIND HOW MANY TIMES; MULTIPLY; SUBTRACT; AND BRING DOWN.

75. PROOF. The same as in Short Division.

76. Divide 14,150 by 115. Quotient $123\frac{5}{115}$.

77. Divide 28,682 by 121. Remainder $\frac{5}{121}$.

78. Divide 2,360,557 by 1,021. Remainder $\frac{5}{1021}$.

79. Divide 17,286 by 1,234. Remainder $\frac{10}{1234}$.

80. Divide 14,797,541 by 12,321. Remainder $\frac{20}{12321}$.

81. A man bought a farm for 14,400 dollars, paying 75 dollars an acre. How many acres did it contain? *A.* 192 acres.

82. Suppose a farm of 192 acres cost 14,400 dollars, how much was it an acre? *A.* 75 dollars.

83. In 1 pound are 16 ounces; how many pounds in 223,365 ounces? *A.* $13,960\frac{5}{16}$ pounds.

84. London contained in 1831 a population of 1,471,405. Now allowing that 13 persons on an average occupy a single house, how many houses then would be required to accommodate all the inhabitants. *A.* 113,185 houses.

85. How much is $\frac{1}{13}$ of 1,932,045? *A.* $148,618\frac{11}{13}$.

86. How much is $\frac{1}{25}$ of 1,840,062? *A.* $73,602\frac{12}{25}$.

87. How much is $\frac{1}{45}$ of 2,500,368? *A.* $55,563\frac{33}{45}$.

88. How much is $\frac{1}{73}$ of 4,210,909? *A.* $57,683\frac{50}{73}$.

89. How much is $\frac{1}{85}$ of 6,301,895? *A.* $74,139\frac{80}{85}$.

90. How much is $\frac{85305}{49}$? *A.* $1,740\frac{45}{49}$.

91. How much is $\frac{122997}{67}$? *A.* $1,835\frac{52}{67}$.

92. How much is $\frac{83456}{75}$? *A.* $1,112\frac{56}{75}$.

93. According to the census of 1830, the entire population of the U. S,[1] was about 12,840,534, and the number of children who never attend school is about $\frac{1}{21}$ of the entire population; what was their number? *A.* 611,454.

XXIV. When the divisor is a composite number.

RULE.

1. *Divide first by one factor and the quotient by the other.*

Q. What is the General Rule? 70, 71. In what cases are two or more figures to be annexed to the remainder? 72. When must the figure you place in the quotient be made larger or smaller? 73. What is the order of proceeding? 74. Proof? 75.

1 U. S., for the United States of America.

2. A prize of 7,200 dollars is to be divided equally among 36 men; what is each man's part?

```
12) 7200
 3)  600
     200
```

Had there been but 12 men, each one's part would have been 600 dollars; but there being 3 times as many, each one's part is only $\frac{1}{3}$ as much.

3. How many times is 30 contained in 1,230, using the factors 3 and 10? *A.* 41.

4. Divide 1,152 by 24, using first the factors 4 and 6; then 3 and 8; and lastly, 2 and 12. *A.* 48.

5. Divide 8,640 by 36. *A.* 240.

6. Divide 2,160 by 144. *A.* 15.

7. Divide 4,320 by 72. *A.* 60.

8. Suppose a man has a bundle of cloth, in which are 4 pieces, containing each 7 yards, (making 28 yards in the bundle;) how many such bundles would 270 yards make?

```
7) 270     yards.
4)  38..4  yards over.
     9..2  pieces over.
```

A. 9 bundles 18 yards, or $9\frac{18}{28}$ bundles.

There are 38 pieces and 4 yards over, and 9 bundles and 2 pieces over. To find how many yards would remain from dividing by 28, multiply the 2 pieces over by the first divisor [7 yards in each piece,] making 14 yards, which added to the 4 yards over at first, makes 18 yards remainder.

9. *Hence to find the true remainder multiply the last remainder by the first divisor and add in the first remainder.*

10. Divide 85,509 by 42, using two factors. *A.* $2{,}035\frac{39}{42}$.

11. Divide 71,252 by 35, using two factors. *A.* $2{,}035\frac{27}{35}$.

12. Divide 81,605 by 108, using two factors. *A.* $755\frac{65}{108}$.

13. Divide 49,823 by 24, using its three factors, 2, 3 and 4; for 2 times 3 are 6, and 4 times 6 are 24.

```
2) 49823
3) 24911..1
4)  8303..2
    2075..3
```

Writing under the 23 the total divisor makes $\frac{23}{24}$.

A. $2{,}075\frac{23}{24}$.

```
Multiply   3, the last remainder.
      by   3, the second divisor,
           9
Add        2, the second remainder.
Multiply  11
      by   2  the first divisor.
          22
Add        1  the first remainder
          23  the true remainder.
```

14. Divide 33,370 by 162, using the factors 9, 6, and 3; for they multiplied together, make the total divisor. *A.* $205\frac{160}{162}$.

15. Divide 2,310,523 by 60, using 3, 4, and 5. *A.* $38508\frac{43}{60}$.

16. Divide 2,310,523 by 60, using 2, 6, and 5. *A.* $38{,}508\frac{43}{60}$.

XXIV. *Q.* What is the rule for dividing by a composite number? 1. How is the true remainder obtained? 9. What are the two factors of 48? Of 108

17. Divide 2,310,523 by 60, using 2, 2, and 15. *A.* 38,508$\frac{43}{60}$.
18. *Hence dividing by either factor first, brings the same result.*
19. Divide 3,707,716 by 120, using 3 factors. *A.* 30,897$\frac{76}{120}$.
20. There are 1,728 solid inches in 1 foot; how many solid feet then in 14,323 solid inches, using 12 as a factor 3 times. *A.* 8$\frac{499}{1728}$.
21. In a pile of wood 8 feet long, 4 feet wide, and 4 feet high, are 128 solid feet or 1 cord; how many cords of wood then in a pile containing 41,726 solid feet? *A.* 325$\frac{126}{128}$ cords.

XXV. When the divisor is 10, 100, 1000, &c.

RULE.

1. *Cut off as many figures from the right-hand of the dividend as there are ciphers in the divisor; the figures cut off are the remainder and the other figures of the dividend the quotient.*

2. At 10 dollars a barrel, how many barrels of flour may be purchased for 369 dollars?

1|0) 3 6 | 9
A. 3 6$\frac{9}{10}$ barrels.

Removing the 9, makes 36 hundred become only, 36; that is, tens become units, hundreds, tens, &c.

3. Divide 875,600,329 by 10. *A.* Remainder $\frac{9}{10}$.
4. Divide 95,013,421 by 100. *A.* Remainder $\frac{21}{100}$.
5. Divide 8,732,509 by 1,000. *A.* Remainder $\frac{509}{1000}$.
6. Divide 340,137 by 10,000. *A.* Remainder $\frac{137}{10000}$.
7. Divide 45,815 by 100,000. *A.* Remainder $\frac{45815}{100000}$.
8. There are 10 mills in every cent; how many cents then in 895,430 mills? *A.* 89,543 cents.
9. Since 100 cents make 1 dollar, how many dollars are there in 9,503,200 cents. *A.* 95,032 dollars.
10. How many dollars in 8,534 cents, and how many cents over? *A.* 85 dollars and 34 cents.
11. In 68,546,035 cents, how many dollars and cents? *A.* 685,460 dollars and 35 cents.

XXVI. When any divisor has ciphers on the right.

RULE.

1. *Cut off the ciphers from the right of the divisor, and an equal number of figures from the right of the dividend.*

2. *Divide the remaining figures as usual, and annex to the right-hand of the remainder for the true remainder, all the figures cut off from the dividend.*

3. Divide 89,952 dollars equally among 500 persons.

XXV. *Q.* When the divisor is 10, 100, &c. what is the rule? 1. What is the effect of removing figures from the right of the dividend? 2. What is $\frac{1}{10}$ of 899 *A.* 89$\frac{9}{10}$. What is $\frac{1}{100}$ of 565? $\frac{1}{100}$ of 836?

XXVI. *Q.* When any divisor has ciphers on the right, what is a concise rule? 1 What is the effect of cutting off two figures? 3. In dividing 89,952 by 500, why is the 52 annexed to the remainder 4? 3.

5|0 0) 8 9 9 | 5 2
A. 1 7 9$\frac{452}{500}$.

Cutting off two figures does in fact divide it by 100, it being one factor of the composite number 500. Then, if we multiply (as in XXIV. 9,) the last remainder 4 by the first divisor 100, it makes 400, to which adding 52, the first remainder, makes 452 ; but by simply annexing the 52 to 4, produces the same effect, hence the rule.

4. Divide 783,456,078 by 2,100. *A.* Remainder $\frac{678}{2100}$.
5. Divide 634,278,975 by 8,000. *A.* Remainder $\frac{6975}{8000}$.
6. Divide 854,267 by 500,000. *A.* Quotient 1$\frac{354267}{500000}$.
7. The annual expense for schools in the United States is about 15,000,000 of dollars, and the number of children about 3,750,000 ; what is the average expense for each child ? *A.* 4 dollars.
8. The number of teachers is about 95,000 : how many scholars then to each teacher ? *A.* 40 nearly.

CONTRACTION OF RULES.

XXVII. 1. To multiply easily by any number from 10 to 20.—*Multiply by the unit figure only of the multiplier, and having removed its product one place further towards the right of the multiplicand, add it to the multiplicand.*

2. Multiply 8,978 by 19.

OPERATION.

```
8 9 7 8  × 1 9
  8 0 8 0 2
-----------
1 7 0 5 8 2 A.
```

```
(3.)   8 9 7 8
           1 9
     ---------
     8 0 8 0 2
     8 9 7 8
   -----------
   1 7 0 5 8 2 A.
```

4. For the figures which are added in both operations are the same ; the results must therefore correspond.
5. Multiply 765,342,001 by 13. *A.* 9,949,446,013.
6. Multiply 678,320,131 by 15. *A.* 10,174,801,965.
7. Multiply 308,954,201 by 18. *A.* 5,561,175,618.
8. Multiply 608,753,108 by 12. *A.* 7,305,037,296.
9. Multiply 735,080,951 by 11. *A.* 8,085,890,461.
10. To multiply by 5.—*Annex a cipher, and divide* 2.
11. For annexing one cipher multiplies it by 10, then 2 times 5 being 10, dividing only by 2, leaves the number increased 5 times.
12. Multiply 6,545 by 5. *A.* 32,725.
13. Multiply, by this rule, 7,521 by 5. *A.* 37,605.
14. To divide by 5.—*Reverse the last process by multiplying by* 2, *and cutting off one figure for a remainder.*

XXVII. *Q.* How can you multiply by 11, 12, 13, &c., up to 20 expeditiously ? 1. Why so ? 4. How can you multiply by 5 easily ? 10. Why ? 11. How divide by 5 ? 14. How many are 5 times 48 ?—times 5 in 240 ?

15. Divide, by this rule, 32,725 by 5. *A.* 6,545.
16. Divide, by this rule, 37,605 by 5. *A.* 7,521.
17. To multiply by 25.—*Annex two ciphers, and divide by* 4.
18. For annexing two ciphers multiplies by 100, and 4 times 25 being 100, dividing only by 4, leaves the number 25 times the greater.
19. Multiply 6,532,405 by 25. *A.* 163,310,125.
20. Multiply 4,230,216 by 25. *A.* 105,755,400.
21. To divide by 25.—*Reverse the last process by multiplying by* 4, *and cutting off two figures for a remainder.*
22. Divide 163,310,125 by 25. *A.* 6,532,405.
23. Divide 105,755,400 by 25. *A.* 4,230,216.
24. To multiply by 125.—*Annex three ciphers, and divide by* 8.
25. For 8 times 125 being 1000, annexing three ciphers and dividing only by 8, leaves the number 125 times the greater.
26. Multiply 6,304,521 by 125. *A.* 788,065,125.
27. Multiply 2,403,450 by 125. *A.* 300,431,250.
28. To divide by 125.—*Reverse the last process by multiplying by* 8, *and cutting off three figures from the right of the product.*
29. Divide 788,065,125 by 125. *A.* 6,304,521.
30. Divide 300,431,250 by 125. *A.* 2,403,450.
31. To multiply by $33\frac{1}{3}$.—*Annex two ciphers, and divide by* 3.
32. For 3 times $33\frac{1}{3}$ being 100, annexing two ciphers multiplies by 100, and dividing only by 3, leaves the number $33\frac{1}{3}$ times the greater.
33. Multiply 65,220 by $33\frac{1}{3}$. *A.* 2,174,000.
34. Multiply 73,410 by $33\frac{1}{3}$. *A.* 2,447,000.
35. To divide by $33\frac{1}{3}$.—*Reverse the last process by multiplying by* 3. *and cutting off two figures.*
36. Divide 2,174,000 by $33\frac{1}{3}$. *A.* 65,220.
37. Divide 2,447,000 by $33\frac{1}{3}$. *A.* 73,410.
38. To multiply by 9, or 99, or 999, &c.—*Annex to the multiplicand as many ciphers as there are 9s, and subtract the multiplicand from it.*
39. For annexing one cipher, for instance, multiplies by 10, and deducting the multiplicand, leaves it 9 times the greater.
40. Multiply 467 by 9, and by 99, and by 9999.

4670	46700	4670000
467	467	467
A. 4203	*A.* 46233	*A.* 4669533

41. Multiply 653,421 by 999. *A.* 652,767,579
42. Multiply 65,342 by 9,999. *A.* 653,354,658.

Q. How is the multiplying by 25 abridged? 17. Why? 18. The dividing by 25 abridged? 21. How many are 25 times 8? 25 in 200? How is the multiplying by 125 abridged? 24. Why? 25. The dividing by 125? 28. Multiply 8 by 125. Divide 1000 by 125. How is the multiplying by $33\frac{1}{3}$ abridged? 31. Why? 32. The dividing by $33\frac{1}{3}$ abridged? 35. How many are $33\frac{1}{3}$ times 15?—times $33\frac{1}{3}$ in 500? What abreviation is there in multiplying by 9, or 99, or 999? 38. Why? 39.

43 Multiply 6,534 by 99,999. *A.* 653,393,466.
44 Multiply 653 by 999,999. *A.* 652,999,347.
45. Multiply 65 by 9,999,999. *A.* 649,999,935.
46. Multiply 6 by 99,999,990. *A.* 599,999,994.

ARITHMETICAL SIGNS.

XXVIII. 1. EQUALITY.[1] The sign = between two numbers shows that the number before it is equal in value to the number after it; as, 100 cents=1 dollar, meaning 100 cents are *equal to* 1 dollar.

2. This sign is two horizontal[2] lines, drawn parallel[3] to each other.

3. ADDITION.* The sign + shows that the number before it is to be

---

XXVIII. *Q.* What is the sign of Equality? 2. What does it show? 1. What do 100 cents and 1 dollar with the sign of Equality between them mean? 1.

---

*PROOF OF ADDITION, SUBTRACTION, MULTIPLICATION, AND DIVISION, by *casting out the* 9s, or by 9s as it is called.

* ADDITION. One 10 contains one 9 and 1 unit; 2 tens or 20, two 9s and 2 units; 3 tens or 30, three 9s and 3 units, and so on, leaving for a remainder each time as many units as there are tens.

Hence if we deduct from any number of tens, as many units as there are tens, the remainder will contain an even number of 9s.

One hundred [100] contains eleven 9s and 1 unit; two hundred [200,] twenty-two 9s and 2 units, and so on, leaving for a remainder each time as many units as there are hundreds.

And universally if from any number of tens, or hundreds, or thousands, &c., there be taken as many units, as there are tens, or hundreds, or thousands, &c., the remainder will contain even 9s.

The number 634, for instance, is made up of 600, 30, and 4. The 600 then contains even 9s, and 6 remainder; the 30, even 9s, and 3 remainder; and the 4 no 9s, and 4 remainder. Now the remainders 6, 3, and 4, are the very numbers that form 634, therefore we derive the following proposition, (4) on which is based (5) the proof of Addition, viz:—

The sum of the figures that compose any number has an excess (6) of 9s equal to the excess of 9s in that number.

And from the nature of Addition, it follows, that the sum of the excesses of 9s in two or more numbers, always has an excess equal to the excess in the sum of those numbers.

6 8 3	8
9 5 6	2
3 4 5	3
1 9 8 4	4

The amount is 1984, and the proof as follows, viz: Adding the figures 6, 8, and 3 together, in the top line, makes 17, or one 9 and 8 over for the excess; reject the one 9 and write down on the right the excess above 9 which is the 8. Do the same with the 9, 5, and 6, rejecting the the 9s and writing down the excess, which is 2. The third line leaves in like manner an excess of 3; next adding these remainders, 3, 2, and 8, makes 13, or one 9 and 4 remainder; reject the 9 and write down the 4 underneath. The bottom line 1, 9, 8, and 4 makes 22, or two 9s and 4 remainder; rejecting the 9s, the remainder is 4, the same as the preceding remainder, and therefore 1984 is the true amount.

RULE. *Add the figures in the uppermost row or line together, reject the 9s contained in their sum, and set the excess directly even with the figures in that row. Do the same with each row and set all the excesses of 9 together in a line, and find their sum; then if the excess of 9s in this sum (found as before,) be equal to the excess of 9s in the total sum, the work may be considered correct.*

It may sometimes be more convenient to reject the 9s while adding: thus, taking 5632 for example, 5 and 6 are 11—one 9 and 2 over; the 2 over and 3 are 5 and 2 are 8

1 EQUALITY, [L. *equalitas.*] Agreement; evenness; uniformity.
2 HORIZONTAL. Relating to the horizon; level; not perpendicular.
3 PARALLEL. A line equally distant through its whole extent from another line
4 PROPOSITION. What is proposed; statement of facts: offer of terms
5 BASED. Founded.
6 EXCESS. What is over; superfluity; remaining.

added to the number after it; as 6+4=10, meaning 6 and 4 added together are equal to 10.

4. This sign is a cross, formed by a horizontal line intersecting[1] a perpendicular[2] one, at right[3] angles,[4] and is read *plus*,[5] which means *more;* thus 6+4=10, means 6 plus 4 are 10.

5. How many are 375+125+100= *A.* 600.

6. How many are 57,563+1,500+1,000,000+42+100+101+5+72? *A.* 1,059,383.

7. SUBTRACTION.* This sign—shows that the number after it is to be subtracted from the number before it; as, 6−4=2, meaning 4 from 6 leaves 2.

8. This sign is a single horizontal line, and is often called *minus*, signifying *less;* thus, 10−3=7, is read 10 minus 3 is 7.

9. How much is 10,000,000−1,001? *A.* 9,998,999.

10. How much is 37,500,209−4,209? *A.* 37,496,000.

11. MULTIPLICATION.† This sign × shows that the number before it and the number after it are to be multiplied into each other; as, 10×6=60, meaning 10 times 6 are 60.

12. This sign is two lines crossing each other in the form of an x

13. How many are 5,320,065×801? *A.* 4,261,372,065.

14. How many are 423×100×200? *A.* 8,460,000.

Q. What is the sign of Addition? 4. How formed? 4. What does it show? 3. How is it read? 4. How much is 6 plus 4? 8 plus 12? What is the sign of Subtraction? 8. What is it often called? 8. What does it show? 7. How much is 10 minus 7? 25 minus 14? What is the sign of Multiplication? 12. What does it show? 11. When 10 and 6 have this sign between them, what do they mean? 11.

* SUBTRACTION. Since the subtrahend and difference added together should equal the minuend, the Proof is the same in principle, as that for Addition.

RULE. *Reject the 9s from the subtrahend and difference, noting the excesses. Add these excesses together, and if the excess of 9s in their sum equal the excess in the minuend, the work is right.*

```
 6 3 4 5 6 . . . 6
 5 2 0 0 1 . . . 8
 1 1 4 5 5 . . . 7
```

The sum of the excesses 8 and 7 are 15, from which rejecting the 9s leaves 6; equaling the excess in the minuend.

† MULTIPLICATION. Since Multiplication is an abbreviation of Addition, it may be proved on the same principle.

RULE. *Multiply the excess of 9s in the multiplicand, by the excess in the multiplier, and if the excess of 9s in this product equal the excess in the total product the work is right.*

```
     6 8 3 4 5 . . . 8
           3 9 . . . 3
   6 1 5 1 0 5 . . . 6
 2 0 5 0 3 5
 2 6 6 5 4 5 5 . . . 6
```

The product of the two excesses is 24, and the excess of its 9s is 6, which is the same as the excess in the product and therefore right.

1 INTERSECTING, [L. *intersecto.*] Cutting or crossing each other.

2 PERPENDICULAR. Hanging in a straight line from any point to the centre of the earth; upright; not level.

3 RIGHT. Straight; lawful; just; most direct. A *square* corner is a *right* angle; a square figure has four right angles; when the four angles, made by two lines crossing each other are equal, each is called a right angle.

4 ANGLE. A corner; the space between two lines that meet.

5 PLUS, from the Latin *plus*, signifying *more.*

6 MINUS, from the Latin *minus*, signifying *less.*

15. How many are $1\times2\times3\times4\times5\times6\times7\times8\times9\times10$? *A* 3,628,800.

16. Division.* The sign $\div$ shows that the number before it is to be divided by the number after it; as, $60\div5=12$, meaning 60 divided by 5 is 12.

17. The sign $\frac{60}{5}$ shows that the number above the line, is to be divided by the number below the line.

18. The first sign is formed by one horizontal line passing between two dots, and the second by writing the divisor under the dividend with a line between.

19. Perform $1,236,000\div5$. *A*. 247,200.

20. Perform $3,756,000\div20$. *A*. 187,800.

21. Perform $\frac{4237500830}{501}$. *A*. $8,458,085\frac{245}{501}$

22. This sign is the proper method of expressing the remainder after division is performed. VI. 1, 2.

23. Perform $750,348\div125$. *A*. $6,002\frac{98}{125}$.

24. Perform $\frac{320658952}{1017}$. *A*. $315,298\frac{886}{1017}$.

25. When we wish merely to indicate there is a remainder, it being not of sufficient importance to be expressed, the sign of Addition is generally adopted;[1] thus 10 mills $\div$ 3 makes 3+.

26. Divide 5,608,354 drams by 117. *A*. 47,934+

27. Divide 7,503,478 gills by 129? *A*. 58,166+

28. When two or more signs occur in succession, each operation is to be performed in the order of the signs.

29. Perform $600+100-150\times20\div11,000=1$.† *A*. 1.

30. Perform $100+100-5+29\div8+6\times11+40\times3+617\times5-1295\div80=100$. *A*. 100.

Q. What are the two signs of Division? 18. What does each show? 16. 17. What two methods are there of indicating a remainder? 22. 25. What do several signs in succession indicate? 28.

---

* Division. From the principles of proof recognized in Addition and Multiplication, we may proceed as follows:—

85) 38998 (458	Divisor,	85, . . . 4 excess.
340	Quotient,	458, . . . 8 excess.
499		32, . . . 5 excess.
425	Remainder,	68, . . . 5 excess.
748		10, . . . 1 excess.
680	Dividend,	38998, . . . 1 excess.
68		

Rule. *Multiply the excess of 9s in the quotient, by the excess in the divisor, and reject the 9s from the product: to which add the excess of 9s in the remainder, and if the result equal the excess of 9s in the dividend, the work is right.*

Note. This property of 9 belongs to it, only because it happens to be 1 less than 10 (the *radix* (2) of the system;) for did we reckon by 11s, then 10 would answer the same purpose; but since any number of 9s always contains an exact number of 3s, we may prove questions as well by casting out the 3s in the manner above, as by casting out the 9s.

† Add 100 to 600, from the amount subtract 150, multiply the remainder by 20, and divide the product by 11,000, the quotient will be 1 the *Answer*.

1 Adopted, [L. *adopto*.] Taken as one's own; selected for use

2 Radix, [L. *radix, a root*.] A primitive word: root.

[illegible] Perform $\frac{100}{50}+\frac{480}{60}\times500\div250-15=5$.*

[illegible] Execute[1] $\frac{225}{3}\times14-50\div4+\frac{250}{25}\times39\div18=563\frac{6}{18}$.

PROBLEMS

XX iX. 1. A PROBLEM is a question proposed for solution; in other words, it is something to be done.

2. PROB. I. The parts of a number being given to find that number.—*Add the several parts together.*

3. If the several parts of a number are 635; 4,008, and 5,025, what is that number? *A.* 9,668.

4. The total value of real estate in the state of New York, in 1831, was 289,457,104 dollars, and of personal estate, 75,258,726 dollars; what number will represent the value of both? *A.* 364,715,830.

5. PROB. II. The sum of two numbers, and one of them being given to find the other.—*Subtract the given number from the given sum.*

6. If 36,085 be the sum of two numbers, one of which is 10,052 what is the other? A. 26,033.

7. What is that number, which, with 25,233 will make 36,085? *A.* 10,852.

8. When the minuend is 6,345 and the subtrahend 3,052, what is the remainder? *A.* 3,293.

9. When the remainder is 3,293 and the minuend 6,345, what is the subtrahend. *A.* 3,052.

10. PROB. III. The difference between two numbers, and the greater of them being given to find the less.—*Subtract one from the other.*

11. What is the smaller number when the greater one is 15,675, and their difference 8,758? *A.* 6,917.

12. When the remainder is 4,080 and the minuend 6,304, what is the subtrahend? *A.* 2,224.

13. North America has about 25,750,000 inhabitants, and the difference between the population of North and South America is about 11,250,000; how many inhabitants has South America? *A.* 14,50●,000.

XXIX. *Q.* What is a Problem? 1. How is a number found from having its parts given? 2. How, from having one of two numbers, and their sum given? 5. How may the less of two numbers be found, from having the greater and their difference given? 10.

---

* Divide 100 by 50, then add the quotient to the quotient of 480 divided by 60; multiply that sum by 500; divide the product by 250 and subtract 15 from the quotient and the remainder will be 5. Answer.

1 EXECUTE. To perform; to finish; to kill.

14. Prob. iv. The difference between two numbers, and the smaller of them being given, to find the greater.—*Add both together.*

15. If the difference between two numbers be 2,340, and the smaller one 1,683, what is the greater number? *A.* 4,023.

16. When the remainder is 5,032 and the subtrahend 4,037, what is the minuend? *A.* 9,069.

17. Suppose a man who was born A. D. 1492, the year in which America was discovered, to live as long as Methuselah, which was 969 years, when would his death happen? A. D. 2,461.

18. France has a population of about 32,000,000, and the rest of Europe about 168,000,000; what then is the entire population of Europe? *A.* 200,000,000.

19. Prob. v. The sum of two or more numbers, and the excess of each above the smallest given, to find those numbers.—*From their whole sum subtract the given excess, and if there be more than two numbers, subtract the sum of the excesses; then divide the remainder into 2 equal parts to find two numbers, into 3 equal parts to find three, and so on; the quotient will be the smallest number; to which, add separately the several excesses at first subtracted for the other numbers required.*

20. What are the two numbers whose sum is 1,824 and the excess of the one above the other 360? *A.* 732; 1,092.

21. If the minuend is 6,800 and the difference between the remainder and subtrahend be 850, what will be the remainder and subtrahend? *A.* 2,975; 3,825.

22. Two men having met on a journey, found by calculation that they both had traveled 1000 miles, and that one had traveled 150 miles more than the other; what distance had each traveled?
A. 425; 575.

23. Suppose that two fat oxen weigh 1950 pounds, and that the difference in their weight is 215 pounds, what is the weight of each?
A. 867½ pounds; 1,082½ pounds.

24. A gentleman gave to both of his sons 65,300 dollars, giving one 4,000 more than the other; what did each receive?
A. 30,650; 34,650.

25. Suppose that John has done in one day 20 sums more than Samuel, and Richard 30 sums more than Samuel, how many did each do, allowing all to have done 260?

Note.—Reckoning what John and Richard both did more than Samuel makes 50; the sum of the excesses then is 50: subtracting and dividing as directed in the rule for three numbers gives 70 for the smallest number, that is, the number which Samuel did. Adding to 70 John's excess over that makes 90, and adding to the same Richard's excess makes 100. *A.* S. 70: J. 90; R. 100.

Q. How may the greater of two numbers be found, from having the smaller and their difference given? 14. When the sum of two or more numbers and their difference are given, how are the numbers themselves found? 19

26. Divide 16,000 dollars so that B may have 300 more than A, and C 400 more than A. *A.* A. 5,100; B. 5,400; C. 5,500.

27. Suppose 97 apples are so divided that James has 20 more than Richard, and Thomas 30 more than James; how many has each?

NOTE.—If James has 20 more than Richard, and Thomas 30 more than James, Thomas evidently has 50 more than Richard; and James and Thomas together have 70 more than Richard; the sum of the excesses here then is 70. *A.* R. 9; J. 29; T. 59.

28. A father has an estate of 600,000 dollars and 3 sons; he gives to the second son 50,000 dollars more than to the youngest, and to the oldest 50,000 more than to the second; what sum did each receive?

NOTE.—The oldest has 50,000 more than the second son and 100,000 more than the youngest; the total excess then is 150,000.
A. 150,000; 200,000; 250,000.

29. Suppose a poor man has labored 4 years for 1,000 dollars, receiving each successive year 50 dollars advance; what sum did he receive each year?

NOTE.—The 2nd year he received 50 more than for the 1st.; the 3rd year 100 more; the 4th year 150 more; then $50 \times 100 + 150 =$ 300, the excess; $1000 - 300 = 700 \div 4$ years$=175$.
A. 175; 225; 275; 325.

30. Divide 1,300 dollars so that B may have 300 more than A, and C 200 less than A; what sum will each receive? The excess is 100. *A.* 400; 700; 200.

31. PROB. VI. The sum of two or more numbers and their rate of increase or decrease being given to find those numbers.—*First find what sum each number would be if the smallest were* 1*; then add them together for the divisor of the given sum, the quotient will be the smallest number; with which proceed as with the* 1 *at first to find the other numbers required.*

32. Divide 900 dollars so that B may have 3 times as much as A, and C 4 times as much as B.

NOTE.—Suppose A has 1, then B will have (3 times $1 =$) 3 and C (4 times $3 =$) 12. Adding 1 and 3 and 12 together makes 16 for the divisor of 1,600. The quotient 100 is A's part, then 3 times $100 =$ 300 B's, and 4 times $300 = 1{,}200$ C's.

A. A's. 100 dollars; B's. 300 dollars; C's. 1,200 dollars.

33. A man bought a sheep, a cow, and a horse, for 165 dollars paying 8 times as much for the cow as for the sheep, and 3 times as much for the horse as for the cow; what price did he pay for each? *A.* 5 dollars for the sheep, 40 for the cow and 120 for the horse.

34. PROB. VII. The product of two or more numbers, and one o.

Q. When the sum of two or more numbers and their rate of increase are given, how are the numbers found? 31. When the product of two or more numbers, and one of them are given, how can the other b[illegible]

them being given to find the other.—*Divide their product by the given number.*

35. If the product of two numbers be 972 and one of them 36, what is the other? *A.* 27.

36. If 375 be a product and 15 a quotient, what is the divisor? *A.* 25.

37. Suppose one of the factors of the composite number 972 is 27, what is the other factor? *A.* 36.

38. Suppose the composite number 11,250 has factors, the product of two of which is 1,875, what is the third factor? *A.* 6.

39. When 18,000 barrels of flour cost 234,000 dollars, what was the cost per barrel? *A.* 13 dollars.

40. PROB. VIII. When the quotient, dividend and remainder are given to find the divisor.—*Divide the difference between the dividend and remainder by the quotient.*

41. Suppose a quotient 65, a dividend 9,526, and a remainder 101; what is the divisor? *A.* 145.

42. A man sold a block of buildings containing several tenements and shops for 30,087 dollars, being on an average for each, 1,002 dollars and 27 dollars besides. How many shops and tenements were there? *A.* 30.

43. A father having 194,000 dollars, gave to each of his sons 20,000 dollars and had 14,000 dollars left. How many sons must he have had? *A.* 9 sons.

44. PROB. IX. To find the cost of several things from having the different prices of each given.—*Add the different prices togther.*

45. A gentleman purchased a farm for 5,360 dollars, a vessel for 18,000 dollars, a span of fine horses for 950 dollars; what was the cost of the whole? *A.* 24,310 dollars.

46. PROB. X. The quantity and the uniform price of each being given to find the cost.—*Multiply the price by the quantity.*

47. What is the cost of 495,000 "morus multicaulis" trees, at $37\frac{1}{2}$ cents apiece? *A.* 18,562,500 cents.

48. PROB. XI. The cost of the whole and the equal price of each being given to find the quantity.—*Divide the cost of the whole by the price of one.*

49. How much hay will 1,836 dollars purchase at 12 dollars a load? *A.* 153 loads.

50. PROB. XII. The quantity and cost being given to find the pric of one.—*Divide the cost by the quantity.*

Q. How is the divisor found from having the quotient, dividend, and remain der given? 40. How is the cost of several things found when the different prices are given? 44. How, when the price is uniform? 46. How is the quantity found when the price of each is uniform? 48. When the quantity and cost are given to find the price of one? 50

51. When a cargo of 8,500 bushels of wheat sells for 17,000 dollars, what is the price per bushel? *A.* 2 dollars.

52. PROB. XIII. When the number of equal parts and the value of one are given to find the value of the whole.—*Multiply the value of one by the whole number.*

53. Suppose a packet is divided into 8 equal parts, and one part sells for 4,500 dollars, what is the whole packet worth at that rate? *A.* 36,000 dollars.

54. If $\frac{1}{7}$ (1-seventh) of a beef creature cost 15 dollars, what would the whole cost? *A.* 105 dollars.

55. There are 105,192 seconds in $\frac{1}{5}$ of a year; how many seconds then in a year? *A.* 525,960 seconds.

56. PROB. XIV. The number of equal parts, and their value being given, to find the value of one.—*Divide the value by the whole number of equal parts.*

57. Suppose the capital of a bank to be 200,000 dollars, and the number of shares 2,000; how much is each share? *A.* 100 dollars.

58. Suppose a packet is divided into eighths, and valued at 36,000 dollars; what is the value of $\frac{1}{8}$? *A.* 4,500 dollars.

MISCELLANEOUS EXAMPLES.

XXX. 1. When one of two numbers is 3,750 and their sum 4,856, what is the other. *A.* 1,106.

2. When the remainder is one million, and the minuend one billion, what is the subtrahend? *A.* 999 million.

3. When the multiplicand is 6,350 and the product 50,800, what is the multiplier? *A.* 8.

4. When the multiplier is 8 and the product 50,800, what is the multiplicand? *A.* 6,350.

5. When the remainder is 52, the quotient 49, and the dividend 17,937, what is the divisor? *A.* 365.

6. When the remainder is $\frac{234}{7001}$, and the dividend 14,023,237, what is the quotient? *A.* 2003.

7. How many times must 814 be added to itself to make 407,000. *A.* 500 times.

8. How many times must 500 be taken from 407,000, to find how many times the former number is contained in the latter? *A.* 814.

9. Suppose a farmer has his live stock distributed as follows, viz: in one pasture 17 oxen, 7 calves, and 4 young horses; in another 510 sheep, 7 calves, and 4 young horses; in his yard 12 cows, 3 horses, 5 young oxen, 14 sheep, 10 colts, 17 turkeys, 15 hens, 25 geese, and 25 ducks; in his barn 22 hens, 14 calves, 11 geese, and 3 ducks, 10 sheep, 5 colts, 4 oxen, 3 cows, and 2 horses. What is the amount

of the whole? *A.* 26 oxen, 28 horses, 15 cows, 28 calves, 534 sheep, 17 turkeys, 36 geese, 28 ducks, and 37 hens.

10. A farmer having a flock of 700 sheep, perceived every time he foddered them, which was twice a day for 30 days, that one of the number was missing; how many sheep had he left at the end of the 30 days? *A.* 640.

11. A grocer bought 12 barrels of flour for 7 dollars a barrel, and 20 barrels for 8 dollars a barrel. What did the whole cost? *A.* 244 dollars.

12. Suppose a man who had 500 dollars, has purchased flour to the amount of 244 dollars; how many more barrels at 8 dollars a barrel, can he buy with the remainder? *A.* 32 barrels.

13. Rufus bought a vest for 3 dollars, a hat for 4 dollars, and for his coat he paid 3 times as much as for both of the other articles; how much did the three articles cost? *A.* 28 dollars.

14. Suppose a poor woman has only 230 cents, with which to purchase calico for a gown, and suppose that it takes 10 yards for a pattern, how high a price can she pay by the yard to just take all the money? *A.* 23 cents.

15. The population of the United States in 1830 was about 12,868,000. Suppose, as has been computed, 1 person to every 400 die annually by intemperance, how many deaths then in the United States may be attributed to this cause alone? *A.* 32,170.

16. Suppose only 30,000 die annually by intemperance, how many is that a month, allowing 12 months to the year? *A.* 2500.

17. What would be the expense of constructing a railroad from Maine to the Oregon Territory, the distance being about 3,000 miles, at the cost of 15,000 dollars per mile? *A.* 45,000,000 dollars.

18. How long a time would it require to travel across the United States, on the above road, at the rate of 20 miles an hour, or 480 miles a day? *A.* $6\frac{120}{480}$ days.

19. Light is supposed to pass from the sun to the earth in about 8 minutes, a distance of 95 millions of miles. How far then does light move in 1 minute? *A.* 11,875,000 miles.

20. The national debt of England was, in 1831, about 3,300,000,000 dollars, and the revenue of Great Britain and Ireland about 300,000,000 annually; suppose this revenue to be applied to the extinction of the debt, how long a time would it require? *A.* 11 years.

21. "Mercury is the smallest and swiftest of the planets, moving at the rate of 111,000 miles every hour." How far then would it move in 24 hours, or 1 day? *A.* 2,664,000 miles. How far in 365 days, or 1 year? *A.* 972,360,000.

22. The earth is about 25,000 miles in circumference; how many days would it take a car, at the rate of 20 miles an hour being 480 miles a day, to move round the earth? *A.* $52\frac{40}{480}$ days.

23. A father divided his property so that his sons had 1,420 dollars

apiece, and his daughters 1,100 dollars apiece; he had 7 sons and 3 daughters. What was the value of the father's estate? *A.* 13,240 dollars

24. A flour merchant sold 108 barrels of flour for 9 dollars a barrel, and gained on it 216 dollars. What price then must he have paid for it by the barrel? *A.* 7 dollars.

25. A purchased of B 500 bushels of wheat, and sold it for 1,475 dollars, which was 475 dollars more than its cost. What must he have paid a bushel for it? *A.* 2 dollars.

26. A merchant purchased 1,400 casks of lime at 3 dollars a barrel, and was obliged from its becoming air slacked, to sell it for 4,000 dollars. What was his loss on the whole? *A.* 200 dollars.

27. How much can a cashier lay up in a year of 365 days whose salary is 1,500 dollars, and daily expenses 3 dollars? *A.* 405.

28. A farmer sold 5 horses which cost him 75 dollars apiece, for 50 dollars advance,[1] and received payment in sheep at 5 dollars a head. How many sheep will pay for the horses? *A.* 85 sheep.

29. Suppose a certain cistern, which will hold 300 gallons, has two pipes; and that every hour 25 gallons run in by one pipe, and 15 gallons run out by the other, how many gallons would stay in every hour? How long a time would be required to fill it? *A.* 10 gallons; 30 hours.

30. Rufus bought a watch for 20 dollars, and paid 4 dollars for repairing it. What must he ask for it to gain 5 dollars? *A.* 29 dollars.

31. A clock strikes 1 time for 1 o'clock, 2 times for 2 o'clock, 3 times for 3 o'clock, and so on to 12 o'clock. How many times then will a clock strike in half a day, or 12 hours? *A.* 78 times.

32. If a clock strikes 78 times in half a day; how many times would it strike in a whole day? *A.* 156 times. How many times in a year of 365 days? *A.* 56,940 times.

33. A farmer sold a grocer 10 bushels of corn at 1 dollar a bushel; 12 barrels of cider at 2 dollars a barrel; he received in payment 3 barrels of flour at 7 dollars a barrel, and the balance in cash. How much money did he receive? *A.* 13 dollars.

34. *A* purchased $\frac{1}{5}$ of a steamboat for 1200 dollars; what did $\frac{2}{5}$ cost? *A.* 2,400. What did the whole boat cost? *A.* 6,000 dollars.

35. If $\frac{1}{8}$ of a manufactory be sold for 8,540 dollars, what would the whole bring at that rate? *A.* 68,320 dollars.

36. How many times can $\frac{1}{3}$ of 1,200 be taken from 1200, and have nothing remain? *A.* 3 times.

37. Suppose a merchant bought 200 barrels of pork of one man, and enough of another man to make 750 barrels, how many barrels did he buy of the second man? *A.* 550 barrels.

38. A farmer has 475 bushels of grain in 2 bins, one holding 125 bushels more than the other; how many bushels does each hold? See XXIX. 19 *A.* 175 bushels; 300 bushels.

1 ADVANCE. Moving forward; additional price; profit.

39. Suppose two persons have a legacy left them of 20,000 dollars, to be so divided that one may have 3,000 more than the other; what is each one's part? Give one 3,000 dollars, then divide the rest equally. *A.* 8,500 dollars; 11,500 dollars.

40. Suppose a man traveled from Albany to Buffalo, a distance of about 363 miles, partly by the canal and partly by the railroad, so that he went 163 miles more on the canal than on the railroad; what distance did he travel on each? *A.* 100 miles; 263 miles.

41. Divide 630 dollars so that B may have 3 times as much as A, and C 5 times as much as A. See XXIX. 31. *A.* 70; 210; 350.

42. Suppose 7 tons of hay are sufficient to keep a calf, a cow, and a horse, through the winter, and that the horse eats 2 times as much as the cow, and the cow 2 times as much as the calf; what quantity is sufficient for each? *A.* 1 ton; 2 tons; 4 tons.

43. It has been estimated, that the population of the globe is about 816 millions, and that every 32 years as many inhabitants as are living at any one time will be dead, and their places supplied by others; how many then must die and be born every year?
A. 25,500,000 persons.

44. At the rate of 25,500,000 persons a year; how many must die and be born every day of the year (=365 days?) *A.* 69,863+. How many every hour of the day (=24 hours?) *A.* 2,911 nearly. How many every minute of the hour (=60 minutes;) or in less time than you are solving[1] the question? *A.* 48+

FEDERAL MONEY.

XXXI. 1. FEDERAL MONEY is the currency or coin of the United States, established by Congress A. D. 1786.

2. The EAGLE, DOLLAR, DIME, CENT, and MILL, are the several denominations of Federal Money.

3. Accounts are kept in dollars and cents. Eagles and dimes are not used at all, the former being expressed in dollars, as tens of dollars; and the latter in cents, as tens of cents.

4. Thus 5 eagles, 4 dollars, 6 dimes, and 5 cents, are read, 54 dollars and 65 cents.

5. The dollar is fixed upon as the unit figure; all inferior denominations are therefore parts of a dollar.

6. Thus the dime is 1-tenth part of a dollar, the cent 1-hundredth part, and the mill 1-thousandth part.

XXXI. *Q.* What is Federal Money? 1. What are its denominations? 2. Repeat the Table of Federal Money? See VII. 1. Which denominations are used in accounts? 3. How are 5 eagles, 4 dollars, 6 dimes, and 5 cents read? 4. How is the value of the different denominations determined? 5. What are the several parts of the inferior denominations? 6.

1 SOLVING, [L. *solvo.*] Loosing; explaining; unfolding; performing.

REDUCTION OF FEDERAL MONEY

XXXII. 1. REDUCTION[1] is the changing of one denomination into another, without altering its value.

2. Thus, 2 dollars into 200 cents ; 40 mills into 4 cents.
3. How many mills are there in 8 cents?—in 27 cents?
4. How many cents in 80 mills?—in 270 mills?
5. Reduce 2 dollars to dimes—to cents—to mills.
6. Reduce 2,000 mills to cents—to dimes—to dollars.
7. Reduce 5,000 cents to dollars—to eagles.
8. Reduce 5 eagles to dollars—to cents.
9. Reduce 25 dollars to cents—2,500 cents to dollars.
10. Reduce 34 dollars to dimes—to cents—to mills.
11. Reduce 34,000 mills to cents—to dimes—to dollars.
12. Reduce 815 cents to dollars and cents.
13. Reduce 8 dollars and 15 cents to cents.
14. *Hence, annexing two ciphers or figures, brings dollars into cents, and annexing three, into mills, and vice versa ;*[2] *that is, cutting off two figures from cents, and three from mills, brings them back again into dollars.*
15. Reduce 27 dollars to cents—to mills.
16. Reduce 27,000 mills to cents—to dollars.
17. Reduce 8 dollars 15 cents 8 mills to mills.
18. Reduce 8,158 mills to cents—to dollars.
19. Dollars and cents are known as such by this sign $, and distinguished one from the other by a *point*, thence called a *Separatrix.*
20. Thus $8.356 means 8 dollars, 35 cents and 6 mills.
21. *Cents then occupy the first two places on the right of dollars, and mills the third place.*
22. Reduce $7.156 to mills.
23. Reduce 7,156 mills to dollars, cents and mills.
24. *Hence, merely removing the separatrix brings dollars, cents and mills into mills, and vice versa.*
25. Reduce $15.756 to cents and mills.
26. Reduce 15,756 mills into dollars, cents and mills.
27. As cents occupy two places, recollect when the cents are less then 10, *to write a cipher in the tens' place, or place of dimes.*
28. Reduce 5 dollars and 6 cents to cents. *A.* 506 cents.
29. Reduce 4 dollars and 2 cents to cents. *A.* 402 cents.

XXXII. *Q.* What is Reduction of Federal Money? 1. How are dollars reduced to cents and mills and the reverse? 14. How many cents in 5 dollars Dollars in 500 cents? Mills in 5 dollars? Dollars in 5,000 mills? How are dollars and cents distinguished the one from the other? 19. What places do cents and mills occupy? 21. What is the direction for supplying vacant places? 27, 31.

---

1 REDUCTION, from *re*, L. *back*, and *duco*, L. *to bring or lead* ; hence it literally mean to bring back; to reduce ; to subjugate.

2 VICE VERSA, L. The terms being exchanged. Thus, the generous should be rich and *vice versa* ; that is, the rich should be generous.

30. Express 3 dollars 7 cents by the signs. *A.* $3.07.

31. *When there are no cents write two ciphers in their place.*

32. Express 9 dollars 5 mills by the signs. *A.* $9.005

33. Express 8 dollars 7 mills by the signs. *A.* 8.007.

34. Reduce 6 dollars and 3 mills to mills. *A.* 6.003 mills.

35. Since Federal Money increases in a tenfold proportion, like whole numbers, its operations are performed in like manner.

ADDITION OF FEDERAL MONEY.

RULE.

XXXIII. 1. *Add dollars to dollars, cents to cents, &c., as in Simple Addition, placing the separatrix directly under the separatrix above.*

2. Add together $8.17; $13.05; 17ct.; 8ct. 5m.; and 8m.

```
$  8 . 1 7
  1 3 . 0 5
      . 1 7
      . 0 8 5
      . 0 0 8
$ 2 1 . 4 8 3  A.
```

NOTE.—d. stands for dimes, ct. for cents, and m. for mills.

Observe the cipher before the 8 cents and 5 mills, and the two ciphers before the 8 mills. See XXXII. 27. 31. Then say, 8 and 5 are 13; set down 3 and carry 1, adding as usual.

3. Add together $36.75; $1.50; $1.25, and $1.43. *A.* $40.93.

4. Add together $5.035, $6.075, $4.127, $13.125. *A.* $28.362.

5. Add together $57, $36.42, $52.01, $6.05; 8 cents and 4 mills, and 9 mills. See the first example No. 2. *A.* $151.573.

6. What is the amount of 8 dollars and 5 cents, $252 and 3 cents, $150 and 5 mills, $17 and 8 mills. *A.* $427.093.

7. A man bought a chaise for $126.18, a watch for $85 dollars and 6 cents; a coach for $850; a hat for $6 and 9 cents; a whip for 62 cents and 5 mills. What did he pay for the whole? *A.* $1,067.955.

8. Find the sum of one dollar and two cents, twenty-five dollars and two mills, $19.09, three dollars and three mills. *A.* $48.115.

9. A man gave an eagle for a coat, five dollars and five dimes for a pair of boots, six dimes and six mills for a pair of gloves, fifty-five mills for blacking his boots. What did he lay out in all? 1 eagle=$10; 5 dimes=50 cents; 55 mills=5 cents 5 mills. *A.* $16.161.

10. Add together $8, $4, 8ct. 3m., 75ct. 19m., 19d. 7m. 3ct., $425 and 1m., 5 eagles and 5 mills. *A.* $489.795.

11. Find the sum of $135, 5ct., 25ct., $18, 5m., 80d., 6 eagles, 3m., 18ct., $9, 2d., 3d. and 3m. *A.* $230.991.

12. Find the sum of 18\frac{1}{4}$, 8\frac{3}{4}$, 5\frac{1}{2}$, 16\frac{3}{4}$, and 25\frac{1}{2}$.

Q. How many cents are 8 dollars 2 cents? 15 dollars 10 cents? Dollars in 802 cents? in 1510 cents? in 8176 cents? in 1000 mills? in 1234 mills? How many dollars, cents and mills be reduced to mills, and the reverse? 24. How are operations in Federal Money performed? 35.

XXXIII. Q. What is the rule for Addition of Federal Money? 1.

NOTE.—$\$\frac{1}{4}$=25 ct.; $\$\frac{1}{2}$=50 ct.; $\$\frac{3}{4}$=75 ct. *A.* \$74.75 or \$74$\frac{3}{4}$

13. A man purchased a plough for \$7$\frac{1}{4}$, a harrow for \$4$\frac{1}{2}$, a load of hay for \$20.50, a yoke of oxen for \$62$\frac{1}{4}$, and a horse for \$150$\frac{1}{2}$. What did the whole cost him? *A.* \$245

14. Find the sum of \$50.37$\frac{1}{2}$, \$65, \$54.20$\frac{1}{2}$, 18ct., 9ct., 45ct., 8m. and \$17.65$\frac{1}{2}$, ($\frac{1}{2}$ct.=5 mills.) *A.* \$187.963.

15. Suppose a traveler's expenses, at various times, were as follows, viz: \$1.25, \$.085, \$2.007, \$6.53, 45ct., 3m. and 9ct.; what was the total amount? *A.* \$11.18.

16. Suppose a father divides an estate equally between his two sons, giving one \$61,537.87$\frac{1}{2}$; what was the value of the estate? *A.* \$123,075.75.

17. Suppose 1-third of a load of hay costs \$5.37$\frac{1}{2}$, what is the load worth at that rate? A. \$16.125.

18. Suppose a ship's cargo to be valued at \$37,567.12$\frac{1}{2}$, and 1-fourth of the ship itself at \$15,000, what is the value of both ship and cargo? *A.* \$97,567.12$\frac{1}{2}$.

SUBTRACTION OF FEDERAL MONEY.

RULE.

XXXIV. 1. *Place the numbers as in Addition, and subtract as in whole numbers.*

2. From \$65.83 take \$39.95; from \$137 take 49 cents; from \$12 take 5 mills.

\$65.83	\$137.00	\$12.000
39.95	.49	.005
A. \$25.88	*A.* \$136.51	*A.* \$11.995

3. Subtract \$15.43 from \$84.91. *A.* \$69.48.

4. Subtract \$3.005 from \$5,650. *A.* \$5,646.995.

5. A gentleman owing \$6,537.50$\frac{1}{2}$, paid \$2,549.655; how much remained unpaid? *A.* \$3,987.85.

6. From \$1,000, take 35 cents 5 mills. *A.* \$999.645.

7. From \$10 take 2 cents 5 mills. *A.* \$9.975.

8. From \$3 take 62$\frac{1}{2}$ cents. *A.* \$2.375.

9. From \$1,000 take 1ct. and 1m. *A.* \$999.989.

10. From 940 dollars take one dime. *A.* \$939.90.

11. From 1 dime take 1ct. and 1m. *A.* \$.089.

12. If a man owes \$3,500.625, and has property worth \$6,500, how much will he have left after paying his debts? *A.* \$2,999.375.

13. A man purchased a barrel of flour for \$8.50, and paid \$3.875, what was there still due? *A.* \$4.625.

14. Subtract 5 dimes from 50 cents. *A.* 0.

XXXIV. *Q.* What is the rule for Subtraction? 1.

15. Subtract 10 eagles from $100. *A.* 0.
16. Subtract 55 mills from 5ct. and 5 m. *A.* 0
17. Subtract $10 from 10 eagles. *A.* $90.
18. If a ship's cargo is valued at $20,875, and the ship at $31,675.31½, how much is the ship worth more than the cargo? *A.* $10,800.31½
19. Suppose 1-fourth of a ship is worth $7,500, and the whole ship at $30,000, what is 3-fourths of the ship worth? *A.* $22,500.
20. If a merchant buys a lot of cotton for $6,500.875, and sells it for $8,215.12, what are his profits? *A.* $1,714.245.

MULTIPLICATION OF FEDERAL MONEY.

RULE.

XXXV. 1 *Multiply as in whole numbers, and call so many figures cents and mills, on the right of the product, as there are figures of cents and mills in the multiplicand or multiplier.*

2. What will 7 yards of cloth cost at $2.50 a yard?

$2.50 7 $17.50 *A.*	7 times $2.50 is the same as $2.50, written down 7 times and added together, which would make $17.50.

	(3.)	(4.)	(5.)
Multiply	$65.01	$151.015	$375 76
by	25	2 001	5.03
	A. $1625.25	*A.* $302,181.015	*A.* $189,007.28

6. Multiply $6,301.25 by 203. *A.* $1,279,153.75.
7. Multiply $420.135 by 652. *A.* $273,928.02.
8. Multiply $1,075.08 by 750. *A.* $806310.
9. At $2.50 a head, what will 7,350 sheep cost? *A.* $18375.
10. What will the winding of 26,750 balls of thread amount to in dollars and cents, at 3 mills each? *A.* $80.25.
11. Suppose a man receive 37½ cents apiece for making 285 pair of slippers, what does he receive for the whole? *A.* $106.875.
12. A merchant bought 200 pieces of calico, each piece containing 40 yards, for 29 cents a yard; what did he pay for the whole? *A.* $2320.
13. What will 17 hogsheads of vinegar, each containing 63 gallons, come to at 13 cents a gallon? *A.* $139.23.
14. A merchant bought 15 barrels of flour for $150, and sold it for 11½ dollars a barrel; what did he make on the whole? *A.* $22.50.
15. A farmer bought 500 sheep for $1100.50, and sold them for $2.75 apiece; what did he gain on the whole? *A.* $274.50.

XXXV. *Q.* Rule for Multiplication? 1. What will 4 yards of cloth cost at 46 cents a yard? At 50 cents a yard? How many dollars will buy 4 yards of broad cloth at $1.50 a yard? At $2.50?

16. A drover bought cows for $27.50 a head, and sold them for $30¼ a head; what profit would he make at that rate on buying and selling 324 cows? Multiply 324 by the profit on one. *A.* $891.

17. Suppose a man's daily income is $3.40, and his daily expenditures $2.16, how much will he have saved at the year's end, or in 365 days? *A.* $452.60.

18. Suppose a merchant pays his clerk $11.50 a month, and $2 a week for his board, what sum does the clerk cost him by the year of 12 months, or 52 weeks? *A.* $242.

19. A sold B 15 bushels of wheat at $2.75 per bushel, and 17 bushels of rye at $1.12½ a bushel, for which B. gave him a hogshead. (63 gallons) of molasses at 35 cents a gallon, and the balance in cash; how much cash must B. pay A. *A.* $38.32½.

20. A farmer gave $3.755 for 1-half a barrel of flour; what was the price by the barrel? *A.* $7.51.

21. When 1-third of an estate sells for $1,652.75, what is the value of the estate? *A.* $4,958.25.

22. If 1-fourth of a ship cost $1,650, and the cargo $18,251.62½ what is the value of both ship and cargo? *A.* $24,851.625.

23. If 1-fifth of a hogshead of molasses cost $5.25, what is a hogshead worth? *A.* $26.25. What are 17 hogsheads worth? *A.* $446.25.

24. A received a legacy[1] of $1,150.125; B 3 times as much, and C 4 times as much as B; what sum did C receive? *A.* $13801½.

25. Suppose a bank has $20,000.50 in specie[2] deposited[3] in its vault,[4] and a circulation[5] of 15 times that amount, what is the amount of its bills in circulation? *A.* $300,007½.

26. Suppose that 1-eighth of a ship's cargo sells for $875.375, and that the ship itself is worth 3 times as much as the entire ca what must be the value of the ship? *A.* $21,009

DIVISION OF FEDERAL MONEY.

XXXVI. 1. When the dividend only consists of Federal Money

RULE.

2. *Divide as in simple numbers, and the quotient will be the answe in the lowest denomination of the dividend, which may then be brough into any other denomination required.*

XXXVI. *Q.* When the dividend is Federal Money, what is the rule fo dividing? 2.

---

1 LEGACY. A bequest; money or property left by will.
2 SPECIE. Coin, copper, silver, or gold, used as money.
3 DEPOSITED. Laid down, lodged in the care of; laid aside.
4 VAULT. A cellar; an arch; a cave; a grave; a receptacle used by banks for the safe keeping of money.
5 CIRCULATION. The act of moving round or in a circle; a series in which things return to the same state; currency

3. Divide $17.50 equally among 7 men.

```
    $   ct.
7 ) 1 7 . 5 0
A. $ 2 . 5 0
```

For $17.50 (by XXXII. 24)=1750 cents÷7=250 cents=$2.50, the same as in the operation.

4. Divide $202.568 by 8. *A.* $25.321.

5. Divide $728.065 by 23. *A.* $31.625.

6. What is the price of 1 pound of sugar, when 20 pounds cos $3? When 8 pounds cost $1?

```
      (5.)                      (6.)
      $   ct.                   $   ct.
 2 0 ) 3 . 0 0             8 ) 1 . 0 0 0
       $ . 1 5=15 cents A.     $ . 1 2 5=12½cents A.
```

7. Hence when there is a remainder in dividing dollars:—*Bring them into cents by annexing two ciphers, and if there be still a remainder; into mills by annexing another cipher.*

8. When 250 bushels of oats cost $80; what is the price per bushel? *A.* 32 cents.

9. A farmer sold 30 bushels of potatoes for $5.85: what was that per bushel? *A.* 19½ cents.

10. A man bought 3 hats for $10.40; what was that apiece?

```
    $.    ct.
 3 ) 1 0 . 4 0 0
A. $ 3 . 4 6 6+.
```

In business all the mills under 5 are rejected, but 5 mills are reckoned as ½ a cent, and all over 5, and not exceeding 10, as 1 cent.

11. A father divided an estate of $30,000 equally among 13 children; what was each one's part? *A.* $2,307.69.+

12. Suppose a man's salary to be $3,650.60 a year of 365 days, what is that for a single day? *A.* $10.+

13. A man buys 36 pounds of sugar for $10.50; what is it a pound? *A.* $.29.+

14. Suppose 1,013 sheep cost $2,532.25, what were they apiece? *A.* $2.499+*but call it* $2.50.

15. Suppose you buy 2 knives for 40 cents, what must you ask apiece for them to gain 10 cents? *A.* 25 cents.

16. A grocer bought 20 barrels of apples for $70; what must he sell them for per barrel, to gain $10. *A.* $4.

17. Suppose a boy gives 50 cents for 3 balls and looses one of them; how much do the 2 left stand him in apiece? *A.* $.25.

18. Suppose a grocer pays $70 for 20 barrels of apples, and after a while 2 barrels become so rotten as to be worthless; what price per barrel will indemnify[1] him? (See No. 10.) *A.* $3.89.+

How do you proceed with the remainder? 7. How are mills to be regarded in the quotient? 10.

---

1 INDEMNIFY. To save from harm or loss, to make good; to reimburse to one what he has lost

19. A merchant bought 5 barrels of flour for $47.50 ; he kept one barrel himself, and sold the rest for what he paid for the whole ; what was that for each barrel? *A.* $11.87½.

20. A father and his 2 sons received a legacy of $30,000. The father had ½ and the remainder was equally divided between the sons; what was each one's part? *A.* Father's $15,000 ; son's $7,500.

21. A boy having $1, purchased 3 knives, which took all the money e had except 25 cents ; what did he pay apiece for them ? *A.* 25 cents.

22. A man having $174.60 purchased 4 cows ; which took all the money he had except $25 ; what did he pay apiece for them ? *A.* $37.40.

XXXVII. 1. When the divisor and dividend both consist of Federal Money.

RULE.

2. *Reduce the divisor and dividend to the lowest denomination mentioned in either, then divide as in simple numbers.*

3. How many hats at $1.50 apiece may be bought for $9 ?

```
 cts.    cts.
1 5 0 ) 9 0 0 ( 6 A.
        9 0 0
```

Reduce $9 and $1.50 each to cents ; then there will be as many hats as there are times 150 in 900.

4. How many barrels of flour may be bought for $300, at $7.50 a barrel ? *A.* 40 barrels.

5. When molasses is 42 cents by the gallon, and $26.46 by the hogshead, how many gallons must a hogshead hold ? *A.* 63 gallons.

6. A certain vessel was owned equally by so many persons that each part was only $1,250 ; what was the number of owners, supposing the vessel cost $62,500 ? *A.* 50 owners.

7. If a farmer received $21 profit from each cow, and from the whole $945, how many cows must he have had ? *A.* 45 cows.

8. When corn is $1.125 per bushel, how many bushels may be bought for $963 ? Bring $963 into mills. *A.* 856 bushels.

9. When horse-keeping costs 87½ cents a day, how long might a horse be kept at that rate for $1,575 ? *A.* 1800 days.

10. How many melons will $5 purchase, at 25 cents apiece ? *A.* 20 melons.

11. How many times greater then in value is $5, than 25 cents ?— than 2 cents 5 mills ? *A.* 20 times ; 200 times.

12. How many times greater is $73 than 40 cents ? *A.* $182\frac{20}{40}$.

13. What number multiplied by 8 mills, will make a product equal in value to $20,000 ? *A.* 2,500,000.

14. Divide $304 by 23 cents ? *A.* $1321\frac{17}{23}$.

XXXVII. *Q.* How do you proceed when the divisor and dividend are both in Federal Money ? How many times greater in value is 8 dollars than 8 cents ?— than 8 mills ?

15. If you pay \$13 a ton for hay, how many tons can you buy for \$240.50? *A.* $18\frac{650}{1300}$ tons.

16. If a district school receives from the state fund \$45 being at the rate of \$1.25 a head; what is the number of scholars? *A.* 36.

17. A merchant invested \$34,500 in cotton, at $11\frac{1}{2}$ cents a pound, and sold it again for $12\frac{1}{2}$ cents a pound; how many pounds did he purchase, and what was the profit on the whole?

NOTE.—First find the quantity bought as before, which multiply by the profit on a single pound. *A.* 300,000 pounds; profit \$3,000.

18. A gentleman invested \$18,000 in trees of the "genuine morus multicaulis" species, paying for each 30 cents, and afterwards sold them for 40 cents apiece; how many trees did he purchase, and how much did he gain on the whole? *A.* 60,000 trees; gain \$6.000

MISCELLANEOUS EXAMPLES

XXXVIII. Add together the following numbers.—

(1.)	(2.)
One dollar and one cent,	One thousand dollars,
One hundred dollars,	One cent and one mill,
Two dollars and two cents,	One dollar and nine mills,
Twenty cents and five mills,	Ten dimes and three cents,
Six dollars and one mill,	One million dollars, one mill,
A. \$109.236	*A.* \$1,001,002.051

(3.)	(4.)
From Eight million dollars	From One billion dollars
Take Forty-five cents.	Take One cent and one mill.
A. \$7,999,999.55	*A.* \$999,999,999.989

5. Multiply six hundred dollars fifteen cents and five mills by sixty-two hundred. *A.* \$3,720,961.

6. If 1,050 be one factor of a certain number, and 1,113 another factor, what is that number. *A.* 1,168,650.

7. If it costs 6 cents for a boy to go once into the museum, how many times could he go in for \$48? How many times could 4 boys go in for the same money? Only $\frac{1}{4}$ as many times as 1 boy.
A. 800 times; 200 times.

8. There are four numbers; the first is 215, the second 401, the third 625, and the fourth as much as all the other three lacking 200 what is the sum of them all? *A.* 2,282.

9. Write down three hundred and fifteen, multiply it by twenty-nine, subtract one hundred and thirty-five from the product, and if you divide the remainder by nine, the quotient will be 1,000.

10. Perform $40,000-500\div20\times25+625=50,000$.

11. Perform $78 \times 6 - 168 \div 60 \times 40 - 200$. *A.* =0.

12. Perform $\frac{10000000}{30}$. *A.* 333,333$\frac{10}{30}$.

13. What is $\frac{1}{5}$ of 100, or the 5th part of 100? *A.* 20.

14. What is the 7th part of 3,567,895? *A.* 509,699$\frac{2}{7}$.

15 What is the 49th part of 1,374,952? *A.* 280,60$\frac{12}{49}$.

16. There are two numbers, the greater of which is 37 times 45 and the less 26 times 19; what is their sum, difference and product? *A.* 2,159; 1,171; 822,510.

17 What number deducted from the 8th part of 200 will leave 10? *A.* 15.

18. If one man receive $500 more than another, and both receive $25,896.60, what sum does each receive? *A.* $12,698.30; $13,198.30.

19. Suppose a stage goes 3 times as fast as a footman, and a rail-road car 4 times as fast as the stage, and that they all go 800 miles; how far does each go? *A.* 50 miles; 150 miles; 600 miles.

20. Suppose a box to contain 275 Grammars and 412 Arithmetics, what will they come to at 33 cents apiece? $226.71

21. A farmer sold a grocer 30 bushels of potatoes for 27 cents a bushel, for the payment of which he took a keg of molasses, containing 8 gallons, at 45 cents a gallon, and the balance in cash; how much money did he receive? *A.* $4.50.

22. A exchanged with B 37 bushels of apples at 45 cents a bushel; 4 bushels of onions at 63 cents a bushel; 200 bushels of corn at $1.08, for 55 yards of cloth at $2.14 a yard, and for the balance he was to receive the cash; what was A's due? *A.* $117.47

23. When oats are 50 cents a bushel, and sugar 10 cents a pound, how many pounds of sugar will 4 bushels of oats purchase? *A.* 20 pounds.

24. When the market price of cheese is 7 cents a pound, and that of salt $1.10 a bushel, how many bushels of salt will 440 pounds of cheese purchase? *A.* 28 bushels.

QUESTIONS INVOLVING FRACTIONS*

PREPARATORY TO THE OPERATIONS WITH COMPOUND NUMBERS

XXXIX. 1. To add halves and quarters.—*Call every 2-quarters (as $\frac{1}{4}$ and $\frac{1}{4}$) $\frac{2}{4}$ or $\frac{1}{2}$; every 2-halves (as $\frac{1}{2}$ and $\frac{1}{2}$) 1 whole, and every 4-quarters (as $\frac{1}{4}$ and $\frac{3}{4}$) 1 whole.*

XXXIX. *Q.* How are halves and quarters added with whole numbers? 1. Why do $\frac{1}{4}$ and $\frac{1}{4}$ make $\frac{1}{2}$? See the Note after 1. How many whole ones in $\frac{1}{2}$ and $\frac{1}{2}$ added together?—in $\frac{1}{4}$ and $\frac{3}{4}$?—in $\frac{1}{2}$ and $\frac{3}{4}$?—in $\frac{1}{4}$, $\frac{1}{2}$ and $\frac{3}{4}$?—in $\frac{1}{2}$, $\frac{3}{4}$, $\frac{1}{4}$?

---

*For the explanation of Fractions, See Part First from VI. to VII.; which the learner would do well to revise before he commences this chapter.

NOTE.—The $\frac{2}{4}$ is called $\frac{1}{2}$ because it means 2 of 4 equal parts, that is $\frac{1}{2}$ of them.

2. How many cents will $\frac{1}{2}$ of a cent, $\frac{3}{4}$ of a cent, $\frac{1}{4}$ of a cent, $\frac{3}{4}$ of a cent, and $\frac{1}{2}$ of a cent make added together? *A.* $2\frac{3}{4}$ cents

3. A man spent $6\frac{1}{4}$ cents for a glass of sarsaparilla, $18\frac{3}{4}$ cents for a water melon, $12\frac{1}{2}$ cents for an inkstand, $37\frac{1}{2}$ cents for a book, what did all these articles amount to? *A.* 75 cents.

4. Rufus bought a ball for $12\frac{1}{2}$ cents, a penknife for $87\frac{1}{2}$ cents, a writing-book for $12\frac{1}{2}$ cents, an Arithmetic for $18\frac{3}{4}$ cents, and a pencil for $6\frac{1}{4}$ cents; what did he pay for the whole? A. \$$1.37\frac{1}{2}$.

5. Suppose a farmer has in one bin $320\frac{1}{2}$ bushels of oats, in an other $62\frac{1}{2}$ bushels, in another $49\frac{3}{4}$ bushels; how many bushels has he in all his bins? *A.* $432\frac{3}{4}$ bushels.

XL. To multiply by halves, thirds, quarters, &c.

1. A father having three sons, promised the youngest 2 cents, the next older son 1 cent, and the oldest $\frac{1}{2}$ a cent for every sum each would do that day. They did 40 sums apiece. How many cents then must he pay each? *A.* 80 cents; 40 cents; 20 cents.

2. To multiply by 2, we take the multiplicand 2 times; by 1, we take it 1 time; and by $\frac{1}{2}$ we take it $\frac{1}{2}$ *a time*, that is, the *half* of it.

3. Hence to multiply by $\frac{1}{2}$, $\frac{1}{3}$, &c.—*Divide the multiplicand by the figure below the line.*

4. For the smaller the multiplier, the smaller will be the product.

```
4 ) 3 6 0 0
         6¼
 ----------
      9 0 0
  2 1 6 0 0
 ----------
$ 2 2 5.0 0
```

5. What will 3,600 yards of ribbon cost, at $6\frac{1}{4}$ cents a yard?

NOTE.—Take $\frac{1}{4}$ of 3,600=900, then multiply by 6 as usual, and add 900 to that product. *A.* 22500 ct.=\$225.

6. Multiply 2 rods by $5\frac{1}{2}$ yards (=1 rod.) *A.* 11 yards.
7. Multiply 26 rods by $5\frac{1}{2}$ yards. *A.* 143 yards.
8. Multiply 2 degrees by $69\frac{1}{2}$ miles (=1 degree.) *A.* 139 miles.
9. Multiply 360 degrees by $69\frac{1}{2}$ miles. *A.* 25,020 miles.
10. Multiply 640 sq. rods by $30\frac{1}{4}$ sq. yards (=1 sq. rod.) *A.* 19,360 sq. yd.
11. Multiply 133 sq. rods by $272\frac{1}{4}$ sq. feet. *A.* $36{,}209\frac{1}{4}$ sq. ft
12. Multiply 60 by $\frac{1}{3}$,—by $5\frac{1}{3}$,—by $12\frac{1}{5}$. *A.* 20 : 320 : 732.
13. Multiply 400 by $8\frac{1}{20}$, by $36\frac{1}{200}$. *A.* 3,220: 14,402.
14. Multiply 100 years by $365\frac{1}{4}$ days. *A.* 36,525 days.
15. To multiply by $\frac{2}{3}$, $\frac{3}{4}$, &c.—*Divide by the figure below the line and multiply the quotient by the upper figure. Or multiply first and divide afterwards, especially if there be a remainder.*

XL. *Q.* How often do we take the multiplicand in multiplying by 2, 1 and $\frac{1}{2}$? See 2. What then is the rule for multiplying by $\frac{1}{2}$, $\frac{1}{3}$, &c.? 3. What is the principle? 4. How is 3,600 multiplied by $6\frac{1}{4}$? 4.

16. How much is $\frac{1}{4}$ of 24? *A.* 6. $\frac{3}{4}$ of 24? *A.* 18. 8 times 24? *A.* 192. How much then is $8\frac{3}{4}$ times 24?

	1st.	2nd.
2 4	4) 2 4	or 2 4
$8\frac{3}{4}$	6	3
1 8	3	4) 7 2
1 9 2	1 8	1 8
2 1 0		

Divide 24 by 4 and multiply by 3; or as the result is the same, multiply first and divide afterwards; making 18 to be added to 192 (=24×8.)

17. Multiply 460 by $6\frac{3}{4}$,—by $16\frac{3}{5}$. *A.* 3,105; 7,636.
18. Multiply 504 by $9\frac{5}{8}$,—by $29\frac{11}{16}$. *A.* 4,851; $14{,}962\frac{8}{16}$.
19. Multiply 370 by $7\frac{5}{9}$,—by $85\frac{36}{37}$. *A.* $2{,}795\frac{5}{9}$; 31,810.
20. How much does $8\frac{7}{10}$ times 1,000 exceed $8\frac{3}{100}$ times 1,000? *A.* 670.

XLI. To divide by halves, thirds, quarters, &c.

1. How many yards of tape may be bought for 4 cents, at 2 cents a yard?—at 1 cent a yard?—at $\frac{1}{2}$ of a cent a yard?—at $\frac{1}{4}$ of a cent a yard? *A.* 2 yd; 4 yd; 8 yd; 16 yd.
2. Hence to divide by $\frac{1}{2}$, $\frac{1}{4}$, &c.—*Multiply by the lower figure.*
3. For the smaller the divisor the greater will be the quotient.
4. How many bushels of oats may be bought for \$1 at $\frac{1}{2}$ of a dollar a bushel?—for \$1,000? *A.* 2 bushels; 2,000 bushels.
5. How many yards of calico at $\frac{1}{5}$ of a dollar a yard, may be bought for \$1?—for \$1,200? *A.* 5 yards; 6,000 yards.
6. How many times greater in value is \$1,800 than \$$\frac{1}{8}$? *A.* 14,400.
7. Since $\frac{3}{4}$ is 3 times as much as $\frac{1}{4}$: $\frac{4}{5}$, 4 times as much as $\frac{1}{5}$, &c. therefore,—
8. To divide by $\frac{2}{3}$, $\frac{3}{4}$, &c.—*Multiply by the lower figure, and divide by the upper one. Or divide first and multiply afterwards, when it can be done without a remainder.*
9. How many bushels of rye may be bought for \$10, at $\frac{5}{8}$ of a dollar a bushel?—for \$2,400? (\$10×8÷5; or 10÷5×8.) *A.* 16 bushels; 3,840 bushels.
10. When rye is $\frac{3}{4}$ of a dollar a bushel, how many bushels may be bought for \$1.50?—for \$300? *A.* 2 bushels; 400 bushels.
11. How many times greater in value is \$6 than $\frac{2}{3}$ of a dollar?—is \$3,000 than $\frac{4}{5}$ of a dollar? *A.* 9 times; 3,750 times.
12. Divide 20,000 by $\frac{3}{20}$,—by $\frac{75}{100}$,—by $\frac{85}{250}$,—by $\frac{500}{4000}$,—by $\frac{4}{5}$.
Answers. $133{,}333\frac{1}{3}$; $26{,}666\frac{50}{75}$; $58{,}823\frac{45}{85}$; 160,000; 25,000.
13. At $\$1\frac{1}{2}$ a bushel, how many bushels of wheat may be bought for 3 dollars?—for 1,200 dollars?

Q. How much is $\frac{1}{4}$ of 24? $\frac{3}{4}$ of 24? How is 24 multiplied by $8\frac{3}{4}$? See 16.

XLI. *Q.* How many yards of cloth at 2 dollars a yard, may be bought for $\frac{1}{2}$ of a dollar? See 2. For $\frac{1}{4}$ of a dollar? See 2. What is the rule? 2. What is the principle? 3. What is the rule for dividing by $\frac{3}{4}$, $\frac{2}{3}$, &c.? 8. How many times is $\frac{5}{8}$ contained in 10? See 9. In 2,400? See 9.

NOTE.—Since 2 halves make 1 whole there are 6-halves in \$3, for \$3 × 2 halves = 6 halves. Again \$1 × 2 halves = 2 halves + 1-half (= $\frac{1}{2}$) 3-halves: which is the divisor of the 6 halves; thus 6 ÷ 3 = 2. So \$1,200 × 2 halves = 2,400 halves to be divided by 3 halves also.
A. 2 bushels; 800 bushels.

14. How many fishhooks at $2\frac{1}{4}$ cents apiece, may be bought for 9 cents? 4-quarters = 1 whole; then 2 cents × 4-quarters = 8-quarters and 1-quarter (= $\frac{1}{4}$) more make 9-quarters the divisor, and 9 cents × 4-quarters = 36-quarters, the dividend. *A.* 4 fishhooks.

15. At \$$2\frac{1}{5}$ a yard, how many yards may be bought \$9,625?

```
$ 2 1/5
   5-fifths.             $ 9 6 2 5
 ----------                      5-fifths.
 1 0-fifths.           -----------
   1-fifth.      1 1 ) 4 8 1 2 5-fifths.
 ----------            -----------
 1 1-fifths.        A.  4 3 7 5 yards.
 ==========             =======
```

16. We multiply the \$2 by 5, and add in the 1 = 11-fifths, and the \$9,625 by 5 = 48,125-fifths.

16. Hence to divide by $5\frac{1}{2}$, $6\frac{1}{4}$, &c.—*Multiply the whole number of the divisor by the figure below the line, and add in the figure above the line for a new divisor.*

17. *Multiply the dividend by the same multiplier for a new dividend, then divide as usual.*

18. At \$$4\frac{3}{5}$ a yard, how many yards of cloth may be bought for \$23? for \$4,600? *A.* 5 yards; 1,000 yards.

19. What is the quotient of 63 divided by $2\frac{5}{8}$? Say 2 × 8 + 5 = 21 the divisor: 63 × 8 = 504, the dividend. *A.* 24.

20. Divide 1,102 by $4\frac{5}{6}$,—by $5\frac{3}{7}$,—by $10\frac{5}{13}$.
A. 228; 203; $106\frac{16}{135}$.

21. There are $5\frac{1}{2}$ yards in every rod; how many rods then in 33 yards?—in 57 yards? *A.* 6 rods; $10\frac{4}{11}$ rods.

22. There are $69\frac{1}{2}$ statute miles in every degree; how many degrees in 400 miles? *A.* $5\frac{105}{139}$ degrees.

23. How many square rods in 24,200 square yards, there being $30\frac{1}{4}$ square yards in every square rod? *A.* 800 square yards.

24. Divide 1,090 square feet by $272\frac{1}{4}$ square feet. *A.* $4\frac{4}{1089}$.

25. How many years are there in 143,830 days allowing $365\frac{1}{4}$ days to the year? *A.* $393\frac{1147}{1461}$ years.

26. How many nails are there in 38 inches, allowing $2\frac{1}{4}$ inches to make 1 nail? $2\frac{1}{4}$ = 9-quarters, the divisor; and 38 × 4 = 152, the dividend: the quotient is $16\frac{8}{9}$ nails, but since the remainder is always like the dividend, the 8 over in dividing is 8-quarters of an inch; then 8-quarters ÷ 4 quarters = 2 inches. *A.* 16 nails and 2 inches.

27. *Hence dividing the remainder, by the multiplier of the dividend; will give a quotient of the same denomination with the given dividend.*

28. Divide 12 yards by $5\frac{1}{2}$ yards (= 1 rod.) *A.* 2 rods 1 yard.

29. Divide 376 yards by $5\frac{1}{2}$ yards (= 1 rod.) *A.* 68 rods 2 yards.

Q. How many pair of boots at $5\frac{1}{2}$ dollars a pair, may be bought for 11 dollars? For 22 dollars? For 33 dollars? What is the rule for dividing thus? 16. 17

30. Divide 73,050 days by $365\frac{1}{4}$ days. *A.* 200 years 8 days.

XLII. 1. To find the cost of articles when the price is an aliquot part of a dollar. See VII. 81, 82, 86.—*Divide the given quantity by the number of aliquot parts which it takes of the price to make a dollar, the quotient will be the cost in dollars. See the Table*, VII. 86.

2. At 50 cents apiece what will 790 cubical blocks cost? 50 ct. = $\$\frac{1}{2}$; divide by 2 because every 2 costs $1. *A.* 395.

3. At $33\frac{1}{3}$ cents apiece, what will 591 inkstands cost? $33\frac{1}{3}$ ct. = $\$\frac{1}{3}$; divide by 3 because every 3 costs $1. *A.* $197.

4. At 25 cents apiece, what will 980 trees cost? 25 ct. = $\$\frac{1}{4}$; divide by 4, because every 4 cost $1. *A.* $245.

5. At $16\frac{2}{3}$ cents apiece, what will 480 mellons cost? $16\frac{2}{3}$ ct. = $\$\frac{1}{6}$; divide by 6 because every 6 cost $1. *A.* $80.

6. At $12\frac{1}{2}$ cents apiece, what will 1,080 books cost? $12\frac{1}{2}$ ct. = $\$\frac{1}{8}$; divide by 8 because every 8 cost $1. *A.* $135.

7. At $6\frac{1}{4}$ cents apiece, what will 960 steel pens cost? $6\frac{1}{4}$ ct. = $\$\frac{1}{16}$; divide by 16 because every 16 cost $1. *A.* $60.

8. At $12\frac{1}{2}$ cents a yard, what will 12 yards of ribbon cost? What 1,565 yards cost? *A.* $1.50.

8) 1 2 . 0 0   8) 1 , 5 6 5 . 0 0 0   *A.* $$195.62\frac{1}{2}$.
$ 1 . 5 0 $ 1 9 5 . 6 2 5

9. *Hence when there is a remainder, annex two ciphers for cents and one for mills.*

10. What will be the cost of the following articles, viz:—
410 bushels of rye at 50 cents a bushel? *A.* $205.
360 bushels of oats at $33\frac{1}{3}$ cents a bushel? *A.* $120.
415 bushels of apples at 25 cents a bushel? *A.* $103.75.
417 yards of calico at 20 cents a yard? *A.* $83.40.
815 yards of shirting at $16\frac{2}{3}$ cents a yard? *A.* $135,833.+
489 quarts of cherries at $12\frac{1}{2}$ cents a quart? *A.* $$61.12\frac{1}{2}$.
853 quarts of cranberries at 10 cents a quart? *A.* $85.30.
353 pounds of cheese at $6\frac{1}{4}$ cents a pound? *A.* $22.06.+
426 pounds of beef at 5 cents a pound? *A.* $21.30.

11. At $2.50 or $\$2\frac{1}{2}$ per yard; what will 4 yards of broadcloth cost? 1,853 yards cost? Multiply by $2\frac{1}{2}$. *A.* $10; $4,632.50.

12. What will be the cost of the following articles, viz.—
201 bushels of rye, at $\$2.12\frac{1}{2}$ or $\$2\frac{1}{8}$ per bushel? *A.* $$427.12\frac{1}{2}$.
640 acres of land at $\$25.06\frac{1}{4}$, or $\$25\frac{1}{16}$ per acre? *A.* $16,040.
315 barrels of flour, at $5.25 or $\$5\frac{1}{4}$ per barrel? *A.* 1,653.75.
941 gallons of oil at $\$1.16\frac{2}{3}$, or $\$1\frac{1}{6}$ per gallon? *A.* $1,097.83.+

XLII. *Q.* What is meant by an aliquot part? See VII. 81. Give several examples. See VII. 82, 83. Repeat the Table of Aliquot parts? See VII. 86. What is the rule for finding the cost when the price is an aliquot part of a dollar? 1. What will be the cost of 96 yards of calico at 50 cents a yard?—at $33\frac{1}{3}$ cents?—at 25 cents?—at $16\frac{2}{3}$ cents?—at $12\frac{1}{2}$ cents?—at $6\frac{1}{4}$ cents?

13 To find the quantity when the price is an aliquot part of a dollar.—*Reverse the last rule, that is, multiply the whole cost by the number of aliquot parts which it takes of the price to make one dollar.*

14. At $12\frac{1}{2}$ cents a pound, how many pounds of sugar may be bought for \$1?—for \$500? $12\frac{1}{2}$ ct. = \$$\frac{1}{8}$, then \$$\frac{8}{8}$ or \$1 will buy 8 pounds and \$500 will buy 8 times 500=4,000 pounds.

A. 8 pounds; 4,000 pounds.

15. When flaxseed is $16\frac{2}{3}$ cents a peck, how many pecks may be bought for \$1?—for \$7,118? *A.* 6 pecks; 42,708 pecks.

16. How many Madeira trees can be bought for \$829, when the price of each is $33\frac{1}{3}$ cents? *A.* 2,487 trees.

17. A gentleman invested \$1,000 in goods of various kinds; what quantity of each did he purchase, taking their several prices from the following memoranda, viz:

For calico \$150, at $12\frac{1}{2}$ cents per yard.	*A.* 1,200 yards.
For gingham \$116, at $16\frac{2}{3}$ cents per yard.	*A.* 696 yards.
For French calico \$50, at 25 cts. per yard.	*A.* 200 yards.
For silk \$200, at 50 cents per yard.	*A.* 400 yards.
For shoes \$100, at $33\frac{1}{3}$ cents a pair.	*A.* 300 pairs.
For gloves \$50, at 20 cents a pair.	*A.* 250 pairs.
For cotton balls \$50, at 5 cents apiece.	*A.* 1,000 balls.
For cotton cloth \$48, at $6\frac{1}{4}$ cents a yard.	*A.* 768 yards.

The balance, or what remains of the \$1,000 after deducting the cost of the above articles, he laid out in linen at 40 cents a yard; how many yards did he buy? *A.* 590 yards.

BILLS OF PARCELS.

XLIII. 1. Bills, or Bills of Parcels, are statements of goods bought and sold, with the particulars of price and quantity, as in the following examples.

(2.) New York, July 26, 1838.

Chauncey Ackley, Esq.,

Bought of John Smith,

20 merino sheep, at \$6 per head.	\$
25 calves, at \$2.$12\frac{1}{2}$ per head.	
200 pounds of cheese, at $6\frac{1}{4}$ cents per pound.	
36 bushels of oats, at $27\frac{3}{4}$ cents per bushel.	
17 bushels of corn at 75 cents per bushel.	
	\$208.$36\frac{1}{2}$.

November 15, 1839, Received payment,

John Smith.

Q. Suppose there is a remainder, how do you proceed? 9. How many yards of cloth may be bought for 10 dollars, at 50 cents a yard?—at $33\frac{1}{3}$ cents?—at 25 cents?—at 20 cents?—at $16\frac{2}{3}$ cents?—at $12\frac{1}{2}$ cents?—at $6\frac{1}{4}$ cents?—at 5 cents? What is the rule? 13.

(3.) Boston, June 6, 1839.

Mr. George Simpson,

Bought of Rufus Paywell,

8 barrels of cider, at $\$2.12\frac{1}{2}$ a barrel. . . . . $

6 bushels of corn, at $\$1.16\frac{2}{3}$ per bushel. . . . .

$24.00

August 8, 1839, Received payment,

Rufus Paywell.

MERCHANT'S BILL.

(4.) Philadelphia, January 1, 1827.

Messrs. Clark & Brothers,

Bought of Peter Rice,

3,800 yards of calico, at $17\frac{3}{4}$ cents a yard. . . . . . $

40 pieces of blue broadcloth, each 37 yards, . . .
at $\$4.62\frac{1}{2}$ per yard. . . . .

400 yards carpeting, at $1.18 per yard. . . . . . .

200 pieces of nankin, each 42 yards, at $.39, . .
per yard. . . . .

$11,267.50

Received payment for Peter Rice,

John Stimpson.

REDUCTION OF COMPOUND NUMBERS

SEE THE TABLES OF MONEY, WEIGHTS, AND MEASURES. VII.

XLIV. 1. How many feet in 36 inches?—inches in 3 feet?

2. How many shillings are there in £5?—pounds in 100 shillings?

3. How many farthings in 1,200 pence?—pence in 4,800 farthings?

4. How many shillings in 1,200 pence?—pounds in 100 shillings?

5. How many days in 2 years?—hours in 730 days?—minutes in 17,520 hours?—seconds in 1,051,200 minutes? years in 63,072,000 seconds?

6. *Hence pounds must be multiplied by what makes a pound, shilling by what makes a shilling, hours by what makes an hour, &c.*

7. This process is called Reduction Descending, because numbers by it are *carried down* to lower denominations.

XLIV. *Q.* How many shillings in 2 pounds?—in 8 pounds? Pence in 5 shillings? Shillings in 60 pence? How many pence in 20 farthings?—in 10 shillings?—in 48 farthings? What is Reduction? XXXII. What are Compound Numbers? IX. 11. Give an example. IX. 12. How are numbers to be multiplied in Reduction? 6. How divided? 8. What is the former process called, and why? 7. What is the latter called, and why? 9. How many shillings in £5. 10s.? Pounds in 80 shillings? How many rods in 5 furlongs 2 rods? Feet in 150 inches? Yards in 37 feet?

8. *Shillings too must be divided by as many shillings as make a pound; minutes, by as many minutes as make an hour, &c.*

9. This process is called REDUCTION ASCENDING, because numbers by it are *carried up* to higher denominations.

10. Reduce £2. 5s. $6\frac{3}{4}$d. to shillings, pence and farthings.

```
  £.  s.  d.  qr.
  2.  5.  6.  3.
      2 0 s. multiplier.
      4 0 s.
        5 s. added
      4 5 s.
      1 2 d. multiplier.
    5 4 0 d.
        6 d. added.
    5 4 6 d.
        4 qr. multiplier.
  2 1 8 4 qr.
        3 qr. added.
  2 1 8 7 qr. Answer.
```

11. Since 20s.=£1; 12d.=1s; 4qr. =1d.; there will be 20 times as many shillings as pounds; 12 times as many pence as shillings, and 4 times as many farthings as pence.

PROOF.

12. How many pounds in 2,187 farthings?

```
   4 qr. ) 2 1 8 7 qr.
  1 2 d. ) 5 4 6 d. 3 qr.
    2 0 s. ) 4 5 s. 6 d. 3 qr.
      Ans. £ 2.5 s. 6 d. 3 qr.
```

13. How many farthings in £4. 10s. 8d. 2qr.?
14. How many pounds in 4,354 farthings?
15. How many quarters in 2 T. 5 cwt. 3 qr?
16. How many tons are there in 183 quarters?
17. How many drams in 3 T. 17 cwt. 3 qr. 17 lb. 8 oz. 5 dr.?
18. How many tons are there in 1,994,885 drams?
19. How many seconds in 1 Y. 51 d. 13 h. 35 m. 40 sec.?
20. How many years are there in 35,991,340 seconds?
21. How many rods in 18 m. 3 fur. 15 rods?
22. How many miles are there in 5,895 rods?

REDUCTION DESCENDING AND ASCENDING.

23. REDUCTION DESCENDING, is reducing numbers from higher denominations to lower ones.

RULE.

24. *Multiply the highest denomination given, by as many of the next lower as make one of that higher, adding in as many of that lower as are in the given sum, and so on.*

25. REDUCTION ASCENDING, is changing numbers from lower denominations to higher ones.

RULE.

26. *Divide the lowest denomination by as many of that as make one of the next higher denomination. Divide that quotient in the same manner, and so on, the last quotient with the several remainders will form the answer.*

Q. How are pounds reduced to farthings? Farthings to pounds? 12. Why multiply in one case by 20, 12 and 4, and in the other case divide by these same numbers? 11. What is Reduction Descending? 23. Rule? Reduction Ascending? 25 Rule? 26.

2?. Reduction Descending and Ascending prove each other.

ENGLISH MONEY.

28. How many farthings are there in £6. 8s. 4d. 2qr.?
29. How many pounds are there in 6,162 farthings?
30. How many farthings are there in £25. 9½d.? (½d.=2 qr.)
31. How many pounds are there in 24,038 farthings?
32. How many times are there 6 pence in 414 pence?
33. How many pence are there in 69 sixpences?
34. How many 6 pences are in £40,000?
35. How many pounds are there in 1,600,000 sixpences?
36. How many pounds are there in $445?*
37. How many dollars are there in £133. 10s.?
38. How many guineas of 21s. each, in £588?
39. How many pounds are there in 560 guineas?
40. How many French crowns at 6 shillings and 8 pence each, are equal in value to 1,161,600 farthings? (6s. 8d.=80d.)
41. How many farthings in 3,630 French crowns?
42. How many half-pence are there in £505. 3s. 11½d.?
43. How many pounds are there in 242,495 half-pence?
44. How many 4½d. pieces are there in 96 guineas?—(4½=18qr.)
45. How many guineas in 5,376 four-pence half-pennies?
46. How many threepences and sixpences in one ninepence, of each an equal number? 6+3=9, then 9÷9=1 of each, *A.*
47. Suppose 630 pence to contain an equal number of sixpences and 3 pences, what is that number? *A.* 70.
48. In £101. 5s. how many guineas and dollars, of each an equal number? *A.* 75.
49. Reduce £32. 16s. to dollars of 8s. each. *A.* $82.
50. Reduce £11. 5s. to dollars of 7s. 6d. each. *A.* $30.
51. Reduce $360 of 4s. 6d. sterling to pounds. *A.* £81.
52. How many dollars will purchase 7,200 bushels of potatoes, at 1s. 6d. per bushel?
53. How many bushels of potatoes may be bought for $1,800, at 1s. 6d. per bushel?
54. When 7,200 bushels of potatoes cost $1,800, what is their price by the bushel?
55. Suppose a merchant imports from England 80,640 yards, of tape, at ½d. per yard, how many pounds will pay for it? *A.* £168.

Q. Proof of both? 27. What is English Money? VII. 2. Repeat the Table. How are pounds brought into sixpences? 34. Dollars into pounds and the reverse? 36. Pounds and shillings into guineas? How many guineas in £2 and 2 shillings? Pounds in 2 guineas? Pounds in $6 and 4 shillings? Dollars in £2? How many 6 pences in 3s. 6d.? How are farthings brought into French crowns of 6s. 8d. each? 40. Guineas into four-pence half-pennies? 45. How are pounds brought into an equal number of 3 pences and 6 pences? How many 6 pences and 9 pences in 2s. 6d.?—in 5s.?

---

* Allow 6 shillings to the dollar when no number is mentioned.

TROY WEIGHT.

56. How many pennyweights in 24 lb. 3 oz. 15 dwt. ?

57. How many pounds are there in 5,835 pennyweights?

58. What is the value of a silver tankard, weighing 4 lb. 8 oz., at $1.15 per ounce?

59. When a silver tankard costs $64.40, at $1.15 per ounce, what will be its weight in pounds?

60. What would be the value of the same tankard at 6 cents a pennyweight? *A.* $67.20

AVOIRDUPOIS WEIGHTS.

61. How many ounces in 5 cwt. 3 qr. 17 lb. 11 oz. ?

62. How many hundred weight in 9,483 ounces?

63. How many drams are there in 3 qr. 15 lb. 10 dr. ?

64. How many quarters are there in 23,050 drams?

65. In 14 hogsheads of sugar, each weighing 10 cwt. 14 lb., how many pounds? *A.* 14,196 pounds.

66. What will 11 cwt. 2 qr. 15 lb. of sugar cost, at $12\frac{1}{2}$ cents per pound? *A.* 145.62\frac{1}{2}$.

67. How many small boxes, each to contain 25 lb. may be filled from 85 hogsheads of tobacco, each weighing 8 cwt. 15 lbs. *A.* 2,771 boxes.

APOTHECARIES' WEIGHT.

68. How many grains in 5 ℔ 5 ℥ 1 ʒ 2 ℈ 13 gr?

69. How many pounds are there in 31,313 grains?

70. How much calomel and aloes are contained in 36 boxes of pills, each box having 20 pills, and each pill 2 grains of calomel and 8 grains of aloes? (2 gr. + 8 gr. = 10 gr.) *A.* 1℔ 3℥.

71. An apothecary having mixed in proper proportion 3 ounces of calomel with 1 pound of aloes, wishes to find how many boxes, each to contain 20 pills, and each pill 2 grains of calomel and 8 grains of aloes, will hold the compound? *A.* 36 boxes.

CLOTH MEASURE.

72. How many nails are there in 750 yds. 1 qr. 1 na. ?

73. How many yards are there in 12,005 nails ?

74. How many flemish ells are there in 1,080 yards ? *A.* 1,440.

75. How many quarters are there in 6 pieces, each containing 20 yards and 1-quarter ? *A.* 486 quarters.

Q. For what is Troy Weight used? VII. 5. Repeat the Table. How are pounds, ounces, &c., brought into grains? How are drams brought into tons? How many pounds in 36 ounces?—in 40 oz.? Ounces in 3 pounds?—in $3\frac{4}{12}$ lb.? For what is Avoirdupois Weight used? Repeat the Table. How many pounds were formerly reckoned for a quarter? (See reference 2, at the bottom of p. 12.) How many hundred weight in 2 tons?—in $2\frac{11}{25}$ tons? Pounds in 2 quarters?—in $5\frac{1}{25}$ quarters? For what is Apothecaries' Weight used? Repeat the Table. How many scruples in 60 grains? Ounces in 5 pounds?—in $11\frac{11}{12}$ pounds? What is the use of Cloth Measure? Repeat the Table. How many yards in 9 qr.? Quarters in $10\frac{3}{4}$ yds.? How are French ells brought into nails? Quarters into yards? Yards into ells Flemish?

76. How many yards of cloth only 2-quarters wide, is equal to 10 yards which is 4-quarters wide? *A.* 20 yards.

77. How many yards 1-qr. wide are equivalent to 50 yards 4 quarters wide? *A.* 200 yards.

78. What will 200 yd. 2 qr. of cloth cost at 25 cents per quarter? *A.* $\$200\frac{1}{2}$. At $12\frac{1}{2}$ cents per quarter? *A.* \$100.25.

DRY MEASURE.

79. In 19,691 quarts how many bushels?

80. In 615 bus. 1 pk. 3 qt. how many quarts?

81. At 40 cents a peck what will 25 bushels 3 pecks of wheat cost? *A.* \$41.20.

82. When rye sells for 20 cents a peck, how many bushels may be bought for \$247.20? *A.* 309 bushels.

83. How many bags will 8,500 bushels of rye fill, allowing each bag to hold 4 bushels 1 peck? *A.* 2,000 bags.

WINE MEASURE.

84. How many pints in 2 hhd. 40 gal. 3 qt. 1 pt.?

85. How many hogsheads are there in 1,335 pints?

86. A tee-totaler found to his sorrow, that he had drank, in all his life, no less than 1 tun of wine; what would it have amounted to at $6\frac{1}{4}$ cents a half gill? *A.* \$1,008.

87. A merchant bought 5 hogsheads of molasses for $12\frac{1}{2}$ cents a quart, and sold it for $6\frac{1}{4}$ cents a pint; did he make or loose? *A.* Neither.

ALE, OR BEER MEASURE.

88. How many pints are there in 1 hhd.? *A.* 432 pints.

89. What will 45 bar. 18 gal. of ale come to at 4 cents a pint?

90. A man having retailed 45 bar. 18 gal. of ale, received for the whole \$524.16; what did he get by the pint?

91. How many $1\frac{1}{2}$ pint bottles can be filled with 3 hogsheads of ale? *A.* 864

LONG MEASURE.

92. How many inches in 100 yd. 2 ft. 5 in.?

93. How many yards are there in 3,629 inches?

94. How many inches in 1 mile? *A.* 63,360.

95. How many barley corns is it round the globe, it being 360 degrees? See XL. 3.

Q. When cloth is only 2-quarters wide, how can you find what quantity 4-quarters wide will equal it? 76. In 15 yards how many Flemish ells?—how many English ells? What is the application of Dry Measure? Repeat the Table. How are bushels reduced to pints? Quarts to chaldrons? How many pints in 3 quarts? Quarts in $10\frac{1}{8}$ pk.? (1 qt. $=\frac{1}{8}$ pk.) Pecks in $12\frac{3}{4}$ bushels? What is the use of Wine Measure? Repeat the Table. How are gills reduced to hogsheads? Pipes to gallons? How many gallons in 24 quarts?—in $1\frac{7}{63}$ hhd.? For what is Ale or Beer Measure used? Repeat the Table. What would a firkin of ale cost at 50 cents a gallon? What would a firkin of beer cost at the same price? For what is Long Measure used? Repeat the Table.

96. How many degrees in 4,755,801,600 barley corns?

97. How many inches is it from Boston to Providence, it being 40 miles? *A.* 2,534,400 inches.

98. Suppose 5 paces to make 1 rod, how many paces will reach round the earth? *A.* 40,032,000 paces.

LAND OR SQUARE MEASURE.

99. How many square rods in 5 square miles?

100. How many square miles in 512,000 sq. rods?

101. In 60 sq. m. 37 A. 17 R. how many sq. poles?

102. In 6,150,600 sq. poles how many square miles?

103. How many square feet in a room 15 feet long and 13 feet wide? (For the rule see VII. 46.) *A.* 195 sq. ft.

104. How many square rods are there in a piece of land 120 rods long and 17 rods wide? *A.* 2,040 sq. rods.

105. How many square acres in 2,040 sq. rods. *A.* 12 A. 3 R.

106. Suppose a road to be 4 rods wide, how many acres will 40 rods in length make?—will 1 mile in length make?—will 10 miles make? *A.* 1 acre; 8 acres; 80 acres.

SOLID, OR CUBIC MEASURE.

107. How many solid inches in 15 cords of wood?

108. How many cords of wood in 3,317,760 solid inches?

109. How many cubic feet of earth will fill a cellar 15 feet long, 12 feet wide, and 8 feet deep? (See VII. 60.) *A.* 1,440.

110. In a pile of wood 20 feet long, 6 feet high, and 4 feet wide, how many cord feet?—how many cords? *A.* 30 C. ft.; $3\frac{6}{8}$ C.

TIME.

111. How many seconds are there in 15 years, 315 days, 23 hours and 57 seconds? *A.* 500,338,857 seconds.

112. How many more seconds in a leapyear than in a common year of 365 days? *A.* 86,400 seconds.

113. Suppose your age to be 15 years, 7 months, 3 weeks, 5 days 17 hours, how many seconds old are you? *A.* 454,698,000 second.

114. How many seconds has it been, since the creation of the world, to the close of the year A. D. 1839, allowing the birth of

---

Q. How are furlongs brought into degrees? Miles into yards? Barley corns into feet? How many miles in 5 leagues? Inches in 5 feet 2 inches? What is the use of Square Measure? Repeat the Table. How are the square contents obtained? 103. How many square yards in a small room 6 feet square? Roods in a piece of land 50 rods long and 2 rods wide? For what is Solid Measure used? Repeat the Table. How are cords brought into solid inches? Feet into inches? How are the solid contents found? 109. How many cord feet of wood in a pile 8 feet long, 4 feet wide, and 2 feet high? How many cord feet are 96 solid feet of wood? For what is the Table of Time used? Repeat the Table. How may centuries be reduced to days? Days to years? Seconds to hours? How many seconds in 120 minutes? Minutes in 1 h. 40 m.?—in $1\frac{30}{60}$ h.?

Christ to have taken place A. M. 4,000, and each year to contain $365\frac{1}{4}$ days?* *A.* 184,264,826,400 seconds.

CIRCULAR MOTION.

115. How many seconds are there in 1 circle?

116. How many circles in 1,296,000 seconds?

117. How many seconds in 11 S. 15° 15′ 15″?

118. How many signs in 1,242,915 seconds?

TABLE OF PARTICULARS.

119. What will be the cost of 420 dozen eggs at $1\frac{1}{2}$ cents for each egg? *A.* $75.60.

120. When buttons are 5 mills apiece, what will 50 dozen cost? *A.* $3. What will 50 gross cost? *A.* $36. What will 50 great gross cost? *A.* $432.

121. Suppose 2 hogs to weigh 40 score and 15 pounds, what is their value at $5\frac{1}{2}$ cents per pound? *A.* $44.825.

122. What will be the expense of 200 reams of paper at 25 cents per quire? *A.* $1,000. At $.015 per sheet? *A.* $1,440.

123. At 7 cents a pound, how many barrels of beef may be bought for $15,050? *A.* 1,075 barrels.

124. At $5\frac{1}{4}$ cents a pound, what will 1 quarter of flour cost? *A.* $$1.31\frac{1}{4}$. What will 196 pounds, or 1 barrel cost? *A.* $10.29.

MISCELLANEOUS

EXAMPLES IN REDUCTION.

XLV. 1. In 80 guineas how many dollars at 6s. each? *A.* $280.

2. In 224 boxes of sugar, each containing 27 lb. how many hundred weight? *A.* 60 cwt. 1 qr. 23 lb.

3. In running 300 miles, how many times will a wheel 9 feet 2 inches in circumference turn round? *A.* 172,800 times.

4. In 172,800 turns of a wheel measuring 9 feet 2 inches in circumference, how many miles will be passed over?

5. How many acres on the surface of the earth, allowing it to contain 197,000,000 square miles? *A.* 126,080,000,000 acres.

6. How many times does a clock tick in a leapyear, supposing it to tick once every second? *A.* 31,622,400 times.

7. How much time will it require, for a man that is worth one million of dollars, to count that number, at the rate of 50 dollars a minute, supposing him to be employed only 10 hours each day? *A.* 33 da. 3 h. 20 m.

Q. For what is Circular Motion used? Repeat the Table. How many degrees in 2 signs? Minutes in 2 degrees?—in $2\frac{2}{60}$ degrees?

---

* The 3 left in dividing by $\frac{1}{4}$ is $\frac{3}{4}$ of a day, or because 24 hours make 1 day, it is $\frac{3}{4}$ of 24 hours, which is 18 hours to be added in when multiplying by 24 hours.

8. Suppose a man to travel 39 miles and 20 rods a day, how long would it take him to travel round the earth, it being about 25,000 miles? *A.* 1 Y. 275 days.

9. How many yards of carpeting will it take to cover the floor of two parlors each 18 feet square? What will be the expense at $2¼ per yard?*

NOTE.—18 feet square means 18 ft. long, and 18 feet wide; therefore 18 ft. × 18ft. × 2 parlors ÷ 9 sq. ft. = 72 sq. yd. *A.* 72 yd.; $153.

10. Suppose a room to be 20 feet square, how many square feet are there in the floor? *A.* 400 sq. ft.

11. How many square feet of plastering over head, in a room 20 feet square? *A.* 400 sq. ft. How many square feet in one side, supposing the room to be 12 feet high? *A.* 240. In the other three sides? *A.* 720 sq. ft.

12. In a room 20 feet long and 16 feet wide, how many square yards of plastering over head? *A.* 35 sq. yd. 5 sq. ft. How many yards of carpeting, 1 yard wide, will cover the floor? *A.* 35 and 5 sq. ft.

13. Suppose the foregoing room to be 10 feet high, how many yards of paper, 1 yard wide, will cover one end? *A.* 17 sq. yd. and 7 sq. ft. How many, the other end? *A.* 17 and 7 sq. ft. How many to cover both sides? *A.* 44 sq. yd. and 4 sq. ft.

14. How many square yards are there in the floor of a church, which is 80 feet long and 67 feet wide? *A.* 595 sq. yd. 5 sq. ft.

15. How many shingles 18 inches long, 4 inches wide, will it take to cover one side of a roof, 45 feet long and 25 feet wide?

NOTE.—In laying shingles, two-thirds of the length of each shingle are overlaid by others; therefore, each shingle must be reckoned as covering only 6 inches in length and four inches in breadth, making 24 square inches; then, 45 × 25 × 144 sq. in. ÷ 24 sq. in. *A.* 6,750.

16. How many shingles 18 inches long and 4 inches wide, will be required to cover one side of a roof 60 feet long and 25 feet wide? *A.* 9,000 shingles. How many for both sides? *A.* 18,000 shingles.

17. On a certain wharf there lies a pile of wood 40 rods long, 6 feet high and 4 feet wide; how many cord feet will it make? *A.* 990 C. ft.

18. How many times will a wheel which is 15 feet 9 inches in circumference, turn round in going 378 feet? *A.* 24. In going from Providence to Norwich, it being 45 miles? *A.* 15,085.+

19. Suppose a farmer rents a plantation of 400 acres, of which no more than 200 are tilled, how many poles are there in the remainder *A.* 32,000 poles.

20. In a lunar month of 27 days, 7 hours, 43 minutes, 5 seconds, how many seconds? *A.* 2,360,585 seconds.

---

* NOTE.—Observe that feet multiplied by feet, make square feet; inches by inches, square inches, &c. Also, that square inches must be divided by square inches; square feet by square feet, &c.

21. How many seconds is it from the birth of our Saviour to Christmas, 1828, allowing the year to contain $365\frac{1}{4}$ days, or 365 days and 6 hours? *A.* 57,687,292,800 seconds.

22. When a person is 21 years old, what is his age in seconds? *A.* 662,709,600 seconds.

23. The wars of Bonaparte caused, as is computed, in 20 years, the deaths of at least 2,103,840 persons; how many would that be for every hour of the 20 years. *A.* 12 persons

---

COMPOUND ADDITION.

XLVI. 1. A man paid 10 shillings for a gallon of oil, 15 shillings for a vest, and 17 shillings for a pair of boots; how many pounds did he pay for the whole? *A.* £2. 2s.

2. How many pounds are 9s., 16s., 19s., and 11s.? *A.* £2.15s.

3. In one lot are 36 roods, in another 57 roods, in another 25 roods, and in the fourth 17 roods; how many acres do all the lots contain? *A.* 33 A. 3 R.

4. *Hence, pounds must be added to pounds, shillings to shillings. miles to miles, &c.*

5. A farmer bought a load of hay for £3. 6s. 5d.; a cow for £4. 6d.; and a horse for £69. 12s.; what did he pay for the whole?

	£.	s.	d.
	3	6	5
	4	0	6
	69	12	0
A.	76	18	11

Write pence under pence, shillings under shillings, &c., supplying vacant places with ciphers, then add up each column as in whole numbers.

6. What is the amount of £1,583. 2s. 4d. 1qr.; £2,036. 10s. $1\frac{1}{2}$d., £806. 4s. 3d.; £456. 1s.? *A.* £4,881. 17s. $8\frac{3}{4}$d.

7. A gentleman purchased four loads of hay weighing as follows, viz.: the first 19cwt. 1qr. 6lb.; the second 17cwt. 10lb.; the third 18cwt. 2qr. 4lb.; and the fourth 22cwt. 3lb.; what was the weight of the whole? *A.* 76cwt. 3qr. 23lb.

8. What is the sum of £116. 12s. $3\frac{1}{4}$d.; £13. 19s. $10\frac{1}{2}$d.; £4. $8\frac{3}{4}$d.; £18. 4s.; £905. 17s. 9d.; £801. 14s.; £9. 2d.?

	£.	s.	d.	qr.
	116	12	3	1
	13	19	10	2
	4	0	8	3
	18	4	0	0
	905	17	9	0
	801	14	0	0
	9	0	2	0
A.	1869	8	9	2

9. The column of farthings makes 6qr. ÷ 4qr. = 1d. 2qr; carry the 1d. to the column of pence The pence make 33d. ÷ 12d. = 2s. 9d.; carry the 2s. to the column of shillings. The shillings make 68s. ÷ 20s. = £3. 8s.; carry the £3 to the pounds, which add as in whole numbers.

XLVI. *Q.* In adding compound numbers how do you proceed? 5. What is to be done when a column of farthings makes 6, for instance? 9

10. *Hence, divide the sum of each column, when it can be done, as in Reduction; write down the remainder and carry the quotient to the next column.*

11. Find the sum of £205. 13s. 4½d.; £211. 15s. 8¼d.; £69. 10s. 3¾.; £49. 6¼d. *A.* £535. 19s. 10¾d.

12. Find the sum of 10 C. 27 Y. 8mo. 1wk. 2d. 17h. 40m. 30sec.; 85 C. 49 Y. 6mo. 6d. 15h. 50m. 50sec.; 65 C. 99 Y. 5mo. 5d. 10h. 27m. 45sec.

	C.	Y.	mo.	wk.	d.	h.	m.	sec.
	10	27	8	1	2	17	40	30
	85	49	6	0	6	15	50	50
	65	99	5	0	5	10	27	45
A.	161	76	7	3	0	19	59	5

The first column makes 125sec. ÷ 60sec. = 2m. 5 sec.; carry the 2 minutes, &c.

13. *When the sum of any column is too small to be divided as above, write down its entire sum and carry none.*

14. Find the sum of 317lb. 11oz. 13dwt. 15gr.; 295lb. 8oz. 2dwt 14gr.; 615lb. 8oz. 16gr.; 819lb. 8gr. *A.* 2,048lb. 3oz. 17dwt. 5gr.

15. Add together £215. 8s. 2¼d.; £425. 6s. 8¾d.; £425. ½d.; £819 2s. 5d. *A.* £1,884. 17s. 4½d.

RECAPITULATION.

16. Compound Addition is the adding of compound numbers of the same kind or general class.

RULE.

17. *Write the same denominations under each other.*

18. *Add up the first right-hand column and divide its sum by as many of that denomination as make 1 of the next greater denomination.*

19. *Write down the remainder and carry the quotient to the next column, proceeding thus to the last column, which add as in whole numbers.*

20. The proof is the same as in Simple Addition.

ENGLISH MONEY.

21. Add together £17. 13s. 11¼d.; £13. 10s. 2½d.; £10. 17s. 3¾d.; £7. 7s. 6½d.; £2. 2s. 3½d.; £18. 17s. 10½d. *A.* £70. 9s. 1½d.

22. Add together £8. 10s. 3½d.; £4. 9s. 8d.; £1. 9s, 1½d.; £2. 8s 7d.; £4. 9s. 6½d.; £8. 5s. 4½d. *A.* £29. 12s. 7d.

TROY WEIGHT.

23. Add together 750lb. 9oz. 17dwt. 29gr.; 450lb. 6oz. 11dwt. 11gr.; 891lb. 7dwt. and 539lb. 3oz. 13dwt. 1gr.
A. 2,631lb. 8oz. 9dwt. 17gr.

24. A goldsmith bought four ingots of silver, the first of which weighed 8lb. 2oz. 12dwt.; the second, 5lb. 4oz. 5dwt.; the third, 6lb.

Q. What is the general direction for such cases? 10. When the sum of any column is less in value than one of the next higher denomination, how do you proceed? 13. What is Compound Addition? 16. What is the Rule? 17, 18, 19 What is the Proof? 20.

10oz. 11dwt.; and the fourth, 6lb. 11oz. 15dwt.; what was the weight of the whole ! *A.* 27lb. 5oz. 3dwt.

AVOIRDUPOIS WEIGHT.

25. Add together 2cwt. 3qr. 19lb. 5oz. 7dr.; 1cwt. 2qr. 16lb. 4oz 6dr.; 3cwt. 1 qr. 15lb. 2oz. 3dr.; 5cwt. 2qr. 12lb. 1oz. 5dr.; 2cwt. 2qr. 14lb. 4dr.; 5cwt. 1qr. 15lb. 2oz. 8dr.
A. 21cwt. 2qr. 17lb. 1dr.

26. A grocer sold four hogsheads of sugar, weighing as follows: the first, 7cwt. 1qr. 14lb.; the second, 5cwt. 2qr. 10lb.; the third, 9cwt. 1qr. 15lb.; the fourth, 7cwt. 1qr. 10lb.; what did the whole weigh ? *A.* 29cwt. 2qr. 24lb.

APOTHECARIES' WEIGHT.

27. Find the sum of 17℔. 5℥. 2ʒ. 1℈. 3gr.; 19℔. 2℥. 7ʒ. 2℈. 17gr.; 65℔. 11℥. 4ʒ. 19gr. and 75℔. 3℥. 3ʒ. 1℈. 8gr.
A. 177℔. 11℥. 2ʒ. 7gr.

28. What is the compound formed from the following ingredients, viz.: 5℔. 2℥. 3ʒ. 1℈. 12gr. of calomel; 3℔. 10℥. 5ʒ. 15gr. of jalap, 7℔. 8℥. 7ʒ. 2℈. 14gr. of rhubarb, and 1℔. 3℥. 2ʒ. 15gr. of the ex tract of colocynth ? *A.* 18℔. 1℥. 2ʒ. 2℈. 16gr.

CLOTH MEASURE.

29. Add together 70yd. 2qr. 1na.; 12yd. 1qr. 1na.; 9yd. 1na., 40yd. 2qr. 1na.; 56yd. 1qr. 1na.; and 48yd. 1qr. 1na. *A.* 237yd. 2na.

30. How many yards are 565yd. 3qr.; 275yd. 3na.; 425yd. 1qr. 1na.; 915yd. 2na.; 617yd. 2qr. 2na., and 719yd. 1qr. 3na.
A. 3,518yd. 1qr. 3na.

LONG MEASURE.

31. How many yards are 617yd. 1ft. 10in.; 810yd. 2ft. 11in.; 6yd. 7in.; 85yd. 2ft. 5in.; and 679yd. 3in. ? *A.* 2,199yd. 2ft.

32. What is the sum of the following distances; 540l. 1m. 3fur. 15rd.; 640l. 7fur. 39rd.; 720l. 2m. 3fur. 20rd. 799l. 39rd.; 560l. 1fur. 17rd.; and 750l. 2m. 6fur. 23rd. *A.* 4,011l. 1 m. 7fur. 33rd.

LAND OR SQUARE MEASURE.

33. Add together 45yd. 8ft. 113in.; 45yd. 3ft. 112in.; 75yd. 8ft 139in.; 49yd. 115in.; and 589yd. 8ft. 90in. *A.* 806yd. 3ft. 137in.

34. Find how many acres are 367A. 2R. 30rd.; 815A. 1R. 16rd., 50A. 2R. 20rd.; and 60A. 2R. 36rd. *A.* 1,294A. 1R. 22rd.

SOLID OR CUBIC MEASURE.

35. Add together 25T. 39ft. 1600in.; 42T. 13ft. 1213in.; 49T 25ft. 895in.; and 60T. 1689in. of round timber.
A. 177T. 30ft. 213in

36. Find how many cords are 189C. 127ft. 1500in.; 3421C. 6ft. 1720in.; 814C. 32ft. 815in.; 617C. 96ft. 1629in.; 915C. 915in., 101C. 120ft. 1700in.; 831C. 16ft. 250in.; and 901C. 113ft. 875in
A. 7,793C. 3ft. 764in.

DRY MEASURE.

37. How many bushels are 715bu. 3pk. 7qt. 1pt.; 695bu. 1pk. 3qt.; 789bu. 2pk. 2qt.; 150bu. 3qt. 1pt.; 167bu. 1qt. 1pt.? *A.* 2,518bu. 1qt. 1pt.

38. Add together 40bu. 2pk. 6qt. 1pt.; 89bu. 1pk. 3qt.; 75bu 2pk. 1qt. 1pt.; 69bu. 2pk. 3qt.; 49bu. 1pk. 2qt. 1pt.; and 65bu 3pk. 1qt. 1pt. *A.* 390bu. 1pk. 2qt.

WINE MEASURE.

39. Add together 38gal. 2qt. 1pt. 2gi.; 16gal. 1qt. 3gi.; 20gal 2qt. 1pt. 1gi.; 18gal. 1qt. 1pt.; 7gal. 1qt. 2gi.; and 30gal. 2qt. 1pt *A.* 132 gallons.

40. Find the sum of the following quantities: 615T. 1p. 1hhd. 62gal. 2qt. 1pt. 3gi.; 700T. 1p. 1hhd. 49gal. 1qt. 1pt. 1gi.; 513T. 61gal. 2qt. 1pt.; 718T. 1p. 1hhd. 22 gal. 2qt. 1pt. 3gi.; and 871T 38gal. 1qt. *A.* 3,420T. 45gal. 2qt. 1pt. 3 gi.

ALE OR BEER MEASURE.

41. Add together 15hhd. 42gal. 2qt. 1pt.; 75hhd. 39gal. 1qt. 1pt., 62hhd. 15gal. 1qt. 1pt.; and 39hhd. 17gal. 1qt. 1pt. *A.* 193hhd. 6gal. 3qt.

42. How many barrels are 17bl. 1kil. 1fir. 8gal. 2qt. 1pt.; 89bl. 1kil. 1fir. 3gal. 1qt. 1pt.; 65bl. 1fir. 6gal. 2qt. 1pt.; and 29bl. 1kil 1fir. 5gal. 3qt. 1pt. *A.* 203bl. 6gal. 2qt.

TIME.

43. What is the sum of 5C. 64Y. 364d. 23h. 40m. 15sec.; 3C. 19Y. 125d. 17h. 39m. 12sec.; 4C. 85Y. 189d. 11h. 13m. 23sec.; 7C. 45Y. 118d. 3h. 25m. 37sec.; 9C. 63Y. 149d. 12h. 12m. 12sec.; and 8C. 8h. 8sec. *A.* 38C. 78Y. 218d. 4h. 10m. 47sec.

44. Find the amount of the following periods of time:—49Y. 11mo. 3wk. 6d. 11h. 59m. 39sec.; 45Y. 8mo. 2wk. 5d. 14m. 42sec.; 65Y. 5mo. 1wk. 3d. 19h. 25m. 11sec.; 40Y. 3mo. 3wk. 2d. 17h. 11m. 4sec.; and 18Y. 1mo. 3wk. 1d. 21h. 8m. 8sec. *A.* 219Y. 7mo. 2wk. 5d. 21h. 58m. 44sec.

CIRCULAR MOTION.

45. Add together 12S. 29°. 59′. 59″; 45S. 15°. 45′. 42″; 65S. 18°. 11′. 40″ and 62S. 13°. 19′. 17″. *A.* 186S. 17°. 16′. 38″.

46. Find the sum of 11S. 29°. 16′. 59″; 20°. 45′. 11″; 8S. 3° 10′. 50″ and 3S. 10°. 6′ 10″. *A.* 24S. 3°. 19′. 10″

COMPOUND SUBTRACTION.

XLVII. 1. Suppose you owe £1 or 20 shillings, and pay 7 shillings, how many shillings remain unpaid? *A.* 13 shillings.

2. Suppose you draw 5 gallons from 1 hogshead of molasses, how much will remain in the hogshead? *A.* 58 gallons.

3. From 1 year take 118 days. *A.* 247 days.

4. *Hence, we must subtract shillings from shillings, pence from pence, days from days, &c.*

£.	s.	d.	qr.
5	13	6	2
3	4	3	1
2	9	3	1

5. From £5. 13s. $6\frac{1}{2}$d. take £3. 4s. $3\frac{1}{4}$d. Begin on the right hand, and say, 1qr. from 2qr. leaves 1qr.; 3d. from 6d. leaves 3d. &c *A.* £2. 9s. $3\frac{1}{4}$d.

6. A horse that cost £17. 13s. $9\frac{3}{4}$d., was sold for only £10. 3s $4\frac{1}{2}$d.; what was the loss upon it? *A.* £7. 10s. $5\frac{1}{4}$.

7. From 8037hhd. 40gal. 2qt 1pt. 3gi., take 7948hhd. 29gal. 2qt. 1pt. 1 gi. *A.* 89hhd. 11gal. 2gi.

8 From £8. 15s. 2d. take £3. 9s. 8d.

£.	s.	d.
8	15	2
3	9	8
5	5	6

Borrow 1s.=12d., which added to 2d. makes 14d.; then say, 8d. from 14d. leaves 6d.; carrying 1s. (borrowed) to 9s.=10s. from 15s.=5s.; £3 from £8 leaves £5.

9. *Hence, we may borrow one of the next higher denomination and add its value in the next lower denomination to the upper figure, then subtract as before and carry the* 1 *to the lower figure in the next column.*

10. From £18. 17s. 6d. take £11. 9s. 8d. *A.* £7. 7s. 10d.

11. From 43 hours 23m. 30sec. take 19h. 5m. 40 sec.

NOTE.—Add 60 sec. to the 30 sec., or we may subtract the 40 sec. from the 60 sec. first, and add the 30 sec. to the remainder; thus, 40 sec. from 60 sec.=20 sec.+30=50 sec. *A.* 24h. 17m. 50sec.

12. From 813C. 102ft. 1000in., take 787C. 35ft. 1727in. *A.* 26C. 66ft. 1001 in.

13. From 812hhd. 23gal. 1qt. take 716hhd. 29 gal. 2 qt.

hhd.	gal.	qt.
812	23	1
716	29	2
95	56	3

Say, 2qt. from 4qt.=2qt.+1qt.=3qt. Next, 1 (to carry) to 29=30gal. from 63 gal.=33+23 =56 gal.; then carry 1 to the 716.

14. From 514hhd. 53gal. 2qt., take 235hhd. 55gal. 3qt. *A.* 278hhd. 60gal. 3qt.

15. From 817m. 4fur. 22rd., take 619m. 6fur. 17rd. *A.* 197m. 6fur. 5rd.

16. From 1 tun take 46 gallons and 2 quarts.

	T.	p.	hhd.	gal.	qt.
	1	0	0	0	0
				46	2
A		1	1	16	2

17. Say, 2 from 4 (qt.) leaves 2. 1 (to carry) to 46 makes 47, which from 63 (gal.) leaves 16; 1 (to carry again) from 2 (hhd.) leaves 1 hhd. &c.

XLVII. *Q.* How do you subtract numbers of different denominations? 4. In subtracting for instance, 8d. from 15s. 2d. &c. how do you proceed? 8. How are 40 seconds subtracted from 30 seconds 23 minutes, &c.? 11. What other method produces a like result? 11.

18. From 785hhd. take 696hhd. 29gal. 2qt. 1pt. 3gi.
A. 88hhd. 33gal. 1qt. 1gi

19. From 87,563 yards take 1 nail. *A.* 87,562yd. 3qr. 3na.

RECAPITULATION.

20. COMPOUND SUBTRACTION is the subtracting of one compound number from another of the same kind or general class.

RULE.

21. *Write the smaller quantity under the greater, with the same denominations under each other; then begin on the right and subtract the numbers in each denomination separately as in Simple Sub traction.*

22. *But when a lower number in any denomination exceeds the one over it, add to the upper number as many units as make one of the next higher denomination, from which subtract as before, and carry* 1 *to the next lower number.*

23. The PROOF is the same as in Simple Subtraction.

ENGLISH MONEY.

24. From £3. 5s. 4d. take £1. 2s. 8d. *A.* £2. 2s. 8d.

25. Suppose a gentleman has £100, and gives £19. 5s. 4½d. for his passage to England; how much will he have left on his arrival there? *A.* £80. 14s. 7½d.

TROY WEIGHT.

26. From 3lb. 5oz. 10dwt. take 1lb. 6oz. 13dwt.
A. 1lb. 10oz. 17dwt.

27. A gentleman has a silver teapot weighing 3lb. 7oz. 5dwt. 22gr., and a silver cup weighing 2lb. 10oz. 13dwt. 15gr.; what is the difference in their weight? *A.* 8oz. 12dwt. 7gr.

AVOIRDUPOIS WEIGHT.

28. From 10T. 15cwt. 1qr. 10lb. take 5T. 17cwt. 2qr. 22lb.
A. 4T. 17cwt. 2qr. 13lb.

29. A merchant bought two hogsheads of sugar, which together weighed 16cwt. 3qr. 17lb. 8oz., and the smaller hogshead weighed 7cwt. 1qr. 20lb. 10oz.; what was the weight of the larger one?
A. 9cwt. 1qr. 21lb. 14oz.

APOTHECARIES' WEIGHT.

30. From 49℔. 3℥. 5ʒ. 1℈. take 11℥. 6ʒ. 2℈. 5gr.
A. 48℔. 3℥. 6ʒ. 1℈. 15gr.

31. Suppose an expectorant[1] to consist of 2℔. 3℥. 7ʒ. 1℈. 15gr. of the mucilage[2] of gum arabic, and 1℔. 9℥. 4ʒ. 2℈. 19gr. of the oxymel of squill; how much is there of one quantity more than of the other? *A.* 6℥. 2ʒ. 1℈. 16gr

Q. What is Compound Subtraction? 20. What the Rule? 21, 22. Proof? 23.

1 EXPECTORANT. That which promotes discharges from the lungs.
2 MUCILAGE. A slimy or viscous body

CLOTH MEASURE.

32. From 810 English ells take 1qr. 1na. *A.* 809E.e. 3qr. 3na.

33. A merchant bought 500yd. 2na. of broadcloth, and sold 412yd 2qr.; how much had he left? *A.* 87yd. 2qr. 2na.

LONG MEASURE.

34. From 19yd. 1ft. 7in. 1b.c. take 6yd. 2ft. 2b.c. *A.* 12yd. 2ft. 6in. 2b.c.

35. Suppose a footman goes 3m. 4fur. 17rd. an hour, and a railroad car 39m. 2fur. 20rd. in the same time; how much does one gain of the other in one hour? *A.* 35m. 6fur. 3rd

LAND OR SQUARE MEASURE.

36. From 657yd. 3ft. 1in. take 398yd. 6in. *A.* 259yd. 2ft. 139in.

37. If from a field containing 40A. 2R. 20rd. there be taken 19A 3R. 30rd., how much will there be left? *A.* 20A. 2R. 30rd.

SOLID OR CUBIC MEASURE.

38. From 17 tons of round timber take 1720 inches. *A.* 16T. 49ft. 8in.

39. Suppose 315C. 68ft. of wood be taken from a pile containing 1000 cords; how many cords will be left? *A.* 684C. 60ft.

DRY MEASURE.

40. Subtract 7bu. 2pk. 6qt. from 12bu. *A.* 4bu. 1pk. 2qt.

41. A farmer having raised 40 bushels of corn, kept 23bu. 2pk. for his own use, and sold the remainder; what quantity did he sell? *A.* 16 bushels 2 pecks.

WINE MEASURE.

42. From 3hhd. 15gal. take 19gal. 3qt. *A.* 2hhd. 58gal. 1qt.

43. A grocer bought 5 hogsheads of molasses, and sold 1hhd 25 gals.; how much had he then on hand? *A.* 3hhd. 38gal

ALE OR BEER MEASURE.

44. From 7bl. 1fir. take 1kil. 3qt. *A.* 6bl. 1kil. 8gal. 1qt.

45. Suppose a brewer has in one cellar 39bl. 1kil. 1fir. 5gal. 1qt of beer, and in another 25bl. 1fir. 6gal.; how much more has he in one cellar than in the other? A. 14bl. 1fir. 8gal. 1qt.

TIME.

46. From 1y. 3mo. 2wk. take 8mo. 3wk. *A.* 6mo. 3wk.

47. Suppose a father's age is 45Y. 6mo. 3wk. 5d., and his son's 22Y. 9mo. 1wk. 6d.; how much does the father's age exceed the son's? *A.* 22Y. 9mo. 1wk. 6d.

CIRCULAR MOTION.

48. From 29S. 8° take 21° 15′ 30″ *A.* 28S. 16°. 44′. 30″

49. The Moon is 5S. 18° 14′ 17″ east of the Sun, and Jupiter 12S. 28° 43′ 45″; how far are the Moon and Jupiter apart? *A.* 7S. 10°. 29′ 28″.

COMPOUND MULTIPLICATION.

XLVIII. 1. At 8 shillings a quarter, how many pounds will purchase 5 quarters of flour? *A.* £2. (=40s.)

2. Suppose it takes 3qr. of a yard of cloth for one vest, how many yards will be required for 12 vests? *A.* 9 yards.

3. How many bushels are 8 times 3 pecks? *A.* 6 bushels.

4. Suppose a bottle to contain 3 quarts of molasses, how many gallons would 9 such bottles hold? *A.* 6gal. 3qt.

5. A ship is valued at £1,976. 5s. $3\frac{1}{4}$d., and her cargo of specie at 3 times as much; how much specie has she on board?

£.	s.	d.	qr.
1976 .	5 .	3 .	1
			3
5928 .	15 .	9 .	3

Say, 3 times 1 qr.=3qr., 3 times 3d.=9d., &c. *A.* £5,928. 15s. $9\frac{3}{4}$d.

6. *Hence we may multiply each denomination separately, as in simple numbers.*

7. Multiply £7,865. 3s. 1d. by 4. *A.* £31,460. 12s. 4d.

8. Multiply 346m. 1fur. 6rd. by 6. *A.* 2,076m. 6fur. 36rd.

9. A merchant bought 5 yards of cloth for £2. 6s. $1\frac{3}{4}$d.; what was the cost of the whole?

£.	s.	d.	qr.
2 .	6 .	1 .	3
			5
11 .	10 .	8 .	3

Say, 5 times 3qr.=15qr. ÷4qr.=3d. 3qr., carry the 3d.: 5 times 1d.=5d.+3d.=8d.: 5 times 6s.=30s.÷20s.=£1. 10s. carry the £1.: and so on. *A.* £11. 10s. $8\frac{3}{4}$d.

10. *Hence, carry after multiplying as in Compound Addition.*

11. Multiply £5. 8s. $1\frac{1}{4}$d. by 6. *A.* £32. 8s. $7\frac{1}{2}$d.

12. Multiply £8. 10s. $6\frac{3}{4}$d. by 8. *A.* £68. 4s. 6d.

13. What is the product of 105T. 1p. 1hhd. 37gal. 3qt. 1pt. 3gi. multiplied by 9?

	T.	p.	hhd.	gal.	qt.	pt.	gi.
	105 .	1 .	1 .	37 .	3 .	1 .	3
							9
A.	953 .	0 .	0 .	26 .	2 .	1 .	3

Say, 9×3gi.=27gi.÷4gi.=6pt. 3gi.; carry the 6pt.: 9 times 1pt.=9pt.+6pt.=15pt.÷2=7qt., 1pt.; carry the 7qt. and so on.

14. Multiply 20l. 2m. 5fur. 15rd. by 10. *A.* 208l. 2m. 5fur. 30rd

15. Multiply 15lb. 5oz. 13dwt. by 11. *A.* 170lb. 2oz. 3dwt.

16. Multiply 17bu. 1pk. 3qt. by 12. *A.* 208bu. 0pk. 4qt.

17. *When the multiplier is a composite number, multiply successively by its factors. See* XVII. 8.

18. What is the product of £2. 10s. $4\frac{1}{2}$d. multiplied by 24? 3×8=24: then multiply first by 3 and that product by 8. *A.* £60. 9s.

XLVIII. *Q.* When oats are 2s. 6d. per bushel, what will be the cost in shillings and pence of 3 bushels?—of 4 bushels?—of 6 bushels?

19. Multiply 5h. 10m. 20sec. by 48. *A.* 248h. 16m.

20. Multiply 5Y. 3mo. 3wk. by 96. *A.* 510 years.

21. *When the multiplier is not a composite number, multiply by the whole of it at once.*

	£ .	s.	d.
	0 .	2 .	6
			9 5
4.	£1 1 .	1 7 .	6

22. What will 95 pair of slippers cost at 2s. 6d. a pair? Say, 95 times 6d.=570d.÷12d.=47s 6d.: 95×2s.=190s.+47s.=237s.÷20= £11. 17s.

23. At 7s. $4\frac{1}{2}$d. per bushel, what will 23 bushels of wheat cost? *A.* £8. 9s. $7\frac{1}{2}$d.

24. Suppose it takes 2gal. 1qt. 1pt. 2gi. to fill a demijohn, how much would be required to fill 19 such vessels? *A.* 46gal. 1qt. 0pt. 2gi.

RECAPITULATION.

25. Compound Multiplication is the multiplying of a compound number by a simple one.

RULE.

26. *Multiply each denomination separately, carrying as in Compound Addition.*

27. Multi[illegible]y £819. 3s. $6\frac{1}{4}$d. by 8. *A.* £6,553. 8s. 2d.

28. Wh[illegible] will be the cost of 73 pair of shoes at 5s. 6d. per pair? *A.* £20. 1s. 6d.

29. [illegible]ultiply 5lb. 10oz. 17dwt. by 9. *A.* 53lb. 1oz. 13dwt.

30 How many pounds will 7 cups weigh, when one weighs 3lb. 5oz [illegible]3dwt. 11gr.? *A.* 24lb. 3oz. 14dwt. 5gr.

3[illegible]. Multiply 3T. 15cwt. 1qr. 15lb. by 13. *A.* 49T. 0cwt. 0qr. 20lb.

32. What is the whole weight of 17 hogsheads of sugar each of which weighs 12cwt. 1qr. 20lb.? *A.* 211cwt. 2qr. 15lb.

33. Multiply 5℔. 3℥. 6ʒ. 1℈. by 27. *A.* 143℔. 6℥. 3ʒ.

34. Suppose a box of pills to contain a compound of aloes and jalap, weighing 1℥. 3ʒ. 1℈.; what quantity would be required to fill 1 dozen boxes? 1 gross of boxes? 1 great gross of boxes? *A.* Dozen, 1℔. 5℥.; gross, 17℔.; great gross, 204℔.

35. Multiply 5yd. 3qr. 1na. by 8. *A.* 46yd. 2qr.

36. Bought 21 pieces of broadcloth, each containing 13yd. 2qr. 3na.; how many yards were there in the whole? *A.* 287yd. 1qr. 3na.

37. Multiply 5m. 2fur. 7rd. by 8. *A.* 42m. 1fur. 16rd.

38. Suppose a man travels 20m. 5fur. 20rd. in one day; how far would he travel in a year at that rate? *A.* 7,550m. 7fur. 20rd.

39. Multiply 8A. 3R. 10rd. by 108. *A.* 951A. 3R.

40. Suppose the floor of a spacious hall to contain 200 sq. yd. 5ft

Q What is Compound Multiplication? 25. What is the Rule? 26

75in.; how many square yards would it contain, if it were 13 times as large? *A.* 2,607yd. 8ft. 111in.

41. Multiply 5T. 1,429in. of hewn timber by 144. *A.* 722T. 39ft. 144in.

42. There are 24 piles of wood, each containing 3 cords and 42 cubic feet; what quantity do all the piles contain? *A.* 79C. 112ft.

43. Multiply 819bu. 3gi. by 11. *A.* 9,009bu. 0pk. 4qt. 0pt. 1gi.

44. How many bushels of oats are there in 6 bins, in each of which are 15 bags, each containing 3bu. 1pk. 6qt. 1pt.? *A.* 310bu. 3pk. 1qt.

45. Multiply 9bl. 1kil. 1fir. 8gal. by 7. *A.* 69bl. 1kil. 1fir. 2gal.

46. How many gallons of "hard cider" can be put into 1,728 bottles, supposing each bottle to hold 3qt. 1pt.? *A.* 1,512 gallons.

47. Multiply 7T. 1p. 1hhd. 20gal. by 6. *A.* 46T. 1p. 1hhd. 57gal.

48. How many hogsheads of water will be sufficient to supply an army of 50,000 men for one day, supposing each man to require 1qt. 1pt. 1gi.? *A.* 322hhd. 26gal. 2qt.

49. Multiply 5C. 89Y. 115d. by 8. *A.* 47C. 14Y. 190d.

50. The sun performs his rotation on his axis in 25d. 14h. 8m.; how many years would he be in performing 450 such revolutions? *A.* 31Y. 200d.

51. Multiply 5S. 29°. 4′ by 96. *A.* 573S. 0° 24′.

52. Suppose one ship is in 5° 15′ 45″ south latitude, and another 4 times farther south; what must be the latitude of the latter? *A.* 21°. 3′ south latitude.

COMPOUND DIVISION.

XLIX. 1. When 5 bushels of wheat cost £2. how many shillings will purchase 1 bushel? (£2=40s.) *A.* 8 shillings.

2. Suppose 8 boys to have gathered 4 bushels of chestnuts, how many pecks will each have if they are divided equally? *A.* 2 pecks.

3. Suppose a ship's cargo, valued at £232. 16s. 8d. to be owned equally by 4 men; what is each one's part?

```
      £.     s.  d.
  4 ) 2 3 2 . 1 6 . 8
A.    £ 5 8 .   4 . 2
```

Divide the £232 first by 4; then the 16s. by 4; lastly the 8d. by 4.

4. *Hence, we may divide each denomination separately, as in simple numbers.*

5. Divide 12816lb. 12oz. 6dr. by 6. *A.* 2,136lb. 2oz. 1dr.

6. Divide 7 pounds of bread equally among 8 soldiers. (See Troy Weight.) *A.* 10oz.; and 4oz. left (undivided).

XLIX. *Q.* How are compound numbers divided? 4.

7. Suppose the 8 soldiers wish to divide the 4 ounces also; how many pennyweights would there be apiece? *A.* 10dwt.

8. *Hence, when there is a remainder, we may reduce it by Reduction Descending, and divide again, and so on.*

9. Divide £118 equally among 6 persons.

	£	s.	d.
6)	1 1 8 .	0 .	0
A.	1 9 .	1 3 .	4

10. Say, £118÷6=19 times and £4 left: £4×20s.=80s.÷6=13 times and 2s. left: 2s.×12d.=24d.÷6=4 times.

11. Divide 43cwt. into 5 equal parts. *A.* 8cwt. 2qr. 10lb.

12. Suppose 203 bushels of corn will fill 5 bins of equal size, what quantity can be put into each bin? *A.* 40bu. 2pk. $3\frac{1}{5}$qt.

13. A father divided 1300 acres of land equally among his 7 sons; what quantity did he give to each? *A.* 185A. 2R. $34\frac{2}{7}$rd.

14. Divide £29. 14s. by 9.

	£.	s.
9)	2 9 .	1 4
A.	3 .	6

15. The £2 left×20s.=40s.+14s.=54s. ÷9=6 times.

16. *Recollect then, when a remainder is brought into the next denomination, to add in the given number of that denomination.*

17. Divide £47. 15s. by 3. *A.* £15. 18s. 4d.

18. Divide 67cwt. 1qr. 10lb. into 5 equal quantities.

	cwt.	qr.	lb.
5)	6 7 .	1 .	1 0
A.	1 3 .	1 .	2 2

The 2cwt. over×4qr.=8qr.+1qr=9qr. ÷5=1 time and the 4qr. over×25lb.= 100lb.+10=110lb.÷5=22 times.

19. Divide 134cwt. 3qr. 19lb. by 6. *A.* 22cwt. 1qr. 24lb.

20. Divide 500Y. 3mo. 18d. by 9. *A.* 55Y. 7mo. 2d.

21. *When the divisor exceeds* 12, *and is a composite number, divide successively by its several factors.* See XXIV. 1.

22. Divide £84. 10s. 6d. by 24. *A.* £3. 10s. $5\frac{1}{4}$d.

23. Divide 155yd. 1qr. 1na. by 35. *A.* 4yd. 1qr. 3na.

24. *When the divisor is not a composite number, we may divide by the whole divisor at once, after the manner of Long Division.*

25. Divide 671 hogsheads 9 gallons by 29.

```
         hhd.  gal.
29 ) 6 7 1 . 9 ( 2 3
     5 8
     ---
       9 1
       8 7
       ---
         4
       6 3
     -----
     2 5 2
         9
29 ) 2 6 1 ( 9
     2 6 1
```

Dividing the 671hhd. by 29, as above directed, leaves 4hhd., which we multiply by 63 gallons, and add in the 9 gallons, making 261 gallons, to be divided by 29, as at first. *A.* 23hhd. 9gal.

Q. When there is a remainder, how do you proceed? 8. Suppose, for instance, in dividing, there is a remainder of £4, how do you proceed with it? 10 How do you proceed with a remainder of 2 shillings? 10.

26. Divide £332. 19s. 9d. by 34. *A.* £9. 15s. 10$\frac{1}{2}$.

27. Divide 120A. 2R. 37rd. by 47. *A.* 2A. 2R. 11rd.

28. Divide 485Y. 180d. 15h. 30m. 45sec. into 105 equal periods of time. *A.* 4Y. 227d. 16h. 8m. 51$\frac{90}{105}$sec.

RECAPITULATION.

29. COMPOUND DIVISION is the dividing of a compound number by a simple one.

RULE.

30. *Begin on the left and divide each denomination separately, as in simple numbers.*

31. *But if there be a remainder, reduce it to the next denomination, to which add the given number in that denomination, then divide as before.*

32. *Each quotient will be of the same name with its dividend; and the several quotients taken together will constitute the required quotient or answer.*

33. Divide £161. 14s. 4d. by 8. *A.* £20. 4s. 3$\frac{1}{2}$.

34. If a man can earn £3. 14s. 5$\frac{1}{2}$d. per week, how much can he earn per day? *A.* 10s. 7d. 2$\frac{4}{7}$qr.

35. Divide 30lb. 7oz. 13dwt. by 9. *A.* 3lb. 4oz. 17dwt.

36. When 7 silver cups weigh 8lb. 9oz., what is the weight of each? *A.* 1lb. 3oz.

37. Divide 205qr. 19lb. 7oz. 2dr. by 10.
A. 20qr. 14lb. 7oz. 1$\frac{8}{10}$dr.

38. Suppose a poor man labors a month for 149lb. 13oz. of pork; how much does he receive each day, on an average, allowing 26 working days to each month? *A.* 5lb. 12$\frac{5}{26}$oz.

39. Divide 853 yd. 2qr. 3na. by 157. *A.* 5yd. 1qr. 3na.

40. If it take 2,700 yards of broadcloth to clothe a regiment of 800 men, what quantity will each man require? *A.* 3yd. 1qr. 2na.

41. If a hogshead of wine costs £45. 8s. 3d., what is it worth by the gallon? *A.* 14s. 5d.

42. Bought 2 dozen (24) silver spoons, which weighed 7lb. 6oz 13dwt.: how much silver did each spoon contain?
A. 3oz. 15dwt. 13gr.

43. Suppose a steamboat, in making 121 trips from Albany to New York, occupies 48d. 17h. 40m.; what will be the average time in which she makes one trip? *A.* 9h. 40m.

44. How far must I travel each day, to accomplish a journey of 1,400 miles 3fur. 10rd. in 51 days? *A.* 27m. 3fur. 26$\frac{44}{51}$rd.

45. Suppose 37 barrels of equal size contain 98bu. 3pk. 2qt. of wheat; what quantity is in each barrel? *A.* 2bu 2pk. 5$\frac{17}{37}$qt.

Q. When there is a remainder, what is to be done with each inferior denomination of the dividend? 16. How do you proceed when the divisor exceeds 12, and is a composite number? 21. How, when it is not a composite number? 24 What is Compound Division? 29. Rule? 30, 31, 32

46. Suppose a king's salary to be £200,000 per annum; what is that per day? *A.* £547. 18s. 10d. $3\frac{145}{365}$qr.

MISCELLANEOUS EXAMPLES.

L. 1. How many farthings are there in £5. 17s. 6d.?

2. How many pounds are there in 5,640 farthings?

3. How many guineas at 28 shillings each, will pay a debt of £25? *A.* 17 guineas 24 shillings.

4. A grocer bought 20 hundred weight of sugar for $112, and sold it for $4\frac{1}{2}$d. per pound; what was the gain? *A.* $13.

5. A merchant in London borrowed £60 and paid at one time £15. 14s. 6d., and at another £20. 3s. $6\frac{1}{2}$d., how much remained unpaid? *A.* £24. 1s. $11\frac{1}{2}$d.

6. From a compound weighing 5℔, an apothecary sold to one man 1℔. 3℥. 5ʒ. 1℈., and to another 3℥. 2℈., how much had he left on hand? *A.* 3℔. 5℥. 2ʒ.

7. A merchant bought 3 hogsheads of sugar, each weighing 8cwt. 1qr. 20lb., and sold five barrels of the same, each weighing 3cwt. 3qr 17lb. How much had he left? *A.* 5cwt. 3qr.

8. If you deduct the days in the months of November and December from the year, how many days will there be left in a leapyear? *A.* 305 days.

9. What is the sum of 30rd. 5yd., 19rd. 4yd., 17rd. 1yd., and 25rd. 4yd.? See XLI. 16, 17. *A.* 93rd. 3yd.

10. Add together 30A. 2R. 39 sq. r. 30 sq. yd., 29A. 1R. 25 sq. r 23 sq. yd., 16A. 3R. 8 sq. r. 15 sq. yd., and 45A. 27 sq. r. 8 sq. yd. *A.* 122A. 21 sq. r. $15\frac{3}{4}$ sq. yd.

11. Add into one sum 500 sq. r. 272 sq. ft., 450 sq. r. 195 sq. ft., 365 sq. rd. 215 sq. ft., and 985 sq. r. 270 sq. ft. *A.* 2,303 sq. r. $135\frac{1}{4}$ sq. ft.

12. If a man travels 25m. 3fur. 15rd. 3yd. a day, for 12 successive days, how far will he go in that time? *A.* 305m. 26rd. 3yd.

13. From 40rd. 2yd. take 17rd. 4yd. Say 4yd. from $5\frac{1}{2}$yd.$=1\frac{1}{2}$yd. $+2$yd.$=3\frac{1}{2}$ and carry 1. *A.* 22rd. $3\frac{1}{2}$yd.

14. Add together 22rd. $3\frac{1}{2}$yd., and 17rd. 4yd. *A.* 40rd. 2yd.

15. Suppose a man travels 305m. 26rd. 3yd. in 12 days; what is the average distance per day? *A.* 25m. 3fur. 15rd. 3yd.

16. How many gallons in 50bl. 25gal. *A.* 1,600

17. How many barrels in 1,600 gallons? *A.* 50bl. 25gal.

18. How many pint, quart and 2 quart bottles, of each an equal number, can be filled with a hogshead of molasses?

NOTE.—4pt. [=2qt.:] 2pt. [=1qt.] and 1pt. make 7 pints; then divide 63 gallons brought into pints by 7 pints. *A.* 72 of each.

19. A merchant has 700 quart, 700 two quart, 700 three quart and

700 gallon bottles, and wishes to know how many hogsheads of wine it will take to fill them? *A.* 27hhd. and 49gal. over.

20. A certain manufacturer employs an equal number of men, boys and girls, to whom he pays daily, as follows, viz: to each man $1, to each boy 50 cents, and to each girl 75 cents, making in all $6 75. How many persons of each class has he in his employ?

A. 300 persons.

21. A merchant has 20 hogsheads of tobacco, each weighing 9cwt. 1qr. 14lb., which he wishes to put into an equal number of small and large boxes, the former to hold $2\frac{1}{2}$lb. and the latter 3 times as much; what number of each must we have? *A.* 1,878 boxes.

22. How many sheets of paper will it take to make an 18mo. book (VII. 80.) which shall contain 288 pages (=144 leaves?) *A.* 8 sheets. How many quires to print an edition of only 96 copies? *A.* 32 quires. How many reams to print an edition of 2,400 copies? *A.* 40 reams.

23. At $3.50 per ream, what will be the expense of paper for printing an edition of 43,200 copies of a 12mo. work, to consist of 192 pages, making the usual allowance of 2 quires of waste paper in each ream? *A.* $2,800.

24. How many years of $365\frac{1}{4}$ days in 49,000 hours?

NOTE.—In $214\frac{3}{4}$ days, the $\frac{3}{4}$ of a day is of course $\frac{3}{4}$ of 24 hours= 18 hours, which added to 16 hours, the first remainder=34h.=1d. 10h. Add the 1 day to the 214 days. *A.* 5Y. 215d. 10h.

25. "A gentleman in Buffalo has just (Feb. 1838) sold all his real estate for $130,000, payable in instalments at the rate of 1 dollar an hour." What period of time has the purchaser allowed him for the payment of the debt, reckoning $365\frac{1}{4}$ days to the year?

A. 14Y. 303d. 4h.

FRACTIONS.

GENERAL PRINCIPLES.

LI. 1. When two numbers are written, one above the other with a line between them, they mean as follows:—

$\frac{1}{2}$ (1-half) means 1 of the 2 equal parts of a unit or any thing.
$\frac{1}{3}$ (1-third) means 1 of the 3 equal parts of a unit or any thing.
$\frac{2}{3}$ (2-thirds) means 2 of the 3 equal parts of a unit or any thing.
$\frac{1}{4}$ (1-fourth) means 1 of the 4 equal parts of a unit or any thing.
$\frac{3}{4}$ (3-fourths) means 3 of the 4 equal parts of a unit or any thing.
$\frac{1}{5}$ (1-fifth) means 1 of the 5 equal parts of a unit or any thing.
$\frac{2}{5}$ (2-fifths) means 2 of the 5 equal parts of a unit or any thing.
$\frac{5}{5}$ (5-fifths) means 5 of the 5 equal parts, that is, the whole of any thing, and so on in respect to any numbers whatever.

2. Then $\frac{2}{2}$, or $\frac{3}{3}$, or $\frac{4}{4}$, or $\frac{5}{5}$, or $\frac{6}{6}$, &c. are each equal to 1 unit.

LI. *Q.* What is meant by $\frac{1}{2}$, $\frac{2}{3}$, $\frac{5}{5}$, &c.? 1. What by $\frac{2}{2}$, or $\frac{3}{3}$, $\frac{4}{4}$, &c.? 2.

3. These expressions are called FRACTIONS (from the Latin *fractio* signifying *broken*,) because they stand for numbers *broken* or divided into parts.

4. The *whole* unit or thing, of which fractions are broken parts, is called an INTEGER (a Latin word signifying *whole*,) in order to distinguish it from fractions.

5. *Fractions then are the expressions for one or more equal parts of a unit or whole number, called an integer.*

6. The number below the line, which shows into how many equal parts the unit or integer is divided, is called a DENOMINATOR[1]; because it gives the *name* or *denomination* to the fraction; as, *halves*, *thirds*, &c.

7. The number above the line, which shows the *number* of parts meant, is, for that reason, called the NUMERATOR.[2] The Numerator and Denominator are called the *Terms* of the Fraction.

8. Thus, in $\frac{1}{2}$, $\frac{3}{4}$, $\frac{5}{6}$, $\frac{7}{8}$, the upper terms, 1, 3, 5 and 7 are the *numerators*, and the lower terms, 2, 4, 6 and 8, are the *denominators*

LII 1. Since the denominator represents *all* the parts of the integer, therefore,—

2. *If we multiply the value of a single part by the denominator, the product will be the entire value of the integer.*

3. When $\frac{1}{10}$ of a bushel of rye costs 12 cents, what will $\frac{10}{10}$ or 1 bushel cost? *A.* $1.20.

4. If $\frac{1}{5}$ of a vessel be valued at $5,000, what is the value of the whole vessel? *A.* $25,000.

5. What is that number of which 36 is $\frac{1}{30}$. *A.* 1,080.

6. 29 is $\frac{1}{50}$ of what number? *A.* 1,450.

7. 75 is $\frac{1}{40}$ of what number? *A.* 3,000.

8. When $\frac{3}{4}$ of a cask of wine sells for $45, what is the whole cask worth at that rate? Find the value of $\frac{1}{4}$ first, by dividing 45 by 3, then multiply the result by 4? *A.* $60.

9. 24 is $\frac{3}{12}$ of what number? The result will be the same, if we multiply by 12 first and divide by 3 afterwards; thus, $24 \times 12 \div 3 = 96$. *A.* 96.

10. HENCE *if we multiply the value of any fraction by its denom-*

Q. What are such expressions called and why? 3. What then are Fractions? 5. What is an Integer and whence its name? 4. What is the figure below the line called, and why? 6. What, the figure above the line, and why? 7. What do both the numerator and denominator form? 7. Which are the numerators and denominators in $\frac{1}{2}$ and $\frac{3}{4}$? 8.

LII. *Q.* How may the value of any integer be ascertained from having its fractional part given? 2. Why so? 1. When you pay 3 dollars for $\frac{1}{5}$ of a ton of hay, what would be the price of a whole ton? When $\frac{4}{7}$ of a hogshead of molasses costs 12 dollars, what is the price of a whole hogshead? What is the rule for it? 10.

---

1 DEMOMINATOR, [L. *denomino.*] He that names
2 NUMERATOR, [L. *numero.*] One that numbers.

nator, and divide the result by its numerator, the product will be the entire value of the integer.

11. If $\frac{7}{11}$ of a ship's cargo be valued at $10,000, what is the value of the entire cargo? *A.* $15,714$\frac{2}{7}$.

12. 509 is $\frac{5}{13}$ of what number? *A.* 1,323$\frac{2}{5}$.

13. The fractional remainders, $\frac{2}{5}$ and $\frac{2}{7}$ above, are, properly speaking, *unexecuted divisions; hence fractions are said to have originated in this manner from Division.*

14. 815 is $\frac{29}{73}$ of what number? *A.* 2,051$\frac{16}{29}$.

15. 940 is $\frac{83}{90}$ of what number? *A.* 1,019$\frac{23}{83}$.

LIII. 1. How many halves are there in 17 dollars? Since 2-halves are equal to 1 dollar, there are 2 times as many halves as there are dollars. *A.* 34 halves=$\frac{34}{2}$.

2. *Hence multiplying any whole number by a given denominator, shows how many parts are to be taken for the numerator.*

3. How many dollars are $\frac{34}{2}$ of a dollar? Evidently as many dollars as there are times 2 in 34, for 2-halves make 1 dollar. *A.* 17 dollars.

4. *Hence dividing the numerator by the denominator, shows what whole number is contained in the fraction.*

5. How many fourths or quarters in $5? Sixths in 118 bushels? Sevenths in 395 barrels?

6. How many dollars in $\frac{20}{4}$ of a dollar? Bushels in $\frac{708}{6}$ of a bushel? Barrels in $\frac{2765}{7}$ of a barrel?

7. Change 10 to a fraction whose denominator shall be 8. *A:* $\frac{80}{8}$. How many units in $\frac{80}{8}$? *A.* 10.

8. Change 625 to a fraction whose denominator shall be 1. How many units are there in $\frac{625}{1}$?

9. Since no number is affected by multiplying or dividing it by 1, therefore,—

10. *Any whole number becomes a fraction by simply writing* 1, *for its denominator.*

11. What fraction, that has 17 for a denominator, is equal to 365? Or to 415? *A.* $\frac{6205}{17}$; $\frac{7055}{17}$.

12. When 1 pound of butter costs $\frac{1}{10}$ of a dollar, how many pounds may be bought for $1? For $365? *A.* 10lb; 3,650lb.

13. When 1 gallon of molasses costs $\frac{1}{5}$ of a dollar, what will be the cost of 5gal.? Of 20gal.? Of 1 tierce? *A.* $1; $4; 8\frac{2}{5}$.

14. How many furlongs are equal to $\frac{495}{8}$ fur.? *A.* 61$\frac{7}{8}$ furlongs.

15. *Hence the value of any fraction, is the quotient arising from dividing the numerator by the denominator.*

Q. 20 is $\frac{3}{5}$ of what number? How did fractions originate? 13.

LIII. *Q.* How many halves are there in 17 dollars, and why? 1. How is it ascertained? 2. In 10 minutes how many fourths?—fifths?—sixths? How many dollars are there in $\frac{34}{2}$ of a dollar, and why? 3. What is the inference? 4. How many furlongs in $\frac{60}{5}$ of a furlong?—in $\frac{121}{10}$ of a furlong? How does any whole number become a fraction? 10 Why so? 9. Give an example.

16. What is the value of $\frac{950}{7}$ of a bushel? *A.* $135\frac{5}{7}$ bushels.

17. What is the value of $\frac{24}{6}$? Of $\frac{415}{1}$? *A.* 4; 415.

18. *Fractions it seems are proper indications of Division, in which the numerator is the dividend, the denominator the divisor, and the quotient the value of the fraction.* XXVIII. 17.

19. What is the value of that fraction, which may be formed by he divisor 21 and the dividend 6,170? *A.* $293\frac{17}{21}$.

20. What is the value of 4,500 divided by 91? *A.* $49\frac{41}{91}$.

21. When the denominator is 18 and the value 25, what is the numerator? *A.* 450.

22. When the numerator is 3,645 and the value 81, what is the denominator? *A.* 45.

23. *When, however, the dividend is less than the divisor, the quotient is the fraction formed by writing the divisor under the dividend.*

24. Divide $1 equally among 4 persons. *A.* $\$\frac{1}{4}$ apiece.

25. Divide 3 by 5. 4 by 9. 723 by 901. *A.* $\frac{3}{5}$; $\frac{4}{9}$; $\frac{723}{901}$.

26. If 20 bushels of wheat be divided equally among 23 poor persons, what will be each one's part? *A.* $\frac{20}{23}$ of a bushel.

LIV. 1. It is plain that every number is divisible into as many equal parts as it contains units,—

2. Thus 8=8 units or 8 equal parts: so 5=5 units or 5 equal parts.

3. Hence if it be asked, what part of 8 is 5, we say $\frac{5}{8}$; because this means, as we have seen, 5 of 8 equal parts.

4. What part of 7 is 3? *A.* 3 of 7 parts, that is, $\frac{3}{7}$.

5. *Hence every number, which is to become a part of another, is properly the numerator of that Fraction, whose denominator is that other number.*

6. What part of 19 is 15? *A.* $\frac{15}{19}$. What part of 10 is 5? *A.* $\frac{5}{10}$. What part of 5 is 10? *A.* $\frac{10}{5}=2$.

7. When hay sells for $10 a load, how many loads may be bought for $10? For $7? *A.* 1 load; $\frac{7}{10}$ of a load.

8. What part of 120 is 40? *A.* $\frac{40}{120}$.

9. What part of 40 is 120? *A.* 3.

10. What part of 3 is 500? *A.* $166\frac{2}{3}$.

11. What part of 500 is 3? *A.* $\frac{3}{500}$.

12. Suppose you owe $23 and pay $15, what part of the debt do you pay, and what part do you still owe? *A.* $\$\frac{15}{23}$; $\$\frac{8}{23}$.

LV. 1. Since fractions, having different numerators but the same

---

Q. What appears to be the value of a fraction? 15. What is the value of $\frac{42}{3}$?—of $\frac{49}{6}$?—of $\frac{103}{10}$? Of what are fractions proper indications? 18. In what particulars do fractions correspond to Division? 18.

LIV. *Q.* Why is 8, for instance, said to have that number of equal parts? 1. What part of 8 is 5, and why? 3. How do you find what part one number is of another? 5. What part of 20 is 3?—is 8?—is 40? Suppose that you owe 60 dollars, and pay 20 dollars; what partof the debt do you pay, and what part do you still owe?

LV. *Q.* How may fractions be added and subtracted? 2. What is the reason for the process? 1.

denominator, express parts, each of equal magnitude or value, it follows,—

2. *That the operations of addition and subtraction of fractions having the same denominator, may be performed by means of the numerators alone, in the same manner as whole numbers.*

3. John has $\frac{1}{12}$ of a dollar, Rufus $\frac{5}{12}$, and Thomas $\frac{3}{12}$; how many twelfths have they all? *A.* $\frac{9}{12}$.

4. Add together $\frac{8}{23}$, $\frac{2}{23}$, $\frac{7}{23}$, and $\frac{6}{23}$. *A.* $\frac{23}{23}=1$.

5. Suppose a man owns $\frac{15}{37}$ of a sloop and sells $\frac{11}{37}$ of it; what par does he still own? *A.* $\frac{4}{37}$.

6. How much does $\frac{115}{808}$ from $\frac{129}{808}$ leave? *A.* $\frac{14}{808}$.

7. A boy having \$1, paid away $\frac{5}{16}$ of it; how many sixteenths had he left? (\$1$=\frac{16}{16}$.) *A.* $\frac{11}{16}$.

8. Subtract $\frac{15}{39}$ from 1 unit ($=\frac{39}{39}$.) *A.* $\frac{24}{39}$.

9. What is the sum of $\frac{15}{39}$ and $\frac{24}{39}$? *A.* 1.

10. Add together \$$\frac{3}{8}$, \$$\frac{5}{8}$, \$$\frac{7}{8}$ and \$$\frac{1}{8}$. *A* \$$\frac{16}{8}=2$.

11. Add together $\frac{37}{49}$, $\frac{11}{49}$, $\frac{50}{49}$, $\frac{3}{49}$, and $\frac{46}{49}$. *A.* 3.

12. What is the sum of $\frac{4}{5}$, $\frac{11}{5}$, $\frac{8}{5}$, $\frac{3}{5}$, $\frac{2}{5}$ and $\frac{1}{5}$? *A.* $5\frac{4}{5}$.

13. How much less than 1 is $\frac{953456}{9000000}$? *A.* $\frac{8046544}{9000000}$.

LVI. 1. Since the greater the number of parts used, the greater must be the value of the fraction, and the reverse, therefore,—

2. *A fraction is as many times greater, as its numerator is made greater; and as many times smaller, as its numerator is made smaller.*

3. *Hence multiplying the numerator multiplies a fraction, and dividing the numerator divides a fraction.*

4. If 1 yard of ribbon costs $\frac{3}{16}$ of a dollar, what will 5 yards cost?
$\frac{3}{16}\times 5=\frac{15}{16}$. *A.* $\frac{15}{16}$.

5. Multiply $\frac{5}{23}$ by 2; by 3; by 4. *A.* $\frac{10}{23}$; $\frac{15}{23}$; $\frac{20}{23}$.

6. If 5 yards of ribbon cost $\frac{15}{16}$ of a dollar, what will 1 yard cost?
A. $\frac{15}{16}\div 5=\frac{3}{16}$ of a dollar

7. Divide $\frac{150}{320}$ by 50; $\frac{1080}{1913}$ by 270; *A.* $\frac{3}{320}$; $\frac{4}{1913}$.

8. Multiply $\frac{7}{315}$ by 45; by 90; by 110. *A.* 1; 2; $2\frac{140}{315}$.

9. If a horse consume in 1 day $\frac{19}{2000}$ of a ton of hay, how much would he consume in 1 week? In 1 month? In 1 year?
A. $\frac{133}{2000}$T.; $\frac{570}{2000}$T.; $3\frac{935}{2000}$T.

10. How many times greater is $\frac{90}{100}$ than $\frac{15}{100}$? $90\div 15=6$ times, the answer; for from No. 1 and 2 above it follows,—

Q. What is the sum of $\frac{9}{23}$, $\frac{8}{23}$, and $\frac{2}{23}$? What is the difference between $\frac{8}{15}$ and the sum of $\frac{4}{15}$ and $\frac{2}{15}$?

LVI. *Q.* On what does the value of a fraction depend? 1. How then may a fraction be made greater or smaller? 2. What is the inference in respect to multiplying or dividing a fraction? 3. Divide $\frac{24}{35}$ by 8; by 24. Multiply $\frac{3}{31}$ by 5; by 6; by 15. If $\frac{20}{23}$ of a dollar will buy 5 dozen of eggs, what is a single dozen worth? How many times is $\frac{5}{81}$ contained in $\frac{60}{81}$? Why divide the 60 by 5? 3.

11. *That when two fractions have the same denominators, one is as many times greater than the other, as the numerator of the one is contained times in the numerator of the other.*

12. Divide $\frac{250}{27}$ by $\frac{5}{27}$; $\frac{1080}{1190}$ by $\frac{216}{1190}$. *A.* 50; 5.

13. Suppose you plant $\frac{4}{32}$ of a bushel of corn on an acre, and that it yield 400 times that quantity; how many bushels will you gather? *A.* 50 bushels.

LVII. 1. Since the greater the number of parts into which any thing is divided, the smaller each part must be, and the reverse, therefore,—

2. *A fraction is as many times greater as its denominator is made smaller, and as many times smaller as its denominator is made greater.*

3. *Hence dividing the denominator multiplies the fraction, and multiplying the denominator divides the fraction.*

4. If a father divides $\frac{1}{2}$ of a dollar equally between his 2 sons, what part of a dollar will each have? $\frac{1}{2}\times2=\frac{1}{4}$. *A.* \$$\frac{1}{4}$.

5. Divide $\frac{1}{2}$ by 2; by 4; by 6; by 8; by 11.
A. $\frac{1}{4}$; $\frac{1}{8}$; $\frac{1}{12}$; $\frac{1}{16}$; $\frac{1}{22}$.

6. When the price of cotton cloth is $\frac{1}{16}$ of a dollar a yard, what will be the cost of 4 yards? Of 8 yards? Of 16 yards? ($\frac{1}{16}\div4=\frac{1}{4}$.)
A. \$$\frac{1}{4}$; \$$\frac{1}{2}$; \$1.

7. Multiply $\frac{3}{960}$ by 8; by 320; by 960. *A.* $\frac{3}{120}$; 1; 3.

8. *Hence suppressing the denominator, multiplies the fraction by that number.*

9. Multiply $\frac{315}{780}$ by 780; $\frac{42}{96}$ by 96. *A.* 315; 42.

10. Divide $\frac{13}{25}$ by 14; by 21; by 45. *A.* $\frac{13}{350}$; $\frac{13}{525}$; $\frac{13}{1125}$.

11. When your board for $\frac{1}{8}$ of a month costs $\frac{15}{8}$ of a dollar, what would the board for 1 month cost? *A.* \$15.

12. We see from the above, that when several numerators are alike, the greatest fraction has the smallest denominator, and the reverse,—

13. Thus $\frac{2}{13}$ is less than $\frac{2}{12}$, or $\frac{2}{11}$, &c.; so $\frac{3}{100}$ is many times smaller than $\frac{3}{4}$.

14. Again, when the denominators are alike, the greatest fraction has the greatest numerator,—

15. Thus $\frac{5}{9}$ is greater than $\frac{4}{9}$ or $\frac{3}{9}$, &c.; so $\frac{99}{102}$ is 99 times greater than $\frac{1}{102}$.

16. We have now two ways for multiplying a fraction, and two ways for dividing it, viz:—

17. *A fraction is multiplied by multiplying its numerator, or by dividing its denominator.*

LVII. *Q.* What effect is produced on a fraction by increasing or decreasing its denominator? 2. Why has it this effect? 1. How then may a fraction be multiplied or divided? 3. Multiply (by dividing the denominator) $\frac{1}{60}$ by 5; by 4; by 12. Divide $\frac{1}{12}$ (by multiplying the denominator) by 5; $\frac{1}{15}$ by 2 $\frac{1}{5}$ by 12. How can you determine which is the greater of two fractions? 12, 14

18. *A fraction is divided by dividing its numerator, or by multiplying its denominator.*

19. Multiply $\frac{80}{400}$ by 5 both ways. *A*. $\frac{400}{400}=1$; $\frac{80}{80}=1$.

20. Divide $\frac{160}{16}$ by 10 both ways. *A*. $\frac{16}{16}=1$; $\frac{160}{160}=1$.

LVIII. 1. Since multiplying the denominator has an opposite effect from multiplying the numerator, and dividing the denominator an opposite effect from dividing the numerator, therefore,—

2. *When both the numerator and denominator are either multiplied or divided by the same number, these operations must compensate or balance each other; that is, have no effect on the value of the fraction.*

3. Find how many half dollars are equal to $\frac{4}{8}$ of a dollar, by dividing each term by 4. *A*. $\$\frac{1}{2}$.

4. Find how many eighths of a dollar are equal to $\frac{1}{2}$, by multiplying each term by 4. *A*. $\$\frac{4}{8}$.

5. That $\frac{4}{8}$ is equal to $\frac{1}{2}$ is obvious from its meaning 4 of 8 parts, which are of course, $\frac{1}{2}$ of the whole.

6. Find what other fractions are equal to $\frac{8}{12}$, by multiplying each term by 3;—by 5;—by 8. *A*. $\frac{24}{36}$; $\frac{40}{60}$; $\frac{64}{96}$.

7. Find what other fractions are equal to $\frac{240}{360}$, by dividing each term by 5;—by 8;—by 120. *A*. $\frac{48}{72}$; $\frac{30}{45}$; $\frac{2}{3}$.

8. Change $\frac{25}{40}$ to eighths by dividing each term by any number that will make the denominator 8. A. $\frac{5}{8}$.

9. Change $\frac{5}{8}$ to fortieths by multiplying each term by any number that will do it.

10. Change $\frac{4}{5}$ to fiftieths, and $\frac{40}{50}$ to fifths.

11. Change $\frac{2}{3}$ to ninetieths, and $\frac{60}{90}$ to thirds.

12. John has $\$\frac{1}{3}\frac{12}{36}$, Rufus $\$\frac{120}{360}$, and Harry $\$\frac{4}{12}$; how many thirds of a dollar has each? *A*$\$\frac{1}{3}$.

13. Reduce $\frac{35}{50}$ to tenths, and $\frac{5}{10}$ to $\frac{1}{2}$. Because the terms in $\frac{1}{2}$ cannot be divided again by any number greater than 1, without a remainder, the fraction is said to be in its *lowest or most simple terms*, and the terms themselves to be *prime to each other*.

14. Reduce $\frac{9}{12}$, $\frac{10}{20}$, and $\frac{150}{750}$ to their lowest terms. *A*. $\frac{3}{4}$; $\frac{1}{2}$; $\frac{1}{5}$.

15. Reduce $\frac{1500}{2000}$ to 200ths; $\frac{150}{200}$ to 20ths; $\frac{15}{20}$ to fourths. *A*. $\frac{3}{4}$.

16. Reduce £$\frac{720}{1200}$ to its lowest terms, by dividing by any number that will divide both terms without a remainder, and these quotients again as before, and so on till the terms become prime to each other. *A*. £$\frac{3}{5}$.

17. Reverse the last process and change £$\frac{3}{5}$ to 1200ths. *A*. £$\frac{720}{1200}$.

Q. Which is the greater fraction, $\frac{2}{17}$ or $\frac{2}{11}$; $\frac{3}{4}$ or $\frac{3}{5}$? What are the two ways for multiplying or dividing a fraction? 17, 18.

LVIII. *Q*. What operations on fractions will produce opposite effects? 1. How may these effects be counteracted? 2. What is the proof that $\frac{4}{8}$ is equal to $\frac{1}{2}$? 5. When are fractions reduced to their lowest terms? 13. How are they reduced to such terms? 16. Reduce $\frac{30}{70}$ and $\frac{500}{1100}$ to their lowest terms.

18. Reduce as above, $\frac{324}{756}$yd. to its lowest terms. *A.* $\frac{3}{7}$yd.

19. Suppose one man buys $\frac{25}{100}$ of a barrel of flour, another $\frac{12}{48}$ of a barrel, a third $\frac{36}{144}$bl., a fourth $\frac{180}{720}$bl., a fifth $\frac{250}{1000}$bl., and the sixth $\frac{375}{1500}$bl.; what part of a barrel has each man? *A.* $\frac{1}{4}$.

20. What is the greatest number that will divide without a remainder both terms in $\frac{120}{180}$, and what are the most simple terms of this fraction? *A.* 60; $\frac{2}{3}$.

21. The 60 in the last example is called the *greatest common divisor* of the terms of the fraction, and by means of it the fraction is reduced at once to its most simple terms.

22. Hence the importance of a rule, by which the greatest common divisor may in all cases be easily ascertained.

LIX. To find the greatest common divisor, or as it is sometimes called, the greatest common measure, of two or more numbers.

1. When a number greater than 1 will divide another without a remainder, it is called a *measure* or *even divisor* of that number.

2. Find by trial, all the even divisors or measures of 12 and 16.
A. Of 12: 2, 3, 4, 12. Of 16: 2, 4, 8, 16.

3. When a number greater than 1 will divide two or more numbers without a remainder, it is called their *common measure* or *common divisor*.

4. Find by trial, all the common divisors of 12 and 16. *A.* 2: 4.

5. The *greatest* number that will divide in this manner two or more numbers, is called their GREATEST COMMON DIVISOR.

6. Find by trial, the greatest common divisor of 24 and 32. Of 175 and 252. *A.* 8; 7.

7. Find the greatest common divisor of 240 and 480. Of 12, 36, and 48. *A.* 240; 12.

8. In these examples the smallest number is the *divisor* sought; whether this be the case with any two numbers is easily ascertained by dividing the greater by the less; thus, taking 125 and 625;—

```
      1 2 5 ) 6 2 5 ( 5
              6 2 5
              ═══════════
Proof.  1 2 5 ) 1 2 5=1
        1 2 5 ) 6 2 5=5
        ═══════════════
```

9. Here 125 is an even divisor of 625, and because 125 can have no greater even divisor than itself; therefore 125 is the greatest com. divisor of 125 and 625. *A.* 125.

10. Find the greatest common divisor of 375 and 2,250. Of 1,817 and 21,804. *A.* 375; 1,817.

LIX. *Q.* What is a measure of a number? 1. What is meant by a common measure or a common divisor? 3. What are the common divisors of 12 and 16? 4. What is meant by the greatest common divisor? 5. What is the greatest common divisor of 125 and 625? How is it ascertained? 8. Why is 12 a common divisor of 72 and 84, rather than the smaller of the given numbers? 12. But why is 12 the greatest common divisor of 72 and 84? 13. What inference is drawn in respect to the common divisor of two numbers, and the difference between these numbers? 14. What is the general rule? 16. When have numbers no common divisor? 17.

11. What is the greatest common divisor of 72 and 84?

```
72)84(1
   72
   12)72(6
      72
Proof. 12)72=6
       12)84=7
```

12. Here the smaller number is not the divisor sought, for 12 remains in dividing; but since 72 is exactly divisible by 12; 84 must be so also, for 84 being 12 more than 72, must contain 12 exactly once more than 72.

13. The number 12 then is a *common* divisor of 72 and 84, and since 84 is only 12 more than 72, it is plain that no number greater than 12, that will divide 72 even, can divide 84 even also, therefore 12 is the greatest common divisor of 72 and 84.

14. *We learn from this illustration, that the greatest common divisor of two numbers, never exceeds their difference.*

15. Since the same reasoning which is employed in example 11 and 12 would apply to any number of successive divisions; therefore, we have the following,—

GENERAL RULE.

16. *Divide the greater number by the less, and that divisor by the remainder, and so on; always dividing the last divisor by the last remainder, till nothing remains; the last divisor is the greatest common divisor required.*

17. When the last divisor is 1, the given numbers are prime to each other, and therefore have no common divisor.

18. Find the greatest common divisor of 495 and 585.

```
495)585(1
    495
     90)495(5
        450
Gr. com. divi. 45)90(2
                  90
```

Proof.
45)495=11
45)585=13

19. The last divisor is 45, and it leaves no remainder; therefore 45 is the greatest com. divisor.

Answer, 45.

20. What is the greatest com. divisor of 356 and 788? *A.* 4.

21. What is the greatest com. divisor of 1,190 and 2,225? *A.* 5.

22. What is the greatest common divisor of 3,760 and 9,024? *A.* 752.

23. When there are more than two numbers—*First find the greatest common divisor of any two of them, then of that common divisor and a third, and so on; the last common divisor will be the greatest common divisor of all the numbers.*

24. What is the greatest common divisor of 54, 126 and 186? The common divisor of 54 and 126 is 18, and of 18 and 186 is 6. *A.* 6.

25. What is the greatest common divisor of 3,672, 5,832 and 1,044? *A* 36.

26. Have 183 and 719 a common divisor? See No. 17.

27. What are the lowest terms of $\frac{4832}{6708}$, and what common divisor will reduce it to those terms by a single operation?

A. $\frac{1208}{1677}$; 4 divisor.

28. Suppose a piece of land lies in the form of a triangle, and that one side is 85 rods in length, another 75 rods, and the other 20 rods; what is the length of the longest chain that will exactly measure each side? *A.* 5 rods long.

29. Suppose a bookseller has an order from A, for 375 Arithmetics; one from B, for 450 Arithmetics, and another from C for 525 Arithmetics, which he would pack in equal boxes, a certain number of which should just hold all the books each man ordered. What is the greatest number of books that he can put into each box?

A. 75 books.

LX. To find the least common multiple of two or more numbers; a process used in reducing fractions to their least common denomination.

1. A COMMON MULTIPLE of two or more numbers is that number which can be divided by each without a remainder.

2. Thus 12 is a common multiple of 3 and 4, for it is divisible by each without a remainder.

3. THE LEAST COMMON MULTIPLE of two or more numbers, is the *least* number that can be divided by each without a remainder.

4. Find by trial, the least common multiple of 3 and 2. *A.* 6.

5. Find by trial, the least common multiple of 4 and 6. Of 6 and 8. Of 9 and 6. *A.* 12; 24; 18.

6. When two or more numbers are multiplied together, they are called *factors*, and their product a *composite number*. XVII. 1.

7. *Hence every composite number is a common multiple of its factors, for it is of course divisible by each factor.*

8. Suppose 12 to be a common multiple, and one of its factors to be 3; what is the other factor? *A.* 4.

9. If 143 be a common multiple, and one of its factors is 11, what is the other factor? *A.* 13.

10. What common multiple may be formed by the factors 30 and 71? *A.* 2,130.

11. A Prime Number is one that is divisible only by itself or unity, as 2, 3, 5, 7, 11, 13, 17, &c.

12. *The product of any two or more prime numbers or factors, is their least common multiple.*

13. What is the least common multiple of the prime numbers 17 and 23? *A.* 391.

LX. *Q.* What is meant by a common multiple? 1. What is the common multiple of 3 and 4? 2. What is meant by the least common multiple? 3. What is the least common multiple of 4 and 6? Of 6 and 3? What are factors? 6. Why is every composite number a common multiple? 7. If 5 be one factor of a common multiple 75, what is the other factor? What is a prime number? 11 Give several examples. What is the least common multiple of 7 and 11?

14. NUMBERS ARE PRIME to each other when they have no common divisor. § VIII. 13.

15. *The product of any two or more numbers prime to each other is their least common multiple.*

16. What is the least common multiple of 2, 9 and 13? *A.* 234.

17. What is the least common multiple of the prime factors 2, 3, 5, 7 and 11? *A.* 2,310.

18. What is the least common multiple of the factors 3, 4, 5 and 7, they being prime to each other? *A.* 420.

```
3 ) 1 2 0
   2 ) 4 0
   2 ) 2 0
   5 ) 1 0
         2
```

19. What are the prime factors of 120?

20. Here the divisors and the last quotient are all prime numbers, and if multiplied together must make, as they do ($3\times2\times2\times5\times2=$) 120; therefore they comprise all the prime factors of 120. *A.* 3, 2, 2, 5, 2.

21. Hence, to find the prime factors of any number—*Divide it successively by any prime number that will divide it without a remainder, till the quotient becomes a prime number, then the several divisors together with the last quotient will become the prime factors required.*

22. What are the prime factors of 2310? *A.* 2, 3, 5, 7, 11.

23. What are the prime factors of 5005? *A.* 5, 7, 11, 13.

24. What is the least common multiple of 2, 3, 5, 7 and 11? Of 5, 7, 11 and 13? *A.* 2310; 5005.

25. What are the prime factors of 6? *A.* 2 and 3.

26. What are the prime factors of 10? *A.* 2 and 5

27. One multiple of 6 and 10 is their product$=60$; and 60 is also a multiple of all the prime factors in both 6 and 10, for $2\times3\times2\times5=60$.

28. But 60 is not the least multiple of 6 and 10, because it has the factor 2 repeated, as $2\times2\times3\times5=60$; therefore we may drop one 2, leaving $2\times3\times5=30$, *which, because it is the product of all the prime factors that are necessary to produce* 6 *and* 10, *is the least common multiple of* 6 *and* 10.

29. When two numbers have one superfluous factor,* it may be excluded by dividing by any prime number that will divide both of them without a remainder; thus, taking 6 and 10 again,—

```
2 ) 6 . 1 0
    3 .   5
```

30. The prime factors $5\times3\times2=30$, the least common multiple as before.

31. *Recollect to divide by a prime number, and to multiply both the divisor and quotients together for the required multiple.*

Q. Why? 12. When are numbers prime to each other? 14. What is the least common multiple of such numbers? 15. What is the direction for finding the prime factors of any number? 21. Why is not 60 the least common multiple of 6 and 10? 28. What then is their least common multiple, and why? 28. How can the superfluous factor be excluded? 29. What is the direction for the process? 31

32. Find the least common multiple of 6 and 20. *A.* 60.

33. Find the least common multiple of 9 and 21. *A.* 63.

34. When the numbers contain more than one common factor, it is plain *that both quotients must be divided successively, as long as they are divisible in this manner; thus, to find the least common multiple of* 120 *and* 210.

5	1 2 0 . 2 1 0
3	2 4 . 4 2
2	8 . 1 4
	4 . 7

35. Here 4, one of the last quotients, though not a prime factor, is nevertheless equal, as a multiplier, to its prime factors 2 and 2; therefore the product of the divisors and quotients being in effect the same as the product of all the prime factors necessary to produce 120 and 210, is their least common multiple.

Then $7\times4\times2\times3\times5=840$ *A.*

36. Find the least common multiple of 96 and 108. *A.* 864

37. Find the least common multiple of 48 and 216. *A.* 432.

38. When there are several numbers and only two are divisible as above, it is evident that the divisor is not a factor of the rest; *these must therefore be written underneath for the next division, thus,—*

5	5 . 9 . 7 . 4 0 . 6 0
2	1 . 9 . 7 . 8 . 1 2
3	1 . 9 . 7 . 4 . 6
2	1 . 3 . 7 . 4 . 2
	1 . 3 . 7 . 2 . 1

39. Observe that every number, which is not divisible by the divisor, is written underneath with the quotients. Then $2\times7\times3\times2\times3\times2\times5=2{,}520$, which, because it contains all the prime factors of 5, 9, 7, 40 and 60, is the least common multiple of these numbers.

Then $2\times7\times3\times2\times3\times2\times5=2520$ *A.*

GENERAL RULE.

40. *Divide by any prime number that will divide two or more of the given numbers without a remainder, and set the quotients, together with the undivided numbers in a line beneath.*

41. *Divide the second line as before, and so on till there is no number greater than* 1, *that will divide two numbers without a remainder; then the divisors and numbers in the last line being multiplied together, will give the least common multiple required.*

42. Find the least common multiple of 5, 18, 9, 4 and 2. *A.* 180.

43. Find the least common multiple of 10, 7, 11, 5 and 8. *A.* 3080

44. Find the least common multiple of 2, 5, 25, 15 and 12. *A.* 300.

45. Find the least common multiple of 2, 3, 4, 5, 6, 12, 24, 30 and 120.

Q. When the numbers contain more than one common factor, how do you proceed? 34. What is the general rule? 40, 41. Why are the undivided numbers arranged with the quotients in a line beneath? 38.

NOTE.—As 120 is exactly divisible by all the other numbers, each is of course a factor of 120, and may therefore be cancelled, leaving 120 as the least common multiple sought. *A.* 120.

46. Hence to abbreviate the process, *cancel every number that will exactly divide any other of the given numbers, and proceed with those that remain as before.*

47. What is the least common multiple of 6, 8, 11, 24, 35, 5, 7 72 and 22? As 24 is divisible by 6 and 8, 35 by 5 and 7, 22 by 11 and 72 by 24, first cancel 6, 8, 5, 7, 11 and 24. *A.* 27720

48. Find the least common multiple of 2, 3, 4, 5, 6, 8, 10, 12, 16 and 20. *A.* 240.

49. Find the least common multiple of 30, 15, 60, 12, 5, 20, 4, 2, 3 and 10. *A.* 60.

50. Suppose a surveyor has one chain 3 rods long, another 4 rods, another 5 rods, and another 6 rods; what is the shortest distance that can be exactly measured by each chain? *A.* 60 rods.

51. There is a circular island, around which A can travel in 5 hours, B in 8 hours, and C in 10 hours. Now suppose they all start together, and go the same way round it; how much time must elapse before they will come together again? *A.* 40 hours

CLASSIFICATION OF VULGAR FRACTIONS

SO CALLED TO DISTINGUISH THEM FROM DECIMAL FRACTIONS.

LXI. 1. A VULGAR OR COMMON FRACTION is one, whose denominator and numerator are both expressed.

2. A PROPER FRACTION is one whose numerator is less than its denominator; consequently its value is less than unity; as, $\frac{1}{2}$, $\frac{2}{3}$, $\frac{4}{7}$, &c.

3. AN IMPROPER FRACTION is one, whose numerator is either equal to, or greater than its denominator; consequently its value is either equal to, or greater than unity; as, $\frac{8}{8}$, $\frac{5}{5}$, $\frac{9}{5}$, $\frac{11}{7}$, &c.

4. A COMPOUND FRACTION is the fraction of a fraction, that is, a part of a part; as, $\frac{3}{4}$ of $\frac{5}{8}$: $\frac{2}{3}$ of $\frac{5}{11}$, &c.

5. A SINGLE or SIMPLE FRACTION has but one numerator and one denominator, and is therefore either Proper or Improper; as, $\frac{2}{5}$ and $\frac{11}{3}$.

6. A MIXED NUMBER is a whole number with a fraction annexed; as, $13\frac{5}{6}$, $8\frac{11}{17}$, &c.

7. A COMPLEX FRACTION is one that has a fraction for its numerator, or for its denominator, or for both its terms; as—

$$\frac{\frac{2}{3}}{7}, \frac{8}{\frac{3}{5}}, \frac{\frac{3}{4}}{\frac{7}{8}}, \frac{5\frac{1}{2}}{11}, \frac{8}{3\frac{2}{5}}, \frac{6\frac{7}{9}}{7\frac{3}{7}}, \&c.$$

Q. How may the process in many cases be shortened? 46. What reason is assigned for it? 45. Note.

Q. LXI. What is a Vulgar Fraction? 1. What is a Proper Fraction? 2. Improper Fraction? 3. Simple Fraction? 5. Compound Fraction? 4. Mixed number? 6. Complex Fraction? 7. How many, and what, appear to be the different kinds of Vulgar Fractions? Give an example of each kind?

REDUCTION OF VULGAR FRACTIONS

LXII. 1. REDUCTION OF FRACTIONS is the process of changing their *forms* without altering their *value*.

CASE I.

To reduce fractions to their lowest terms.

RULE.

1. *Divide both the terms of the fraction by any number that will divide them without a remainder, and the quotients again as before, and so on, till no number greater than 1 will divide them.** LVIII. 1, 2.

2. *Or divide both terms by their greatest common divisor.* LVIII. 21.

3. Reduce $\frac{840}{1120}$ to its lowest terms by both methods.

7) 5)
8)$\frac{840}{1120}=\frac{105}{140}=\frac{15}{20}=\frac{3}{4}$ (Or gr. com. div. 280) $\frac{840}{1120}=\frac{3}{4}$ *Ans.*

4. Reduce $\frac{544}{1088}$ of a barrel to its lowest terms. *A.* $\frac{1}{2}$.

5. Reduce $\frac{379}{1137}$ of a dollar to its lowest terms. *A.* $\frac{1}{3}$.

6. Reduce $\frac{4052}{5065}$ of a tun to its lowest terms. *A.* $\frac{4}{5}$.

7. Suppose a merchant has several remnants of cloth, one containing $\frac{4}{16}$ of a yard, another $\frac{60}{240}$, another $\frac{317}{1268}$, another $\frac{5190}{20760}$, and another $\frac{187}{748}$; how many quarters of a yard in each? *A.* $\frac{1}{4}$yd.

CASE II.

To reduce a whole number to an improper fraction having a given denominator.

RULE.

1. *Multiply the whole number by the denominator for the numerator.* LIII. 1, 2.

2. Reduce 85 to an improper fraction whose denominator is 8. Thus, 85×8=680. *A.* $\frac{680}{8}$.

3. Reduce 2,439 to a fraction whose denominator is 7. *A.* $\frac{17073}{7}$.

4. How many ninths are there in 41?—in 208?—in 207?—in 423?
A. $\frac{369}{9}$; $\frac{1872}{9}$; $\frac{1863}{9}$; $\frac{3807}{9}$.

5. At $\frac{1}{8}$ of a dollar a yard, how many yards of ribbon may be bought for $1?—for $27?—for $3,106? *A.* 8yd.; 216yd.; 24,848yd.

6. Change 295 to halves—to thirds—to fourths—to fifths—to sixths—to sevenths. *A.* $\frac{590}{2}$; $\frac{885}{3}$; $\frac{1180}{4}$; $\frac{1475}{5}$; $\frac{1770}{6}$; $\frac{2065}{7}$.

CASE I. *Q.* How are fractions reduced to their lowest terms? 1. How can they be reduced by one operation in division? 2. On what principle is the rule based? LVIII. 21. Reduce to their lowest terms $\frac{9}{12}$, $\frac{8}{12}$, and $\frac{24}{48}$.

CASE II. *Q.* How is a whole number reduced to an improper fraction with a given denominator? 1. What is the reason for the rule? LIII. 1, 2. A man having 50 dollars, spent it in as many days as that sum contains fifths; how many days was he in spending it? What fraction may be formed with 20 and a denominator? 9.

---

* The following rules are useful in finding the common divisors of both terms
A number ending in an even number or 0, is divisible by 2.
A number ending in 5 or 0, is divisible by 5.
A number ending in 0 or 00, &c. is divisible by 10 or 100, &c.
A number is divisible by 3 or 9, when the sum of its figures is divisible by 3 or 9.
A number is divisible by 6, when the right hand figure is even, and the sum of the digits is divisible by 6.
A number is divisible by 12 when it is divisible by 4 and 3.
A number is divisible by 4 when its two right-hand digits are divisible by 4

7. Suppose a man gives $\frac{1}{13}$ of a bushel of rye for 1 pound of sugar; how many pounds of sugar may be bought for 1 bushel of rye?—for 53 bushels?—for 216 bushels? *A.* 13lb.; 689lb.; 2,808lb.

CASE III.

To reduce a mixed number to an improper fraction.

RULE.

1. *Multiply the whole number by the denominator, and to the product add the numerator for a new numerator.*

2. For, whatever number of parts the whole number may make, it is plain that the fraction will make as many more such parts as are indicated by its numerator.

$$\begin{array}{r} 208\tfrac{3}{7} \\ 7 \\ \hline 1459 \\ \hline 7 \end{array}$$

3. How many sevenths are there in $208\frac{3}{7}$ weeks? In multiplying 208 by 7-sevenths, add in the 3-sevenths thus: 7 times 8 are 56 and 3 are 59, &c. *A.* $\frac{1459}{7}$.

4. Reduce $115\frac{3}{11}$ to an improper fraction. *A.* $\frac{1268}{11}$.

5. Reduce $986\frac{75}{83}$ to an improper fraction. *A.* $\frac{81913}{83}$.

6. Change $210\frac{3}{5}$ to fifths; $342\frac{5}{6}$ to sixths; $425\frac{2}{3}$ to thirds; $305\frac{4}{7}$ to sevenths. *A.* $\frac{1053}{5}$; $\frac{2057}{6}$; $\frac{1277}{3}$; $\frac{2139}{7}$.

7. If a horse eat 1 bushel of oats in $\frac{1}{7}$ of a week, how many bushels will he eat in 1 week?—in $1\frac{5}{7}$ weeks?—in $4\frac{5}{7}$ weeks?—in $219\frac{2}{7}$ weeks? *A.* 7; 12; 33; 1,535.

8. At $\frac{1}{8}$ of a dollar a yard, how many yards of cloth may be bought for $618\frac{3}{8}$ dollars? *A.* 4,947 yards.

CASE IV.

To reduce an improper fraction to a whole or mixed number.

RULE.

1. *Divide the numerator by the denominator.* LIII. 15.

2. A man by saving $\frac{1}{16}$ of a dollar a day, saved in 33 days $\frac{33}{16}$; how many dollars is that? *A.* $\$2\frac{1}{16}$.

3. Reduce $\frac{853}{19}$ to a mixed number. *A.* $44\frac{17}{19}$.

4. Reduce $\frac{372}{89}$ to a mixed number. *A.* $4\frac{16}{89}$.

5. Reduce $\frac{828}{23}$ to a whole number. *A.* 36.

6. If a man spend daily $\frac{1}{8}$ of a dollar, how much will he spend in 8 days?—in 365 days? *A.* \$1; $\$45\frac{5}{8}$.

7. In $\frac{348}{60}$ of an hour, how many hours? *A.* $5\frac{48}{60}=5\frac{4}{5}$ hours.

8. If a steamboat sail 1 mile in $\frac{1}{15}$ of an hour, how long will it be in performing a trip of 205 miles? *A.* $13\frac{2}{3}$ hours.

CASE V.

To reduce a compound fraction to a simple one.

RULE.

1. *Multiply the numerators together for a new numerator, and the denominators together for a new denominator.*

CASE III. Q. How is a mixed number reduced to an improper fraction? 1. Why add in the numerator? 2. How is $208\frac{3}{7}$ reduced to an improper fraction? 3. Suppose the toll at a certain gate is $\frac{1}{5}$ of a dollar for a sulkey; how many times can it pass for $7\frac{3}{5}$ dollars?

CASE IV. Q. How is an improper fraction reduced to a whole or mixed number? 1. Why divide by the denominator? LIII. 15.

2. A man, owning $\frac{3}{7}$ of a vessel, sold $\frac{4}{5}$ of his share; what part of the vessel did he sell?

3. He sold $\frac{4}{5}$ of $\frac{3}{7}$ of the whole vessel. To get $\frac{1}{5}$ of any number we divide by 5; but to divide a fraction we may (by LVII. 3.) multiply the denominator, thus: $\frac{3}{7\times5}=\frac{3}{35}$; then $\frac{4}{5}$ of $\frac{3}{7}$ would be 4 times as much, which (by LVI. 3.) is $\frac{3\times4}{35}=\frac{12}{35}$. Therefore, $\frac{4}{5}$ of $\frac{3}{7}=\frac{4}{5}\times\frac{3}{7}=\frac{12}{35}$, Answer.

4. Reduce $\frac{4}{11}$ of $\frac{315}{417}$ to a simple fraction. *A.* $\frac{1260}{4587}$.

5. Reduce $\frac{3}{5}$ of $\frac{109}{211}$ to a simple fraction. *A.* $\frac{327}{1055}$.

6. A, having $\frac{3}{4}$ of a grist mill, sold $\frac{2}{3}$ of his part to B, who sold $\frac{4}{9}$ of his part to C; what part of the mill does C own? *A.* $\frac{2}{9}$.

7. How much is $\frac{1}{2}$ of $\frac{4}{5}$ of $\frac{7}{8}$ of $\frac{3}{11}$? *A.* $\frac{84}{880}=\frac{21}{220}$.

8. How much is $\frac{3}{4}$ of $\frac{2}{3}$ of $\frac{5}{7}$ of $\frac{5}{6}$? *A.* $\frac{25}{84}$.

9. When a whole or mixed number occurs, *reduce it first to an improper fraction, then proceed as before.*

10. What is $\frac{5}{8}$ of 20? $(20=\frac{20}{1}.)$ *A.* $\frac{100}{8}=\frac{25}{2}=12\frac{1}{2}$.

11. What is $\frac{3}{8}$ of $\frac{5}{6}$ of $\frac{2}{3}$ of 1,000? *A.* $208\frac{1}{3}$

12. What is $\frac{2}{5}$ of $40\frac{5}{8}$ yards? $(40\frac{5}{8}=\frac{325}{8}.)$ *A.* $16\frac{1}{4}$ yards.

13. What is $\frac{3}{4}$ of $\frac{2}{3}$ of $\frac{4}{7}$ of $20\frac{2}{5}$ gallons? *A.* $5\frac{29}{35}$ gallons.

14. A having $208\frac{3}{4}$ hogsheads of molasses, sold $\frac{3}{5}$ of it to B, who sold $\frac{2}{3}$ of what he bought to C, who sold $\frac{1}{4}$ of what he bought to D. How many gallons did each purchaser buy?

A. B $125\frac{1}{4}$ gal.; C $83\frac{1}{2}$ gal.; D $20\frac{7}{8}$ gal.

15. *When any two opposite terms have a common divisor, use their quotients in their stead; and when they are alike, cancel both, which is called cancelling equal terms.*

16. For, the effect is that of dividing both terms of the product by the same number, which (by LVIII. 2.) does not alter the value.

17. What is $\frac{3}{5}$ of $\frac{5}{8}$? Cancel the 5s. *A.* $\frac{3}{8}$

18. What is $\frac{2}{3}$ of $\frac{5}{7}$ of $\frac{375}{419}$ of a hogshead? *A.* $\frac{1250}{2933}$hhd.

19. What is $\frac{2}{3}$ of $\frac{5}{7}$ of $\frac{3}{5}$ of $5\frac{1}{2}$ pints? *A.* $1\frac{4}{7}$ pints.

20. What is $\frac{11}{38}$ of $\frac{57}{119}$? The greatest common divisor of 38 and 57 is 19; therefore $\frac{11}{38}$ of $\frac{57}{119}=\frac{11}{2}\times\frac{3}{119}$. *A.* $\frac{33}{238}$.

21. What is $\frac{318}{35}$ of $\frac{245}{318}$ of $\frac{4208}{3111}$? $(=\frac{7}{1}\times\frac{4208}{3111}.)$ *A.* $\frac{29456}{3111}$

22. What is the value of $\frac{5}{9}$ of $\frac{9}{5}$? *A.* $\frac{5}{5}$ or $\frac{9}{9}$ or 1.

23. Employ both modes of abbreviating in the following, viz: How much is $\frac{3}{4}$ of $\frac{4}{3}$ of $\frac{7}{49}$ of $\frac{40}{49}$ of $\frac{1}{2}$? *A.* $\frac{1}{14}$.

24. A ship's cargo was valued at \$22,000; $\frac{3}{11}$ of which in distress of weather was thrown overboard; what part of the cargo did that man lose who owned $\frac{4}{5}$ of it? What was the value of his loss? (LII. 10.) *A.* $\frac{12}{55}$; \$4,800.

25. Suppose a boy can do a job of work in $3\frac{2}{9}$ days, and that a man

---

Q. What whole or mixed numbers are equal to $\frac{108}{9}$?—to $\frac{110}{11}$?—to $\frac{115}{12}$?

CASE V. *Q.* What is the rule for reducing compound fractions to simple ones? 1. Why is $\frac{4}{5}$ of $\frac{3}{7}$ equal to $\frac{12}{35}$? 3. What is to be done when a whole or mixed number occurs? 9. How much is $\frac{1}{4}$ of $2\frac{1}{5}$? When can the terms be reduced or cancelled? 15. Why? 16. How much is $\frac{1}{3}$ of $\frac{3}{4}$?—of $\frac{4}{5}$?

can do the same in $\frac{3}{7}$ of the time; how many days would the man be in doing it? *A.* $1\frac{8}{21}$.

26. A father at his decease gave $\frac{3}{5}$ of his estate, which was valued at \$20,000, to his wife, who at her decease gave $\frac{5}{8}$ of her portion to her daughter. What part of the father's estate did the daughter receive, and what was its value? *A.* $\frac{3}{8}$=\$7,500.

27. Suppose a man pays for $\frac{5}{16}$ of a ship \$14,000, and for $\frac{3}{5}$ of its cargo \$20,000, and subsequently gives $\frac{3}{4}$ of all his interest in both ship and cargo to his son,—

What is the son's part of the ship and its value? *A.* $\frac{15}{64}$=\$10,500.

What, the son's part of the cargo and its value? *A.* $\frac{9}{20}$=\$15,000.

What is the entire value of both ship and cargo? *A.* \$78,133$\frac{1}{3}$.

CASE VI.

To change one fraction for another of equal value, having a given numerator.

RULE.

1. *Multiply the numerator of the required fraction, by the denominator of the given fraction; and divide the product by the numerator of the same fraction, for the required denominator.*

2. Reduce $\frac{4}{5}$ to an equal fraction whose numerator shall be 12.

```
   1 2  given numer.
     5
4 ) 6 0  A. 12/15.
   1 5  required denom.
```

3. If the $\frac{4}{5}$ were $\frac{1}{5}$, then the denominator of every equal fraction would exceed its numerator 5 times=5×12=60, but $\frac{4}{5}$ being 4 times as much, the 60 must be decreased on that account 4 times= 60÷4=15, the denominator sought.

4. Reduce $\frac{8}{13}$ to an equal fraction whose numerator is 24 *A.* $\frac{24}{39}$.

5. What fraction, having 20 for its numerator, is equal to $\frac{2}{3}$?—to $\frac{3}{4}$?—to $\frac{6}{7}$? *A.* $\frac{20}{30}$; $\frac{20}{26\frac{2}{3}}$; $\frac{20}{23\frac{2}{6}}$.

6. Suppose a company of 105 men purchase $\frac{7}{8}$ of a bank, into how many equal shares must the whole capital of the bank be divided that each purchaser may own one share? *A.* $\frac{105}{120}$.

CASE VII.

To change one fraction for another of equal value, having a given denominator.

RULE.

1. *Multiply the given denominator by the numerator of the given fraction, and divide the product by the denominator of the same fraction.*

2. Reduce $\frac{4}{5}$ to an equal fraction whose denominator shall be 15.

CASE VI. *Q.* What is the rule for changing one fraction for another with a iven numerator? 1. In reducing $\frac{4}{5}$ to an equal fraction, that has 12 for its umerator, how do you proceed? 3. What is the reason for the operation? 3.

CASE VII. *Q.* What is the rule for finding that fraction whose denomination eing known will equal a given fraction? 1. How many fifteenths are $\frac{4}{5}$? Why o you multiply 15 by 4 and divide by 5? 3.

$\begin{array}{r} 15 \\ \underline{4} \\ 5\,)\,\overline{60} \\ \underline{12} \end{array}$	3. If the $\frac{4}{5}$ were $\frac{1}{5}$, then the numerator of every equal fraction must be 5 times smaller than its denominator, that is, $15\div5=3$; but $\frac{4}{5}$ being 4 times as much as $\frac{1}{5}$, the 3 must be increased on that account 4 times$=4\times3=12$, the numerator sought; or *multiply first by 4 and divide by 5 afterwards.* *A.* $\frac{12}{15}$.

4. Reduce $\frac{5}{8}$ to a fraction whose denom. shall be 400. *A.* $\frac{250}{400}$.
5. How many sixteenths in £$\frac{2400}{4800}$? *A.* £$\frac{8}{16}$.
6. How many thirds in $\frac{11}{16}$ of a pint? *A.* $\dfrac{2\frac{1}{16}\text{ pt.}}{3}$

7. Suppose one man has $\frac{25}{100}$ of a barrel of flour; another $\frac{120}{480}$; a hird $\frac{36}{144}$; a fourth $\frac{271}{1084}$; a fifth $\frac{119}{476}$; and a sixth $\frac{197}{788}$: what frac ion, whose denominator is 4, will express each one's part? *A.* $\frac{1}{4}$.

CASE VIII.

To Reduce fractions to a common denominator.

RULE.

1. *Multiply both the numerator and denominator of each fraction by the denominators of all the other fractions.*

2. For both terms being multiplied by the same numbers, the value of every fraction (by LVIII. 2.) remains the same; and the denominator of each fraction must be a common one, for it is in each instance, the product of the same numbers.

3. Reduce $\frac{3}{4}$, $\frac{2}{5}$ and $\frac{7}{8}$ to a common denominator.

$\frac{3}{4}\times5$ denom.$=\frac{15}{20}\times8$ denom.$=\frac{120}{160}$. *A.*
$\frac{2}{5}\times4$ denom.$=\frac{8}{20}\times8$ denom.$=\frac{64}{160}$. *A.*
$\frac{7}{8}\times5$ denom.$=\frac{35}{40}\times4$ denom.$=\frac{140}{160}$. *A.*

4. Since the reduction of each fraction involves the multiplication of the same denominators, we need multiply them together only once; but recollect to multiply the numerators as before, according to the following

GENERAL RULE.

5. *If the numbers are not all single fractions reduce them to such first, then multiply each numerator by all the denominators except its own, for a new numerator; and all the denominators together for a new denominator.*

6. Reduce $\frac{2}{3}$, $\frac{5}{6}$, and $\frac{4}{7}$ to a com. denominator. *A.* $\frac{84}{126}$; $\frac{105}{126}$; $\frac{72}{126}$.	$2\times6\times7=84$. Then $\frac{2}{3}=\frac{84}{126}$. $5\times3\times7=105$. Then $\frac{5}{6}=\frac{105}{126}$. $4\times6\times3=72$. Then $\frac{6}{7}=\frac{72}{126}$. $3\times6+7=126$

7. Reduce $\frac{3}{11}$, $\frac{3}{5}$, and $\frac{2}{3}$ to a com. denom'r. *A.* $\frac{45}{165}$; $\frac{99}{165}$; $\frac{110}{165}$.
8. Reduce $\frac{5}{17}$, $\frac{1}{2}$ and $\frac{4}{5}$ to a com. denom'r. *A.* $\frac{50}{170}$; $\frac{85}{170}$; $\frac{136}{170}$.
9. Reduce $\frac{1}{4}$ of $\frac{2}{7}$ and $\frac{2}{3}$ of 11 to a com. denom'r. *A.* $\frac{6}{84}$; $\frac{616}{84}$.

CASE VIII. Q. Reduce to a common denominator, $\frac{2}{3}$, and $\frac{3}{4}$;—$\frac{1}{2}$ and : What is the rule? 1. What is the illustration of the rule? 2. How may th process be shortened and why? 4. What is the general rule for it? 5.

10. Reduce the fractional parts of $14\frac{3}{4}$ barrels, $25\frac{2}{3}$ barrels, $17\frac{4}{5}$ barrels, and $18\frac{3}{10}$ barrels, to a common denominator.
A. $14\frac{450}{600}$; $25\frac{400}{600}$; $17\frac{480}{600}$; $18\frac{180}{600}$.

11. Suppose A sells $\frac{1}{2}$ of a hogshead of molasses to B; who sells $\frac{2}{3}$ of his part to C; who sells $\frac{3}{5}$ of his part to D; what fractions of a hogshead will express each one's part, and have their denominators all alike? *A.* B $\frac{15}{30}$; C $\frac{10}{30}$; D $\frac{6}{30}$.

CASE IX.

To find the least common denominator.

RULE.

1. *Having reduced the numbers as before, find the least common multiple of all the denominators, for a common denominator.*

2. *Then divide this com. denom. by the denom. of each fraction, and multiply the quotient by the numerator for a new numerator.*

3. For the common denominator becomes, from being a product of all the given denominators, a common multiple, of which each of said denominators is a factor.

4. Therefore the least common denominator must be the least common multiple of the given denominators.

5. Reduce $\frac{3}{4}$, $\frac{5}{6}$, $\frac{2}{3}$, $\frac{5}{8}$, to their least common denominator.

6. The least common multiple of the denominators 4, 6, 3, and 8, is 24, which is the denominator sought, (see LX. 40); therefore,—

Com. denom. $24 \div 4 \times 3 = 18$, new numer.; then $\frac{3}{4} = \frac{18}{24}$, *A.*
Com. denom. $24 \div 6 \times 5 = 20$, new numer.; then $\frac{5}{6} = \frac{20}{24}$, *A.*
Com. denom. $24 \div 3 \times 2 = 16$, new numer.; then $\frac{2}{3} = \frac{16}{24}$, *A.*
Com. denom. $24 \div 8 \times 5 = 15$, new numer.; then $\frac{5}{8} = \frac{15}{24}$, *A.*

7. Reduce £$\frac{2}{3}$, £$\frac{5}{6}$, and £$\frac{7}{12}$, to their least common denominator.
A. £$\frac{8}{12}$; £$\frac{10}{12}$; £$\frac{7}{12}$.

8. Reduce \$$\frac{3}{4}$, \$$\frac{5}{6}$, and \$$\frac{11}{16}$, to their least common denominator.
A. \$$\frac{36}{48}$; \$$\frac{40}{48}$; \$$\frac{33}{48}$.

9. Reduce $\frac{1}{2}$ of $\frac{2}{3}$, $\frac{2}{3}$ of 8 and $4\frac{1}{2}$, to their least common denominator. *A.* $\frac{2}{6}$; $\frac{32}{6}$; $\frac{27}{6}$.

10. Reduce $\frac{2}{3}$, $\frac{1}{3}$, $\frac{1}{2}$, $\frac{1}{6}$, $\frac{1}{2}$ of $\frac{2}{3}$ and $105\frac{2}{3}$, to their least common denominator. *A.* $\frac{4}{6}$; $\frac{2}{6}$; $\frac{3}{6}$; $\frac{1}{6}$; $\frac{2}{6}$; $\frac{634}{6}$.

11. Suppose A owns $\frac{2}{5}$ of a brick block, B $\frac{3}{10}$, C $\frac{1}{4}$, and D $\frac{1}{20}$; what other fractions having the least common denominator will express each man's part? *A.* $\frac{8}{20}$; $\frac{6}{20}$; $\frac{5}{20}$; $\frac{1}{20}$

CASE X.

To reduce a complex fraction to a simple one.

RULE.

1. *If both terms be not single fractions, reduce them to such first then to a common or least common denominator; which strike out*

Q. How do you proceed with mixed numbers? 10. What is a common denominator for the fractions connected with $6\frac{1}{2}$ and $5\frac{5}{6}$?

CASE IX. Q. What is the rule for finding the least common denominator? 1, 2. Why is the least common multiple the least common denominator? 3 What is the least common denominator for $\frac{2}{3}$ and $\frac{1}{6}$?—for $\frac{2}{3}$ of $\frac{3}{4}$ and $\frac{1}{4}$?

from both terms, and the numerators alone as they stand will form the terms of the simple fraction required.

2. *Or, having multiplied, as above, the numerator of each term by the denominator of the opposite term, reject the denominators.*

3. For cancelling equal terms has no effect on the value of the fraction. See Case v. 15, 16.

4. Reduce $\frac{\frac{3}{7}}{\frac{9}{11}}$ of a hogshead to a simple fraction.

5. Reducing the $\frac{3}{7}$ and $\frac{9}{11}$ to a com. denom. makes $\frac{33}{77}$ and $\frac{63}{77}$; then cancelling the 77 in each fraction, we have 33 left for a numerator and 63 for a denominator, all of which may be indicated thus: $\frac{\frac{3}{7}}{\frac{9}{11}}=\frac{\frac{33}{77}}{\frac{63}{77}}=\frac{33}{63}$. Or by rule 2. $\frac{\frac{3}{7}}{\frac{9}{11}}=\frac{3\times 11}{9\times 7}=\frac{33}{63}$. *A.* $\frac{33}{63}=\frac{11}{21}$.

6. Reduce $\frac{\frac{2}{3}}{\frac{5}{7}}$ to a simple fraction. *A.* $\frac{14}{15}$.

7. Reduce $\frac{\frac{3}{8}}{\frac{5}{6}}$ to a simple fraction. *A.* $\frac{18}{40}=\frac{9}{20}$.

8. Reduce $\frac{5}{\frac{7}{8}}$ to a simple fraction. $(5=\frac{5}{1}.)$ *A.* $\frac{40}{7}=5\frac{5}{7}$.

9. Reduce $\frac{34\frac{5}{7}}{84}$ to a simple fraction. $(34\frac{5}{7}=\frac{243}{7}.)$ *A.* $\frac{243}{588}$.

10. Reduce $\frac{44}{147\frac{5}{9}}$ to a simple fraction. *A.* $\frac{396}{1328}=\frac{99}{332}$.

11. Reduce $\frac{247}{\frac{5}{7}}$ to a simple fraction. *A.* $\frac{1729}{5}=345\frac{4}{5}$.

12. Reduce $\frac{394\frac{74}{99}}{894\frac{547}{719}}$ to a simple fraction. *A.* $\frac{28098520}{63689967}$.

13. Reduce $\frac{\frac{2}{3}\text{ of }\frac{3}{4}}{8}$ to a simple fraction. $(\frac{2}{3}\text{ of }\frac{3}{4}=\frac{6}{12}.)$ *A.* $\frac{6}{96}=\frac{1}{16}$.

14. Reduce $\frac{\frac{3}{4}\text{ of }20\frac{1}{2}}{\frac{2}{3}\text{ of }45\frac{5}{6}}$ to a simple fraction. *A.* $\frac{2214}{4400}=\frac{1107}{2200}$.

15. When the numerator is $\frac{5}{16}$ and the denominator $\frac{3}{8}$, what is the value of the fraction, expressed in its simplest terms? *A.* $\frac{5}{6}$.

16. When the dividend is $\frac{17}{18}$, and the divisor $\frac{11}{12}$, what is the quotient? *A.* $\frac{34}{33}$.

17. When tape is $\frac{1}{32}$ of a dollar a yard, how many yards may be bought for $\frac{7}{8}$ of a dollar? *A.* 28 yards.

18. When $\frac{2}{5}$ of 8 is a dividend, and $16\frac{3}{4}$ a divisor, what single fraction will express the quotient? *A.* $\frac{64}{335}$.

19. A grocer has $45\frac{5}{8}$ gallons of cider, which he wishes to put into bottles, each to hold $\frac{5}{16}$ of a gallon; how many bottles must he get? *A.* $12\frac{1}{6}$ dozen bottles.

CASE X. *Q.* What is the rule for reducing a complex fraction to a simple one? 1, 2. On what principle are the denominators dropped? 3.

CASE XI.

To find what part one number is of another, which is called *finding their* RATIO.

RULE.

1. *Make that number which is to become the part, the numerator of a fraction, and the other number, the denominator, that is, always divide the second by the first.* LIV.

2. What part of 20 is 5? *A.* $\frac{5}{20}=\frac{1}{4}$.
3. What part of 5 is 20? *A.* $\frac{20}{5}=4$ times 5.
4. What part of 400 is 50?—is 200? *A.* $\frac{1}{8}$; $\frac{1}{2}$.
5. What is the ratio of 23 to 253? *A.* 11.
6. What is the ratio of 253 to 23? *A.* $\frac{23}{253}=\frac{1}{11}$.
7. What is the ratio of 832 to 624? *A.* $\frac{3}{4}$.
8. What is the ratio of 624 to 832? *A.* $1\frac{1}{3}$.
9. What part of 625 is the sum of 175 and 225? *A.* $\frac{400}{625}=\frac{16}{25}$.
10. A and B bought a barrel of flour for $11; A paid $6 and B $5; what part of the whole did each pay? A $\frac{6}{11}$; B $\frac{5}{11}$.
11. Suppose A puts into trade $100, B $200, and C $500; what part of the whole capital did each man advance? *A.* A $\frac{1}{8}$; B $\frac{2}{8}$; C $\frac{5}{8}$.
12. Three men form a copartnership, A advancing $1,000, B $1,500, and C $2,500; what is each one's part of the capital? *A.* A's $\frac{1}{5}$; B's $\frac{3}{10}$; C's $\frac{1}{2}$.
13. Suppose in the last question, that they gained the first year $2,000, which they are to share in proportion to the sum each advanced; how many dollars did each gain? ($\frac{1}{5}$ of $2,000=$400, &c.) *A.* A's $400; B's $600; C's $1,000.
14. Next suppose that the same company lost the second year $1,700; how many dollars did each lose? *A.* A's $340; B's $510; C's $850.
15. A, B and C traded together and gained $3,000, which they agreed to share in proportion to each one's capital. A advanced $2,000, B $3,000, and C $1,000; how many dollars did each man receive for his profits? *A.* A's $1,000; B's $1,500; C's $500.
16. What part of $3\frac{3}{4}$ gallons is $2\frac{1}{2}$ gallons? Reduce the complex fraction to a single one by Case x. *A.* $\frac{2}{3}$.
17. A gentleman having $205\frac{1}{5}$ barrels of flour, sold $76\frac{19}{20}$ barrels; what part of the whole did he sell? *A.* $\frac{3}{8}$.

CASE XII.

To find what fraction or part, one quantity is of another of the same kind, but of different denominations.

RULE.

1. *Reduce the given quantities to the lowest denomination mentioned in either quantity; then proceed to find the part as in the last case.*

CASE XI. *Q.* What part of 12 is 8? What part of 45 is 15? What is the rule? 1. What is the ratio of 20 to 10?—of 10 to 20?—of 40 to 120?

2 What part of \$5 is 50 cents? (\$5=500ct.) *A.* $\$\frac{50}{500}=\frac{1}{10}$.
3. What part of £1 is 2s. 6d.? *A.* $\frac{1}{8}$.
4. What part of £2. 3s. 4d. is 4s. 2d.? *A.* $\frac{5}{52}$.
5. Reduce 16s. 8d. to the fraction of £1. *A.* £$\frac{5}{6}$.
6. Reduce 18s. to the fraction of a guinea (28s). *A.* $\frac{9}{14}$.
7. What part of £1 is 1s. 8d.?—is 2s. 6d.?—is 3s. 4d.?—is 5s.?—is 6s. 8d.?—is 10s.?—is 12s. 6d.?—is 15s.? is 17s. 6d.?
A. $\frac{1}{12}$; $\frac{1}{8}$; $\frac{1}{6}$; $\frac{1}{4}$; $\frac{1}{3}$; $\frac{1}{2}$; $\frac{5}{8}$; $\frac{3}{4}$; $\frac{7}{8}$.
8. What part of 15 hours is 30 minutes? *A.* $\frac{1}{30}$.
9. What fraction of 3cwt. is 3qr.? *A.* $\frac{1}{4}$.
10. What part of 7 miles 4 furlongs is 1m. 2fur.? *A.* $\frac{1}{6}$.
11 What part of \$1 (=6s.) is 2s. 6d.?—is 4s. 6d.? *A.* $\frac{5}{12}$; \$$\frac{3}{4}$.
12. What part of \$1 is 1d.?—is 1ct.?—is 1m.?
A. \$$\frac{1}{10}$; \$$\frac{1}{100}$; \$$\frac{1}{1000}$.
13. What part of 1 month is 1 day?—is 5d.?—is 10d.?—is 15d.?—is 20d.?—is 25d.? *A.* $\frac{1}{30}$mo.; $\frac{1}{6}$mo.; $\frac{1}{3}$mo.; $\frac{1}{2}$mo.; $\frac{2}{3}$mo.; $\frac{5}{6}$mo.
14. What part of 2 yards is 3qr. 3na. *A.* $\frac{15}{32}$.
15. What part of 5 bushels is 3pk. 7qt. 1 pt.? *A.* $\frac{63}{320}$.
16. What part of £3. 15s. 6d. is £1. 5s. 2d.? *A.* $\frac{1}{3}$.
17. Reduce 3qr. 18lb. 12oz. to the fraction of 1cwt. *A.* $\frac{15}{16}$.
18. What part of 15Y. 10mo. 1wk. 5d. 2h. 30m. 30sec. is 6Y 4mo. 4d. 20h. 12m. 12sec.? *A.* $\frac{2}{5}$
19. Reduce 7cwt. 2qr. 12lb. 8oz. to the fraction of a ton.
A. $\frac{61}{160}$T.
20. What part of 4d. $2\frac{3}{4}$qr. is 1d. $1\frac{5}{8}$qr.? *A.* $\frac{3}{10}$.
21. A merchant bought 2cwt. 2qr. 18lb. $9\frac{6}{7}$dr. of sugar, and sold 2qr. 17lb. $2\frac{13}{28}$dr.; what part of the whole did he sell? *A.* $\frac{1}{4}$.

CASE XIII.

To reduce the fraction of any given quantity to whole or compound numbers.

RULE.

1. *Multiply the given quantity or its equivalent in the next lower denomination, by the numerator, and divide the product by the denominator; proceeding with the remainder, if there be any, as in Compound Division.*

2. For dividing by the denominator first would show the value of one equal part, and multiplying the quotient by the numerator would show the value of all the parts meant; a process the same in effect as that described in the rule.

CASE XII. Q. What part of 1 dollar is 75 cents?—is 50 cents?—is 40 cents?—is 30 cents?—is 25 cents?—is $16\frac{2}{3}$ cents?—is $12\frac{1}{2}$ cents?—is $6\frac{1}{4}$ cents? What is the rule? 1. What part of £1 is 5s.?—is 2s. 6d.?—is 6s. 8d.?—is 13s. 4d.?—is 15s.?

CASE XIII. Q. How many cents are there in $\frac{4}{5}$ of a dollar?—shillings in $\frac{3}{5}$ of £1. What is the rule? 1. Why multiply by the numerator and divide by the denominator? 2. How many seconds are there in $\frac{5}{9}$ of a minute?—furlongs in $\frac{3}{4}$ of a mile?—feet in $\frac{2}{3}$ of a yard?

3. How much is $\frac{3}{4}$ of a day? 1 day=24 hours; then 24 hours × 3÷4=18. *A.* 18 hours.

4. What is the value of $\frac{3}{4}$ of a shilling? *A.* 9 pence.

5. What is the value of $\frac{3}{5}$ of £1? *A.* 12 shillings.

6. What is the value of $\frac{2}{3}$ of £1? *A.* 13s. 4d.

7. Suppose a railroad car goes 10 miles in $\frac{2}{5}$ of an hour, how many minutes is it in going that distance? *A.* 24 minutes.

8. What is the value of $\frac{1}{25}$ of 5 dollars? *A.* 20 cents.

9. How much is £$\frac{1}{12}$?—£$\frac{1}{8}$?—£$\frac{1}{6}$—£$\frac{1}{4}$?—£$\frac{1}{3}$?—£$\frac{7}{8}$?

Answers: 1s. 8d.; 2s. 6d.; 3s. 4d.; 5s.; 6s. 8d.; 17s. 6d.

10. How many shillings and how many cents are equal to $\frac{5}{12}$ of a dollar? *A.* 2s. 6d.=41$\frac{2}{3}$ cents.

11. How many days are equal to $\frac{1}{5}$ of a month?—to $\frac{5}{6}$ of a month? to $\frac{3}{4}$ of a month? *A.* 6d.; 25d.; 22$\frac{1}{2}$d.

12. How much is $\frac{1}{6}$ of 7 miles 4 furlongs? *A.* 1m. 2fur.

13. How much is $\frac{15}{32}$ of 2 yards? *A.* 3qr. 3na.

14. How much is $\frac{63}{320}$ of 5 bushels? *A.* 3pk. 7qt. 1pt.

15. Suppose a grocer buys $\frac{5}{8}$ of a hogshead of molasses; how many gallons, quarts, &c. does he buy? *A.* 39gal. 1qt. 1pt.

16. What is the value of $\frac{5}{7}$ of a pound avoirdupois? *A.* 11oz. 6$\frac{6}{7}$dr.

17. What is the value of $\frac{9}{13}$ of a day? *A.* 16h. 36m. 55$\frac{5}{13}$sec.

18. Suppose you resided in one place 4Y. 3mo. 3d., and in another place $\frac{1}{7}$ as long; what period of time did you spend in the latter place? *A.* 7mo. 9d.

19. What is the value of $\frac{2}{7}$ of 2hhd. 27gal. 2qt. 1pt. 3gi.? *A* 43gal. 3qt. 1pt. 1$\frac{3}{7}$gi.

CASE XIV.

To reduce a fraction of one denomination to an equivalent fraction of another denomination.

RULE.

1. *Multiply and divide the fraction according to the principles of Reduction of whole numbers, but recollect—*

2. *That a fraction is multiplied by multiplying its numerator, or by dividing its denominator.* LVII. 17.

3. *Also, that a fraction is divided by dividing its numerator, or by multiplying its denominator.* LVII. 18.

4. Reduce $\frac{1}{1920}$ of a pound to the fraction of a farthing.

Thus: $£\frac{1 \times 20s. \times 12d. \times 4qr.}{1920 \ldots\ldots\ldots\ldots} = \frac{960}{1920}qr. = \frac{1}{2}qr.$ *A.* $\frac{1}{2}$qr.

Or thus: $£\frac{1 \ldots\ldots\ldots\ldots}{1920 \div 20s. \div 12d \div 4qr.} = \frac{1}{2}qr.$ *A.* $\frac{1}{2}$qr.

CASE XIV. Q. What part of £1 is $\frac{1}{2}$ of a shilling? What part of a gallon is $\frac{1}{8}$ of a quart? What is the rule? 1. How is the fraction multiplied or divided? 2, 3 How is $\frac{1}{1920}$ of a £ reduced to the fraction of a farthing? 4

5. Reduce $\frac{960}{1920}$ $(=\frac{1}{2})$ of a farthing to the fraction of a pound.

Thus: $\frac{960\div 4\text{qr.}\div 12\text{d.}\div 20\text{s.}}{1920 \ldots\ldots\ldots} = £\frac{1}{1920}$. A. $£\frac{1}{1920}$

Or thus: $\frac{960 \ldots\ldots\ldots\ldots}{1920\times 4\text{qr.}\times 12\text{d.}\div 20\text{s.}} = \frac{960}{1843200} = £\frac{1}{1920}$. A. $£\frac{1}{1920}$

Or thus: $\frac{960}{1920} = \frac{1 \ldots\ldots\ldots\ldots}{2\times 4\text{qr.}\times 12\text{d.}\times 20\text{s.}} = £\frac{1}{1920}$ A. $£\frac{1}{1920}$

6. Reduce $\frac{1}{240}$ of a pound to the fraction of a shilling.
7. Reduce $\frac{1}{12}$ of a shilling to the fraction of a pound.
8. Reduce $\frac{3}{50000}$ of an eagle to the fraction of a mill.
9. Reduce $\frac{3}{5}$ of a mill to the fraction of an eagle.
10. Reduce $\frac{5}{10752}$ of a guinea to the fraction of a farthing.
11. Reduce $\frac{5}{8}$ of a farthing to the fraction of a guinea.
12. Reduce $\frac{4}{5}$ of a pound to the fraction of a guinea.
13. Reduce $\frac{4}{7}$ of a guinea to the fraction of a pound.
14. Reduce $\frac{1}{1441}$ of a day to the fraction of a minute.
15. Reduce $\frac{1440}{1441}$ of a minute to the fraction of a day.
16. Reduce $\frac{5}{171}$ of a bushel to the fraction of a quart.
17. Reduce $\frac{160}{171}$ of a quart to the fraction of a bushel.
18. Suppose one man has $\frac{1}{480}$ of a pound, another $\frac{1}{24}$ of a shilling, and another $\frac{1}{2}$ of a penny; what fraction of a dollar has each.

A. $\$\frac{1}{144}$.

19. Reduce $\frac{1}{3}$ of a pound to the fraction of a crown at 6s. 8d. each

A. $\frac{80}{80}$=1 crown

20. What fraction of a pound is $\frac{1}{20}$ of $\frac{3}{5}$ of a hundred weight?

A. $\frac{300}{100}$lb.=3 pounds.

21. Suppose you owe $\frac{3}{7}$ of a guinea, (28s.) and pay $\frac{1}{2}$ of the debt, what part of a pound do you still owe? A. $£\frac{3}{10}$.

22. If A buys $\frac{1}{40}$ of 5 hogsheads of molasses, and sells B. $\frac{4}{50}$ of it, what fraction of a gallon does B buy? A. $\frac{63}{100}$ of a gallon.

23. Suppose $\frac{2}{3}$ of $\frac{5}{6}$ of a pound is the numerator of a fraction, and $\frac{3}{10}$ of £2$\frac{1}{2}$ the denominator; what fraction of an eagle will express the same value? A. $\frac{20}{81}$ of an eagle.

CASE XV.

To find the integer from having a fractional part given.

RULE.

1. *Multiply the integer by the denominator, and divide the product by the numerator.* LII. 10.
2. Find that number $\frac{3}{13}$ of which is 205. A. 888$\frac{1}{3}$.
3. What number is that $\frac{4}{11}$ of which is 20,000? A. 55,000.
4. Suppose a merchant sells exactly 11hhd. 49gal. 3qt. 1pt. 3gi. of molasses, it being $\frac{3}{20}$ of all he has on hand; what quantity had he at first? A. 78hhd. 39gal. 1pt.

Q. How is $\frac{960}{1920}$ $(=\frac{1}{2})$ of a farthing reduced to the fraction of a pound? 8, 9.

CASE XV. Q. 40 is $\frac{1}{3}$ of what number? If 12 be $\frac{3}{4}$ of a certain number, what is that number? What is the rule? 1. When a man pays 8 dollars for $\frac{2}{3}$ of a barrel of flour, what is a barrel worth at that rate?

5. Find that sum $\frac{5}{9}$ of which is £3. 5s. $6\frac{1}{4}$d. *A.* £5. 17s. $11\frac{1}{4}$d.

6. The postage of a letter for 400 miles is $\frac{3}{16}$ of a dollar; how far, at that rate, can a letter be carried for one dollar? *A.* $2{,}133\frac{1}{3}$ miles.

7. 503 is $\frac{3}{4}$ of $\frac{5}{11}$ of what number? *A.* $1475\frac{7}{15}$.

8. It is estimated that nearly 25,000,000 persons die annually being $\frac{1}{32}$ of the population of the earth; what, according to that esti mate, must be the whole population of the globe? *A.* 800,000,000

ADDITION OF FRACTIONS.

GENERAL RULE.

LXIII. 1. *Reduce complex and compound fractions to single ones, and all to a common or least common denominator; over which write the sum of the numerators.* LV. 1, 2.

2. This sum may often be reduced either to lower terms, or to a whole or mixed number.

3. Add together £$\frac{3}{5}$, £$\frac{4}{5}$ £$\frac{2}{5}$ and £$\frac{1}{5}$. *A.* £$\frac{10}{5}$=£2.

4. Add together $\frac{181}{941}$, $\frac{367}{941}$, $\frac{417}{941}$, and $\frac{811}{941}$. *A.* $1\frac{835}{941}$.

5. To add mixed numbers.—*Find the sum of the fractional parts first, and if it contain a whole number, add it to the sum of the other whole numbers.*

6. *Or, reduce the mixed numbers first to improper fractions, then add them by the general rule.*

7. Add together $\$1{,}203\frac{17}{18}$, $\$9{,}406\frac{13}{18}$, and $\$8{,}906\frac{15}{18}$.

\$ 1 2 0 3	$\frac{17}{18}$
\$ 9 4 0 6	$\frac{13}{18}$
\$ 8 9 0 6	$\frac{15}{18}$
\$ 1 9 5 1 7	$\frac{1}{2}$

8. The fractions make $\$\frac{45}{18}=\$2\frac{9}{18}=\$2\frac{1}{2}$; carry the \$2 to the column of dollars. *A.* $19{,}517\frac{1}{2}$.

9. Find the sum of $675\frac{21}{35}$, $891\frac{34}{35}$, and $915\frac{32}{35}$. *A.* $2{,}483\frac{17}{35}$.

10. Suppose that one pile of wood contains $136\frac{11}{64}$ cords, another $956\frac{53}{64}$ cords, and another $452\frac{31}{64}$ cords; how many cords are there in all the piles? *A.* $1{,}545\frac{31}{64}$ cords.

11. Add together $\frac{2}{3}$, $\frac{3}{4}$, $\frac{1}{2}$, and $\frac{4}{5}$. Reduce them first to their least common denominator. *A.* $2\frac{43}{60}$.

12. Find the sum of $46\frac{2}{5}$, $89\frac{3}{4}$ and $40\ \frac{7}{8}$. *A.* $177\frac{1}{40}$.

13. Add together $\$18\frac{3}{4}$, $\$5\frac{1}{2}$, $\$7\frac{5}{16}$, $\$8\frac{2}{3}$, and $\$9\frac{7}{8}$. *A.* $\$50\frac{5}{48}$.

14. What is the sum of $789\frac{4}{7}$ years, $817\frac{2}{9}$ years, $316\frac{5}{6}$ years, and $216\frac{1}{2}$ years? *A.* $2{,}140\frac{8}{63}$.

15. Find the sum of $\frac{3}{4}$, $\frac{2}{3}$ of $\frac{5}{6}$, and $\frac{3}{5}$ of $5\frac{1}{2}$. Reduce the fractions to single ones first. *A.* $4\frac{109}{180}$.

LXIII. *Q.* What is the general rule for adding fractions? 1. What may oftentimes be done with the result? 2. What is the sum of $\frac{9}{11}$, $\frac{8}{11}$, $\frac{7}{11}$, and $\frac{6}{11}$? What is the sum of $8\frac{4}{5}$, $9\frac{2}{5}$, and $10\frac{1}{5}$? What is the rule for the last example? 5, 6 What is the sum of $\frac{3}{4}$ and $\frac{2}{3}$? See 11.

16. What is the sum of $819\frac{3}{4}$ barrels, $\frac{1}{5}$ of $\frac{5}{8}$ barrels, $409\frac{3}{3}$ barrels, and $\frac{1}{2}$ of $5\frac{1}{2}$ barrels? *A.* 1,232 barrels.

17. Add together $\frac{\frac{3}{4}}{8}$, $\frac{6}{\frac{3}{7}}$, and $\frac{\frac{4}{5}}{\frac{8}{15}}$. Reduce them first to single fractions, then proceed as before. *A.* $15\frac{19}{32}$.

18. Add together the complex fractions that may be formed by dividing 2 by $\frac{3}{5}$, $\frac{3}{4}$ by 4, $\frac{2}{3}$ by $\frac{4}{5}$, and $2\frac{1}{3}$ by 3. *A.* $5\frac{19}{144}$.

19. To add fractions of different denominations.—*First find the value of each in compound numbers, then add them as in Compound Addition.*

20. Or, first reduce the fractions of different denominations to those of the same, then add them by the general rule.

21. Add together $\frac{1}{6}$ of a pound and $\frac{2}{3}$ of a shilling. £$\frac{1}{6}$=3s. 4d.—$\frac{2}{3}$s.=8d. Then 3s. 4d.+8d.=4s. *A.* 4s.

22. Add together £$\frac{3}{7}$ and $\frac{7}{24}$ of a shilling. *A.* 8s. 10d. $1\frac{3}{7}$qr.

23. Find the sum of $\frac{6}{11}$ of a ton, $\frac{5}{7}$ of a hundred weight, and $\frac{7}{8}$ of a pound. *A.* 11cwt. 2qr. 13lb. 3oz. $6\frac{34}{77}$dr.

24. Find the sum of $\frac{3}{5}$ of a league, $\frac{5}{11}$ of a yard, and $\frac{3}{4}$ of a foot.
A. 1m. 6fur. 16rd. 2ft. 1in. $1\frac{1}{11}$b.c.

25. Add together $\frac{3}{5}$ of an ounce and $\frac{2}{7}$ of an ounce. *A.* $14\frac{6}{35}$dr.

26. A man labored $10\frac{3}{5}$ hours in one day, $9\frac{3}{11}$ hours in another, and $11\frac{5}{7}$ hours in anotner; how many hours did he labor in all?
A. 31h. 35m. $13\frac{19}{77}$sec

SUBTRACTION OF FRACTIONS.

GENERAL RULE.

LXIV. 1. *Reduce complex and compound fractions to singte ones, and all to a common or least common denominator, over which write the difference of the numerators.* LV. 1, 2.

2. From $\frac{5}{16}$ of a dollar take $\frac{3}{16}$ of a dollar. *A.* \$$\frac{2}{16}$=$\frac{1}{8}$.
3. From $\frac{8567}{9539}$ take $\frac{2759}{9539}$. *A.* $\frac{5808}{9539}$.
4. From $\frac{7531}{9328}$ take $\frac{3795}{9328}$. *A.* $\frac{3736}{9328}$.
5. From $\frac{5}{9}$ take $\frac{4}{21}$. Reduce them first to their least common denominator. *A.* $\frac{23}{63}$.
6. From $\frac{2}{3}$ take $\frac{5}{24}$. *A.* $\frac{11}{24}$
7. From $\frac{4}{5}$ take $\frac{6}{11}$. *A.* $\frac{14}{55}$
8. A bought $\frac{11}{16}$ of a load of hay, and B. $\frac{2}{3}$ of a load; how much dıd one buy more than the other? *A.* A $\frac{1}{48}$ the most.

Q. How are fractions of different denominations added? 19, 20. What is the sum of £$\frac{2}{5}$ and $\frac{3}{4}$ of a shilling? What is the sum of $\frac{3}{4}$ of a mile and $\frac{39}{40}$ of a furlong?

LXIV. *Q.* From $\frac{3}{4}$ take $\frac{2}{3}$. From $\frac{4}{9}$ take $\frac{1}{3}$. What is the rule? 1. From $\frac{2}{3}$ of $\frac{3}{8}$ of a dollar take $\frac{1}{12}$ of a dollar.

9. From $\frac{2}{3}$ of $5\frac{1}{2}$ take $\frac{7}{8}$. *A.* $\frac{67}{24}=2\frac{19}{24}$.

10. From $4\frac{2}{3}$ take $\frac{5}{6}$ of 3. *A.* $2\frac{1}{6}$.

11. A having bought a quantity of sugar, sold $\frac{2}{3}$ of it to B, who sold $\frac{3}{5}$ of what he bought to C; what part had B left? *A.* $\frac{4}{15}$.

12. From $\frac{\frac{2}{3}}{\frac{3}{4}}$ take $\frac{5}{6}$. From $\frac{8}{\frac{2}{5}}$ take $\frac{\frac{3}{8}}{2}$. *A.* $19\frac{13}{16}$; $\frac{1}{18}$.

13. To subtract a mixed number from a whole number.—*Write the difference between the numerator and denominator over the denominator, and carry* 1 *to the whole number.*

14. For, since the denominator expresses all the parts of 1 unit, the process is the same in pinciple, as borrowing and carrying in simple numbers.

15. From \$ 8 1 5
Take \$ 3 6 $\frac{5}{16}$
A. \$ 7 7 8 $\frac{11}{16}$

Say, 5 from 16=11, that is, $\frac{11}{16}$ and carry 1. For $\frac{5}{16}$ from $\frac{16}{16}$ (=1 unit borrowed from 815) leaves $\frac{11}{16}$ and 1 to carry as before.

16. From £5,075 take £2,536$\frac{15}{20}$. *A.* £2,538$\frac{14}{20}$.

17. From 487 years take $259\frac{81}{97}$ years. *A.* $227\frac{16}{97}$ years.

18. From 1 league take $\frac{25}{41}$ of a league. *A.* $\frac{16}{41}$ league.

19. How much less than unity is $\frac{3999}{4000}$? *A.* $\frac{1}{4000}$.

20. From 1 take the sum of $\frac{3}{7}$ and $\frac{1}{7}$. *A.* $\frac{3}{7}$.

21. From 1 take the sum of $\frac{2}{3}$ and $\frac{5}{18}$. *A.* $\frac{1}{18}$.

22. What fraction added to the sum of $\frac{2}{5}$ and $\frac{4}{30}$ will make a unit? *A.* $\frac{14}{30}$.

23. Suppose a pole standing so that $\frac{1}{5}$ of it is in the mud, $\frac{3}{8}$ in the water, and the rest above water; what part is above water? *A.* $\frac{17}{40}$.

24. From 1 take the sum of $\frac{3}{11}$ and $\frac{3}{5}$. *A.* $\frac{7}{55}$.

25. A gentleman spent $\frac{1}{5}$ of his life at school, $\frac{1}{4}$ in England, $\frac{1}{3}$ in America, and the rest in France. What part of his life did he spend in the latter country? *A.* $\frac{13}{60}$.

26. To subtract one mixed number from another.—*Having reduced the fractions to a common denominator, subtract as at first, but if the lower numerator exceed the upper one, subtract it from the common denominator and add the difference to the upper numerator, carrying one as before.*

27 From £ 9 6 $\frac{11}{16}$
Take £ 1 8 $\frac{5}{16}$
£ 7 8 $\frac{6}{16}=78\frac{3}{8}$, Answer.

Say, 5 from 11=6, that is, $\frac{6}{16}$, or $\frac{5}{16}$ from $\frac{11}{16}=\frac{6}{16}=\frac{3}{8}$.

28. From $608\frac{29}{108}$ take $504\frac{13}{108}$. *A.* $104\frac{16}{108}=104\frac{4}{27}$.

29. A merchant owing \$$405\frac{7}{8}$, paid \$$293\frac{5}{8}$; how many dollars re main unpaid? *A.* $112\frac{1}{4}$.

Q. How is a mixed number subtracted from a whole number? 13. Why is the numerator to be subtracted from the denominator? 14. From 4 gallons take $2\frac{3}{10}$ gallons. From £1 take £$\frac{3}{47}$. How much does unity exceed $\frac{7}{217}$? How is one mixed number subtracted from another? 26. A man having $4\frac{3}{8}$ yards of broadcloth, sold $2\frac{1}{4}$ yards; how many yards had he left? From $4\frac{2}{3}$ take $1\frac{5}{8}$.

30. From £ 1 5 0 $\frac{2}{7}$
Take £ 7 5 $\frac{6}{7}$
Ans. £ 7 4 $\frac{3}{7}$

Say, 6 from 7=1+2=3, that is, $\frac{3}{7}$, or $\frac{6}{7}$ from $\frac{7}{7}=\frac{1}{7}+\frac{2}{7}=\frac{3}{7}$. Carry 1 to the 5.

31. From $150\frac{31}{151}$ take $107\frac{43}{151}$. *A.* $42\frac{139}{151}$.

32. From $4\frac{2}{3}$ take $3\frac{3}{4}$. Reduce the fractional parts to the least common denominator first. *A.* $\frac{11}{12}$.

33. From the sum of $\$45\frac{3}{5}$ and $\$62\frac{5}{8}$ take $\$49\frac{3}{16}$. *A.* $\$59\frac{3}{80}$.

34. A merchant purchased $21\frac{1}{5}$ barrels of flour of one man, and $13\frac{3}{10}$ barrels of another; how much will he have left after he has sold $30\frac{2}{3}$ barrels? *A.* $3\frac{5}{6}$ barrels.

35. To subtract fractions of different integers.—*First find their separate values, then subtract as in compound numbers.*

36. From $\frac{4}{5}$ of a day take $\frac{3}{4}$ of an hour. *A.* 18h. 27m.

37. From £$\frac{3}{7}$ take $\frac{2}{9}$ of a shilling. *A.* 8s. 4d. $\frac{16}{21}$ qr.

38. When a man has traveled $\frac{4}{7}$ of the distance from New York to Philadelphia, it being 90 miles, how far has he still to travel?
A. 38m. 4fur. 22rd. 4yd. 2ft. 1in. $2\frac{1}{7}$b.c

MULTIPLICATION OF FRACTIONS.

LXV. 1. To multiply a fraction by a whole number.—*Multiply the numerator, or rather divide the denominator when it can be done without a remainder.* LVII. 17.

2. If you pay $\frac{1}{8}$ of a dollar for a yard of ribbon, what will 2 yards cost?—4 yards cost?—8 yards cost? *A.* $\$\frac{1}{4}$; $\$\frac{1}{2}$; \$1.

3. Multiply $\frac{3}{960}$ by 10;—by 4;—by 8. *A.* $\frac{3}{96}$; $\frac{3}{240}$; $\frac{3}{120}$.

4. Multiply $\frac{2}{107}$ by 3;—by 9;—by 7. *A.* $\frac{6}{107}$; $\frac{18}{107}$; $\frac{14}{107}$.

5. All fractional answers should be expressed in their lowest terms.

6. Multiply $\frac{7}{990}$ by 4;—by 12;—by 14. *A.* $\frac{14}{495}$; $\frac{14}{165}$; $\frac{49}{495}$.

7. Multiply $\frac{3}{8}$ by 8;—by 6;—by 4. *A.* 3; $2\frac{1}{4}$; $1\frac{1}{2}$.

8. If one horse consume $\frac{4}{21}$ of a ton of hay in a month, how much will 7 horses consume in the same time? *A.* $1\frac{1}{3}$ ton.

9. To multiply a whole number by a fraction.—*Multiply by the numerator and divide by the denominator, or divide first when it can be done without a remainder.*

10. This rule proceeds on the same general principle as that for whole numbers, viz.

11. *That as many times as the multiplier is made smaller, so many times the product is made smaller.*

Q. How are fractions of different integers subtracted? 35. From $\frac{1}{7}$ of a day take $\frac{3}{4}$ of a minute. From 4 dollars take $\frac{3}{4}$ of a dollar. From $\frac{1}{2}$ of $\frac{2}{3}$ of 1cwt take $\frac{3}{7}$ of 1qr.

LXV. Q. How is a fraction multiplied by a whole number? 1. Multiply $\frac{3}{20}$ by 5;—$\frac{4}{21}$ by 5;—$\frac{7}{320}$ by 320. How is a whole number multiplied by a fraction? 9. On what principle is the rule based? 11.

12. When the multiplier is 1, for instance, the product is the same as the multiplicand.

13. But in multiplying by $\frac{1}{8}$, for example, the multiplier is 8 times less than unity; consequently the product must be 8 times less than the multiplicand.

14. In multiplying by $\frac{3}{8}$, the multiplier is 3 times greater than $\frac{1}{8}$, consequently the product must be made 3 times greater than that of $\frac{1}{8}$

15. Multiply 64 by 1;—by $\frac{1}{2}$;—by $\frac{1}{4}$;—by $\frac{1}{8}$;—by $\frac{1}{16}$;—by $\frac{1}{32}$;—by $\frac{1}{64}$;—and by $\frac{2}{64}$;—by $\frac{4}{64}$;—by $\frac{8}{64}$;—by $\frac{16}{64}$;—by $\frac{32}{64}$;—by $\frac{64}{64}$.

16. Answers. 64; 32; 16; 8; 4; 2; 1; 2; 4; 8; 16; 32; 64.

17. If a laborer receive 20 dollars a month, and saves $\frac{3}{10}$ of it, how many dollars does he save? *A.* \$6.

18. Multiply 43 by $\frac{3}{8}$;—by $\frac{4}{5}$;—by $\frac{2}{3}$. *A.* $16\frac{1}{8}$; $34\frac{2}{5}$; $28\frac{2}{3}$.

19. Multiply 101 by $\frac{3}{50}$;—319 by $\frac{4}{9}$. *A.* $6\frac{3}{50}$; $141\frac{7}{9}$.

20. Since either factor in multiplication may be made the multiplier without affecting the result, therefore,

21. *If more convenient, multiply the fraction by the whole number instead of the whole number by the fraction.*

22. How much is $\frac{5}{12}$ of 6?—6 times $\frac{5}{12}$? *A.* $2\frac{1}{2}$ each.

23. How much is $\frac{201}{366}$ of 180, or 180 times $\frac{201}{360}$? *A.* $100\frac{1}{2}$.

24. To multiply one fraction by another.—*Multiply the numerators together for a new numerator, and the denominators together for a new denominator.* LXII. CASE V. 1.

25. Multiply $\frac{250}{360}$ by $\frac{115}{341}$. *A.* $\frac{2875}{12276}$.

26. Multiply $\frac{5036}{6003}$ by $\frac{85}{96}$. *A.* $\frac{42806}{54027}$.

27. How much is $\frac{3}{5}$ of 400? *A.* 240.

28. How much is $\frac{4}{5}$ of 208? *A.* $166\frac{2}{5}$.

29. How much is $\frac{3}{4}$ of $5\frac{1}{2}$? *A.* $4\frac{1}{8}$.

30. Multiply $\frac{3}{4}$, $\frac{5}{6}$, $\frac{2}{3}$ and $\frac{4}{5}$ together. *A.* $\frac{1}{3}$. LXII.—CASE V. 15.

31. A merchant had $\frac{3}{4}$ of a yard of cloth in one remnant, and in another only $\frac{2}{3}$ as much; what part of a yard was there in the smaller piece? *A.* $\frac{1}{2}$ of a yard.

32. To multiply a mixed number by a whole number, and vice versa.—*Multiply the whole number first by the fraction of the mixed number, then by the whole number connected with the fraction, and add the two products together.*

```
       8
       9 2/3
       -----
       5 1/3
      72
Ans.  77 1/3
```

33. Multiply 8 by $9\frac{2}{3}$, or $9\frac{2}{3}$ by 8. First find $\frac{2}{3}$ of $8=5\frac{1}{3}$, &c. Or, $9\frac{2}{3}=\frac{29}{3}\times 8=\frac{232}{3}=77\frac{1}{3}$, Answer: that is, reduce first to an improper fraction, then multiply as before.

Q. Why should any product be less than the multiplicand? 13. Why is the product of $\frac{3}{8}$ greater than $\frac{1}{8}$? 14. What is the product of 12 multiplied by $\frac{3}{4}$?—by $\frac{5}{6}$?—by $\frac{2}{3}$? Why is $\frac{5}{12}$ of 6 the same as 6 times $\frac{5}{12}$? 20, 21. How is one fraction multiplied by another? 24. Multiply $\frac{2}{3}$ by $\frac{4}{5}$;—$\frac{7}{8}$ by $\frac{2}{3}$;—$\frac{1}{2}$ of $\frac{3}{4}$ by $\frac{4}{5}$;—by $5\frac{1}{2}$;—by $\frac{3}{4}$. How much is $9\frac{2}{3}$ times 8? What is the rule? 32

34. Multiply 92 yards by $27\frac{5}{9}$. *A.* $2{,}535\frac{1}{9}$ yards.

35 Multiply $8\frac{3}{4}$cwt. by 5. *A.* $43\frac{3}{4}$cwt.

36. Suppose 45 hogsheads of molasses contain each $37\frac{3}{8}$ gallons, how many gallons do all the hogsheads contain? *A.* $1{,}681\frac{7}{8}$gal.

37. Reduce 241 rods to yards. *A.* $1{,}325\frac{1}{2}$ yards.

38. Reduce 65 barrels to gallons. *A.* $2{,}047\frac{1}{2}$ gallons.

39. Reduce 40 sq. rods to sq. feet. *A.* 10,890 sq. feet.

40. Reduce $34\frac{2}{11}$ miles to furlongs. *A.* $273\frac{5}{11}$ furlongs.

41. Reduce $45\frac{3}{7}$ minutes to seconds. *A.* $2{,}725\frac{5}{7}$ seconds.

42. In some cases it is more convenient *to multiply the fraction first by the whole number as above.* (1.)

43. Multiply 4 0 8 3 2 $\frac{5}{8}$
by 9 Thus: $\frac{5}{8}\times 9=\frac{45}{8}=5\frac{5}{8}$; carry the 5.
Answer, 3 6 7 4 9 3 $\frac{5}{8}$

44. Multiply $2{,}517{,}937\frac{115}{198}$ by 98. *A.* $246{,}757{,}882\frac{91}{99}$.

45. Multiply $8{,}300{,}675\frac{31}{726}$ by 22. *A.* $182{,}614{,}850\frac{31}{33}$.

46. When both terms are mixed numbers.—*Reduce them to improper fractions, then multiply as above.* (24.)

47. How many yards are there in $5\frac{3}{4}$ rods? *A.* $31\frac{5}{8}$ yards.

48. How many square yards are there in a carpet which is $5\frac{1}{2}$ yards square? *A.* $30\frac{1}{4}$ sq. yards.

49. How many square rods in a piece of land which has four sides, each $25\frac{3}{8}$ rods long? *A.* $643\frac{57}{64}$.

50. How many square feet are there in a room $16\frac{1}{2}$ feet square? *A.* $272\frac{1}{4}$ sq. feet.

51. How many solid feet are there in a block, the length, breadth and depth of which are each $3\frac{2}{3}$ feet? *A.* $49\frac{8}{27}$ solid feet.

52. To multiply complex fractions.—*Reduce them first to single terms, then proceed as above directed.* (24.)

53. Multiply $\dfrac{\frac{6}{11}}{9}$ by 100. *A.* $6\frac{2}{33}$.

54. Multiply $\dfrac{8}{\frac{2}{3}}$ by 200. *A.* 2,400.

55. Multiply $\frac{1}{3}$ of $\dfrac{5\frac{1}{2}}{9}$ by 3. *A.* $\frac{11}{18}$.

56. Suppose A, who owns $\frac{3}{8}$ of a ship, divides his interest into 10 equal parts; what fraction of the whole ship will express $\frac{2}{3}$ of each of these equal parts? *A.* $\frac{1}{40}$.

Q. What is the cost of $4\frac{2}{5}$ yards of broadcloth at $10 a yard? How can this process be varied? 42. Multiply in this manner $2\frac{3}{4}$ by 5. How are complex fractions multiplied? 52.

DIVISION OF FRACTIONS

LXVI. 1. Division of Fractions, like that of whole numbers, proceeds on the same general principle, viz.

2. *That as many times as the divisor is made smaller, so many times the quotient is made greater, and the reverse.*

3. When the divisor is unity, the quotient is of course the same as the dividend.

4. But when the divisor is any number of times less than unity, the quotient is that number of times greater than the dividend.

5. For instance, there are four times as many quarters of an apple in a basket as there are whole apples.

6. When the divisor and dividend are alike, the quotient is of course unity.

7. But when the divisor is any number of times greater than the dividend, the quotient is that number of times less than unity.

8. For though the dividend cannot, strictly speaking, be said to *contain* the divisor, it may nevertheless be divided into any assignable number of equal parts, as,—

9. For example, 1 cannot contain either 2, 3, or 4, but it may be divided into either 2, 3, or 4 equal parts, as $\frac{1}{2}$, $\frac{1}{3}$, $\frac{1}{4}$.

10. To divide a fraction by a whole number.—*Divide the numerator or multiply the denominator.* LVII. 17, 18.

11. Recollect to reduce, in all instances, complex and compound fractions to single ones, and mixed numbers to improper fractions, unless otherwise directed.

12. When 3 bushels of oats cost $\frac{3}{8}$ of a dollar, how many eighths will buy one bushel? *A.* $\$\frac{1}{8}$.

13. Divide $\frac{5}{8}$ by 5;—by 7;—by 11. *A.* $\frac{1}{8}$; $\frac{5}{56}$; $\frac{5}{88}$.

14. Divide $\frac{675}{1821}$ by 27, and $\frac{199}{398}$ by 45. *A.* $\frac{25}{1821}$; $\frac{1}{90}$.

15. Divide $\frac{3}{5}$ of $5\frac{1}{2}$ by 3;—by 33. *A.* $1\frac{1}{10}$; $\frac{1}{10}$.

16. Divide $\dfrac{\frac{8}{9}}{4}$ by 6, and $\dfrac{4\frac{3}{8}}{5}$ by 10. *A.* $\frac{1}{27}$; $\frac{7}{80}$.

17. A man having a tierce of molasses containing $42\frac{1}{5}$ gallons, sold $\frac{3}{4}$ of it to a grocer, who retailed it out in equal portions to 20 different persons. What part of the hogshead did each of the 20 persons buy? *A.* $\frac{211}{8400}$ hhd.

18. To divide a whole number by a fraction.—*Multiply by the fraction inverted, that is, multiply by the denominator and divide the result by the numerator.*

LXVI. *Q.* On what principle is Division of Fractions based? 2. Why should the quotient ever be greater than the dividend? 3, 4. Give an example. 5 When is the quotient less than unity, and why? 6, 7. But can the greater contain the less? 8. Give an example. 9. How is a fraction divided by whole number? 10. What preparation is often necessary? 11. Divide $\frac{6}{7}$ by 8;—by 4;—by 3;—by 10. Divide $\frac{1}{8}$ of $\frac{15}{16}$ by 5;—by 15. How is a whole [illegible] divided by a fraction? 18

19. For instance, when we divide by $\frac{1}{8}$ the divisor is 8 times less than unity; consequently, the quotient must be 8 times greater than the dividend. (4.)

20. But when we divide by $\frac{3}{8}$, the divisor is 3 times greater than $\frac{1}{8}$; consequently, the quotient must be made 3 times smaller. (1, 2.)

21. At $\frac{1}{5}$ of a dollar a yard, how many yards of calico may be bought for $1?—for $5?—for $17?—for $400?

A. 5yd.; 25yd.; 85yd.; 2,000yd.

22. Divide 563,015 yards into pieces containing only $\frac{7}{30}$ of a yard each. *A.* 2,412,921$\frac{3}{7}$ pieces.

23. Divide 908,070 by $\frac{81}{83}$. *A.* 930,491$\frac{13}{27}$.

24. Divide 8,450 by $\frac{1}{5}$ of $3\frac{1}{2}$. *A.* 12,071$\frac{3}{7}$.

25. Divide 8,307 by $\frac{2\frac{1}{2}}{3\frac{3}{4}}$. *A.* 12,460$\frac{1}{2}$.

26. How many times does 85 exceed $\frac{3}{4}$? *A.* 113$\frac{1}{3}$ times.

27. Divide 100 by 2;—by 1;—by $\frac{1}{2}$;—by $\frac{1}{4}$;—by $\frac{1}{8}$;—by $\frac{1}{16}$;—by $\frac{1}{32}$. *A.* 50; 100; 200; 400; 800; 1600; 3200.

28. Reduce 221 yards to rods. ($5\frac{1}{2}=\frac{11}{2}$.) *A.* 40$\frac{2}{11}$ rods.

29. Reduce 13,945 degrees to statute miles. *A.* 200$\frac{90}{139}$.

30. Reduce 69,563 sq. yards to rods. If we divide the remainder by the denominator, (which is the multiplier of the dividend,) the quotient will be of the same denomination with the dividend.

A. 2,299 sq. rd. 18$\frac{1}{4}$ sq. yd.

31. Reduce 1,229 sq. yards to sq. rods. *A.* 40 sq. rd. 19 sq. yd.

32. Reduce 30,052 gallons to barrels. *A.* 954 bl. 1 gal.

33. How many years, of 365$\frac{1}{4}$ days each, are there in 28,567 days?

A. 78 Y. 77$\frac{1}{2}$d.

34. If one bushel of apples will make $\frac{3}{17}$ of a barrel of cider, how many bushels will be required to make 250 barrels? *A.* 1,416$\frac{2}{3}$.

35. To divide one fraction by another.—*Invert the divisor, then multiply the upper terms together for a new numerator, and the lower terms together for a new denominator.*

36. The reason for this rule is the same, in reality, as that for the preceding one.

37. For, multiplying the numerator of the dividend by the denominator of the divisor multiplies the dividend by that number. (LVI. 2, 3.)

38. And multiplying the denominator of the dividend by the numerator of the divisor divides the dividend by that number. (LVII. 2, 3.)

39. When a yard of ribbon costs $\frac{3}{16}$ of a dollar, how many yards will $\frac{4}{5}$ of a dollar purchase?

40. The divisor $\frac{3}{16}$ inverted$=\frac{16\times4}{3\times5}=\frac{64}{15}=4\frac{4}{15}$. *A.* 4$\frac{4}{15}$ yd.

41. Divide $\frac{5}{9}$ by $\frac{3}{11}$;—$\frac{2}{3}$ by $\frac{3}{4}$. *A.* 2$\frac{1}{27}$; $\frac{8}{9}$.

Q. Illustrate the rule by the divisors $\frac{1}{8}$ and $\frac{3}{8}$. 19, 20. Divide 40 by $\frac{3}{4}$;—by $\frac{2}{5}$. When wood is 6 dollars a cord, what is $\frac{2}{3}$ of a cord worth? How is one fraction divided by another? 35. What is the proof that this rule is the same in principle as the last? 37, 38. Divide $\frac{2}{3}$ by $\frac{2}{11}$;—$\frac{3}{10}$ by $\frac{2}{5}$. How may this process be abbreviated? 50.

42. Divide $\frac{3}{8}$ by $\frac{2}{5}$;—$\frac{2}{5}$ by $\frac{3}{7}$. *A.* $\frac{15}{16}$; $\frac{14}{15}$.

43. Divide $5\frac{1}{2}$ yards into pieces containing each $\frac{3}{7}$ of a yard. *A.* $12\frac{5}{6}$ pieces.

44. If a barrel of molasses leak out $\frac{2}{3}$ of a gallon a week, how many weeks, at that rate, must elapse before the whole will have leaked out? *A.* $47\frac{1}{4}$ weeks.

45. Divide $30\frac{1}{4}$ by $5\frac{1}{2}$, and $272\frac{1}{4}$ by $16\frac{1}{2}$. *A.* $5\frac{1}{2}$; $16\frac{1}{2}$.

46. How many bottles, each holding $2\frac{1}{10}$ pints, may be filled with $503\frac{3}{5}$ pints? *A.* $239\frac{17}{21}$.

47. How many measures, each holding $2\frac{5}{7}$ pints of grain, may be filled with 5 bushels 3 pecks and $1\frac{2}{3}$ pints of grain. *A.* $136\frac{11}{57}$.

48. How many paces, each equal to $\frac{1}{2}$ of $7\frac{1}{11}$ feet, are there in 1 mile? $1,489\frac{3}{13}$ paces.

49. Divide $\frac{12}{35}$ by $\frac{6}{7}$. Thus: $\frac{12\times7}{35\times6}=\frac{84}{210}=\frac{2}{5}$; that is, the $\frac{12}{35}$ is multiplied by 7 and divided by 6, for multiplying the denominator divides the fraction (LVII. 3); but a fraction may be multiplied by 7 by dividing its denominator (LVII. 3), and divided by 4 by dividing its numerator (LVI. 3); thus: $\frac{12\div6}{35\div7}=\frac{2}{5}$, as before. *A.* $\frac{2}{5}$.

50. Hence, to abbreviate the process of dividing one fraction by another—*Divide the numerator of the dividend by the numerator of the divisor, and the denominator by the denominator, if it can be done without a remainder, otherwise the quotient will be a complex fraction*

51. Divide $\frac{120}{6307}$ by $\frac{120}{901}$. Thus: $\frac{120\ \div120}{6307\ \div901}=\frac{1}{7}$. *A.* $\frac{1}{7}$.

52. Divide $\frac{8}{21}$ by $\frac{4}{7}$;—$\frac{231}{924}$ by $\frac{3}{11}$. *A.* $\frac{2}{3}$; $\frac{77}{84}$.

53 At $\frac{3}{16}$ of a dollar a yard, how many yards of cloth may be bought for $\frac{15}{16}$ of a dollar? $\frac{15\div3}{16\div16}=\frac{5}{1}=5$. Or by No. 35, thus: $\frac{15\times16}{16\times3}$, which by cancelling equal terms becomes $\frac{15}{3}$, that is, $15\div3=5$, as before. *A.* 5 yards.

54. Hence, when fractions have a common denominator, the process may be still more abbreviated, thus:—*Reject the common denominator, then divide as in whole numbers.*

55. How many times greater is $\frac{4}{5}$ than $\frac{1}{5}$? *A.* 4 times.

56. How many times is $\frac{2}{20}$ contained in $\frac{18}{20}$? *A.* 9 times.

57. Divide $\frac{180}{361}$ by $\frac{36}{361}$, and $\frac{150}{211}$ by $\frac{49}{211}$. *A.* 5; $3\frac{3}{49}$.

58. If a man perform $\frac{3}{37}$ of his journey in half a day, how many half days will he be in performing $\frac{36}{37}$ of the journey? *A.* 12.

59. In the same manner we might divide all fractions, *if we reduce*

Q. Divide $\frac{12}{35}$ by $\frac{6}{7}$. See 49. How and when may the process be still more abbreviated? 54. What is the quotient of $\frac{15}{16}$ divided by $\frac{3}{16}$? See 53. Why not use the denominators at all? Can all fractions be divided in the same manner? 59. How then does this process differ from the general rule? 60.

them first to a common denominator; for example, $\frac{3}{4}$ by $\frac{5}{8}$. Thus $\frac{3}{4} \div \frac{5}{8} = \frac{24}{32} \div \frac{20}{32} = \frac{24}{20} = 24 \div 20 = 1\frac{1}{5}$. *A.* $1\frac{1}{5}$.

60. This process, however, involves no new principle, for it is in reality the same as that of the general rule. For, by inverting the divisor, the same terms are multiplied together in both cases.

61. Thus, taking the last example, $\frac{3 \times 8}{4 \times 5} = \frac{24}{20}$.

62. When the divisor is a whole number, and the dividend consists of a large integer and a fraction—*Divide each separately, and if the whole number leave a remainder, multiply it by the denominator of the fraction, and add the product to the numerator, for a new numerator, which divide as before.*

63. For every unit of the whole number is equal to as many parts of the fraction as are indicated by the denominator.

$$
\begin{array}{r|l}
3 & 4\ 6\ 8\ 2\ 4\ 8 \\
4 & 1\ 5\ 6\ 0\ 8\ 2\ \frac{2}{3} \\
2 & 3\ 9\ 0\ 2\ 0\ \frac{2}{3} \\
5 & 1\ 9\ 5\ 1\ 0\ \frac{1}{3} \\
6 & 3\ 9\ 0\ 2\ \frac{1}{15} \\
A. & 6\ 5\ 0\ \frac{31}{90}
\end{array}
$$

64. Divide 468,248 by 3, 4, 2, 5, and 6. *These operations involve all the variations that can possibly take place in dividing a fraction by an integer.*

The 1st rem. is $\frac{2}{3}$: 2nd, $2\frac{2}{3} = \frac{8}{3} \div 4 = \frac{2}{3}$: 3d, $\frac{2}{3} \div 2 = \frac{1}{3}$: 4th, $\frac{1}{3} \div 5 = \frac{1}{15}$: 5th, $2\frac{1}{15} = \frac{31}{15} \div 6 = \frac{31}{90}$.

65. Divide 47,325,737 by 6, 5, 4, 3, and 2. *A.* $65,730\frac{137}{720}$.
66. Divide 76,543,210 by 8, 7, 5, 4, and 3. *A.* $22,780\frac{241}{336}$.
67. Divide 37,754,276 by 5, 4, 7, 6, and 5. *A.* $8,989\frac{17}{150}$.
68. Divide 76,864,207 by 8, 7, 5, 5, and 4. *A.* $13,725\frac{601}{800}$.
69. Divide 42,680,960 by 12, 2, 6, 2, and 7. *A.* $21,171\frac{1}{9}$.
70. Divide 98,765,432 by 9, 8, 7, 6, 5, 4, 3, and 2. *A.* $272\frac{7759}{45360}$.
71. Find by the preceding rules what number multiplied by $\frac{3}{4}$ will make $15\frac{3}{4}$. *A.* 21.
72. What part of 108 is $\frac{5}{12}$ of an unit? *A.* $\frac{5}{1296}$.
73. What number is that which, if multiplied by $\frac{5}{8}$ of $\frac{3}{7}$ of $15\frac{1}{8}$, will produce only $\frac{5}{8}$ of an unit? *A.* $\frac{56}{363}$.

MISCELLANEOUS EXAMPLES.

LXVII. 1. A gentleman has two sons; the age of the elder added to his, make 126 years, and the age of the younger son is equal to the difference between the age of the father and the elder son. Now if the father be 80 years of age, how old is each of his sons?
A. 34 years, and 46 years.

2. What number is that, from which, if a twelfth part of 1,728 be,

---

Q. How may a mixed number be divided without any reduction, by a whole number? 62. Why is the remainder, if there be any multiplied by the denominator of the fractional part? 63

deducted, and the remainder increased by the ninety-fifth part of 82175, the sum will be 1185? *A.* 464.

3. What number divided by 1185 will give 497 for the quotient, and leave just a fifth part of the divisor remaining? *A.* 589,182.

4. A merchant, having bought at one time 25bl. 27gal. 2qt. 1pt 2gi. of molasses, and at another 5 times as much, sold $\frac{1}{4}$ of the whole; how much has he on hand, unsold?

A. 116bl. 14gal. 1qt. 3gi.

5. "At the Clinton works in Scotland, a sheet of paper has been recently [1839] produced, which, though but 50 inches wide, measures a mile and a half in length." How many square feet does this mammoth sheet contain? How many smaller sheets $1\frac{1}{4}$ feet long and 1 foot wide will it make? How many quires will these sheets make? What will be their value at $4\frac{1}{2}$ per ream?

A. 33,000 sq. ft.; 26,400 sheets; 1,100 quires; $247\frac{1}{2}$.

6. How many acres of land are there in a piece 80 rods long and 36 rods wide? *A.* 18A.—In a piece $80\frac{3}{4}$ rods long and $36\frac{7}{8}$ rods wide? *A.* 18A. 2R. $17\frac{21}{32}$rd.

7. If a *talent* of silver be worth £357. 11s. $10\frac{1}{2}$d. sterling, what is the value of a *shekel*, of which 300 make a *talent*; and what is the weight of a *talent*, a *shekel* weighing 9dwt. 3gr.

A. £1. 3s. $10\frac{3}{40}$d.; 136oz. 17dwt. 12gr.

8. If London was built 1108 years before Christ's nativity, how many hours is it since, to Christmas, 1835, allowing $365\frac{1}{4}$ days to the year? *A.* 25,798,338 hours.

9. A father gave $\frac{2}{7}$ of his property to his daughter, $\frac{4}{7}$ to his son, and the rest, being $1,200, to his nephew; what was the value of the father's estate? *A.* $8,400.

10. "The tail of the comet of 1811 was no less than one hundred and thirty-two millions of miles in length. Now, allowing the earth to be 25,000 miles in circumference, and the tail of the comet a bandage, how many times would it enwrap the earth?" *A.* 5,280 times.

11. A general distributed £307. 17s. among 4 captains, 5 lieutenants, and 60 common soldiers; to every lieutenant he gave twice as much as to a common soldier, and to every captain three times as much as to a lieutenant; what did each receive?

A. Soldier £3. $5\frac{1}{2}$s.; lieut. £6. 11s.; capt. £19. 13s.

12. What number is that from which if you subtract $\frac{1}{11}$ of $\frac{5}{9}$ of a unit, and to the remainder add $\frac{3}{5}$ of $\frac{7}{8}$ of a unit, the sum will be 9?

A. $8\frac{2081}{3960}$.

13. Suppose a quotient to be $3\frac{1}{2}$ times $\frac{2}{11}$, and a dividend $5\frac{1}{2}$ time $\frac{9}{13}$, what will be the divisor? *A.* $5\frac{179}{182}$.

14. A merchant invested $8,300 in a certain bank, being just $\frac{5}{115}$ of its capital; what was the capital of the bank? *A.* $190,900.

15. A merchant gave $$1,956\frac{3}{4}$ for $\frac{4}{7}$ of a sloop, and $\frac{2}{5}$ of the value of the sloop for its entire cargo; what was the estimated value of both sloop and cargo? *A.* $$4,794.03\frac{3}{4}$.

16. Suppose a man's family expenses are $\$730\frac{8}{35}$ annually, being only $\frac{3}{8}$ of his profits in trade; how much then can he save every year? *A.* $\$1,217\frac{1}{21}$.

17. Suppose a carriage wheel to be $15\frac{5}{24}$ inches in circumference, how many times would it turn round in going $367\frac{3}{8}$ miles?

A. $1,530,534\frac{42}{73}$ times.

18. Divide 12 sq. rd. 30 sq. yd. 6 sq. ft. by 7, and multiply 1 sq. rd. $25\frac{25}{28}$ sq. yd. $123\frac{3}{7}$ sq. in. by 7.

19. Suppose a pile of wood to be $200\frac{1}{4}$ feet long, $6\frac{2}{5}$ feet high, and $4\frac{5}{8}$ feet wide; how many solid feet does the pile contain, and how many cords? *A.* $6,194\frac{2}{5}$ s. ft.; 48C. $50\frac{2}{5}$ s. ft.

20. How many cubic feet in a box $9\frac{3}{5}$ inches long, $8\frac{3}{4}$ feet high, and 5 feet wide? *A.* $411\frac{1}{4}$ s. ft.

21. Suppose a room to be 10 feet between floors and $18\frac{1}{4}$ feet square, what will be the expense of plastering it at 8 cents per square yard? There are 4 sides each 10 ft. by $18\frac{1}{4}$, and the surface over head is $18\frac{1}{4}$ by $18\frac{1}{4}$. *A.* $\$9.44\frac{17}{18}$.

22. A wealthy merchant, on retiring from business, invested $\frac{7}{18}$ of all his property in banks, $\frac{4}{9}$ in private loans, $\frac{11}{72}$ in real estate, and the rest, which was $3,000, he reserved for repairs on his estate; how much must he have accumulated? *A.* $216,000.

23. Suppose a man buys $\frac{3}{4}$ of a ship, which is valued at $63,000, and divides it equally among his sons, giving to each $\frac{3}{21}$ of his part; how many sons has he, and how many dollars does each receive?

A. 7 sons; $6,750 each.

24. Divide $\frac{15}{16}$ of a roll of broadcloth, which contains 32 yards, into equal pieces, each to contain $\frac{2}{11}$ of the whole roll.

A. $5\frac{5}{32}$ pieces.

25. Suppose two boys, having bought a kite together, one paying $\frac{1}{4}$ of a dollar and the other $\frac{7}{8}$ of a dollar, sell it for $\frac{5}{8}$ of a dollar more than they paid for it; what did they pay for the kite? *A.* $\$1\frac{1}{8}$.

What did they get for the kite? *A.* $\$1\frac{3}{4}$.

What is each one's part of the kite? *A.* $\frac{2}{9}$; $\frac{7}{9}$.

What is each one's share of the profit? *A.* $\$\frac{10}{72}$; $\$\frac{35}{72}$.

What is each one's share of what it sold for? *A.* $\$\frac{14}{36}$; $\$1\frac{13}{36}$.

26. If $\frac{7}{9}$ of a ship valued at $20,000 be worth $\frac{5}{7}$ of her cargo, what is the value of both ship and cargo? *A.* $\$41,777\frac{7}{9}$.

27. A person left $\frac{2}{5}$ of his property to A, $\frac{3}{10}$ to B, $\frac{1}{8}$ to C, $\frac{1}{20}$ to D, $\frac{1}{40}$ to E, $\frac{1}{50}$ to F, and the rest, which was $800, to his executor; what was the value of the whole property, and of each person's share?

A. A's $4,000; B's $3,000; C's $1,250; D's $500; E's $250; F's $200

DECIMAL FRACTIONS.

LXVIII. 1. A DECIMAL FRACTION[1] differs from a vulgar fraction only in respect to its denominator, being uniformly either 10 or 100 or 1000, &c., and therefore it need not be, and seldom is, expressed.

2. The numerator then is written alone with a point before it, to distinguish it from whole numbers; this point is thence called a separatrix, and sometimes the *decimal point.*

3. Thus .3 is $\frac{3}{10}$; .34 is $\frac{34}{100}$; .345 is $\frac{345}{1000}$; .3456 is $\frac{3456}{10000}$.

4. UNITY then in decimals is first divided into 10 equal parts, which are therefore called TENTHS.

5. The TENTH is divided into 10 other equal parts, making 100 equal parts of unity, which are thence called HUNDREDTHS.

6. The HUNDREDTH is divided into 10 other equal parts, making 1000 equal parts of unity, which are thence called THOUSANDTHS; and so on, as in the following

TABLE I.

10-tenths - - - - - - - - - -	make 1 unit.
10-hundredths - - - - - - - -	make 1 tenth.
10-thousandths - - - - - - -	make 1 hundredth.
10-ten-thousandths - - - - - -	make 1 thousandth.
10-hundred-thousandths - - - -	make 1 ten-thousandth.
10-millionths - - - - - - - -	make 1 hundred-thousandth.

7. Since one decimal figure has for its denominator 1 with one cipher, as $.5=\frac{5}{10}$; two decimal figures, 1 with two ciphers, as $.25=\frac{25}{100}$; three decimal figures, 1 with three ciphers, as $.125=\frac{125}{1000}$, and so on; therefore,

8. A DECIMAL FRACTION is that fraction whose denominator is always understood to be a unit, or 1, with as many ciphers annexed as the given decimal has places of figures.

9. Thus, .8 is $\frac{8}{10}$; .08 is $\frac{8}{100}$; .35 is $\frac{35}{100}$; .0125 is $\frac{125}{10000}$.

10. When the numerator has not so many decimal places as the denominator has ciphers, we must prefix ciphers enough to the numerator to make as many.

11. Thus $\frac{5}{100}$ is written .05; $\frac{45}{10000}=.0045$; $\frac{6}{10000}$ .0006.

12. Since $.5=\frac{5}{10}$, $.05=\frac{5}{100}$, $.005=\frac{5}{1000}$, then .05 is 10 times less in value than .5, and .005 is 10 times less than .05:

LXVIII. Q. How does a decimal fraction differ from a vulgar one? 1. How is the numerator written? 2. What does .3, .34, .345, .3456, with the point before each number, mean? 3. How is unity divided and sub-divided in decimals? 4, 5, 6. Repeat the table of these divisions. What is the denominator of one, two, or three decimal figures? 7. What then is a Decimal Fraction? 8. What is the denominator for .8? See 9. What is the denominator for .08?—for .35?—for .0125? 9. When are ciphers to be prefixed to the numerator? 10. How do you write decimally $\frac{5}{100}$, or $\frac{45}{10000}$, or $\frac{6}{10000}$?

1 A DECIMAL FRACTION is so called from the Latin word *decimus*, signifying *tenth* because it increases and decreases in a tenfold proportion

13. For 10 times $\frac{5}{1000}$ is $\frac{50}{1000}=\frac{5}{100}$, and 10 times $\frac{5}{100}=\frac{50}{100}=\frac{5}{10}$

14. *Hence, a cipher placed on the left of any decimal decreases its value in a tenfold proportion, by removing it farther from the separatrix or decimal point.*

15. But a cipher on the right of a decimal merely changes its name without altering its value.

16. For .5 is $\frac{5}{10}=\frac{1}{2}$; so .50 is $\frac{50}{100}=\frac{1}{2}$, and .500 is $\frac{500}{1000}=\frac{1}{2}$, but the first is read 5-tenths, the second, 50-hundredths, the third, 500-thousandths.

17. *Decimals then increase from the right to the left like whole numbers, and of course their decrease from right to left is in the same proportion.*

18. *Hence every removal of any figure one place further towards the right decreases its value in a tenfold proportion.*

19. Thus 555.555 is really 500, 50, 5, $\frac{5}{10}$, $\frac{5}{100}$, $\frac{5}{1000}$.

20. Our system of notation, then, which begins in whole numbers, is carried out by means of decimals, so as to embrace as many places below units as above or beyond them, even millions and millionths, billions and billionths, as in the following

TABLE II.

BILLIONS.	HUNDRED MILLIONS.	TEN MILLIONS.	MILLIONS.	HUNDRED THOUSANDS.	TEN THOUSANDS.	THOUSANDS	HUNDREDS.	TENS.	UNITS.	Separatrix.	TENTHS.	HUNDREDTHS.	THOUSANDTHS.	TEN-THOUSANDTHS	HUNDRED-THOUSANDTHS.	MILLIONTHS.	TEN-MILLIONTHS.	HUNDRED-MILLIONTHS	BILLIONTHS.
5	5	5	5	5	5	5	5	5	5	.	5	5	5	5	5	5	5	5	5

A s c e n d i n g. ☜ ☞ *D e s c e n d i n g.*

21. The first decimal figure is 5 tenths; the second is 5 hundredths, or the first two 55 hundredths; the third is 5 thousandths, or the first three 555 thousandths; the fourth is 5 ten-thousandths, or the first four 5555 ten thousandths, &c.

RULE FOR NUMERATING AND READING DECIMALS.

22. *Begin on the left and say, tenths, hundredths, thousandths, ten-thousandths, &c., as in the table.*

Q. What is the difference in value between .5 and .05?—.05 and .005? (See 12.) What is the effect of the cipher? 14. What is the use of a cipher on the left of a decimal? 15. How is a decimal figure affected by changing its place? 18 Why? 17. What is the value of each figure in 555.555? 19. Describe our system of notation. 20. Repeat the decimal part of the table beginning with "tenths" and ending with "billionths." Suppose each decimal place to be filled with the figure 5, what would be the value of the first 5?—of the second?—third?—fourth? &c. 21. What is the rule for numerating decimals? 22

23. *Then begin on the left and read, giving each figure the value assigned it in numerating.*

24. *Or numerate and read the entire decimal, as if it were a whole number, giving the name of the last right hand place to the whole.*

25. Write on the slate the decimal figures expressing the following numbers, to be numerated and read at recitation.

26. Five tenths.

27. Seventy-six hundredths.

28. Nine tenths and two hundredths.

29. Three hundred and twenty-one thousandths.

30. Five tenths, two hundredths, and six thousandths.

31. Six tenths, two hundredths, three thousandths, and one ten thousandth.

32. Six thousand nine hundred and fifteen ten-thousandth.

33. Six tenths, 1 ten thousandth, and four millionths.

34. NOTE.—Supply all vacant places with ciphers.

35. Three tenths, five thousandths, and two millionths.

36. One hundred and one thousandths.

37. To express 5 hundredths, which has one vacant place, viz. tenths, we prefix 1 cipher [.05]; to express 5 millionths, which has five vacant places, we prefix 5 ciphers [.000005], and so on to any extent.

38. Hence, to express any number of hundredths or thousandths, &c.—*Prefix as many ciphers as there are vacant places between it and the separatrix.*

39. Write on the slate and recite as before the following numbers.

40. Seven hundredths.

41. Forty-five ten-thousandths.

42. Six hundred thousandths and one millionth.

43. Fifteen hundred-thousandths and fifteen billionths.

44. One thousandth, one millionth, and one billionth.

45. Nine hundredths, nine thousandths, and nine billionths

46. Three hundred and sixty-five millionths.

47. One hundred and twenty-five trillionths.

48. When a whole number has a decimal annexed, they form a mixed decimal fraction, and may be read like decimals, giving the name of the last decimal figure to both.

49. Thus 45.2 is $45\frac{2}{10}$ or $\frac{452}{10}$, that is, 452 tenths.

50. So 5.62 is 562 hundredths, and 3.005 is 3005 thousandths.

51. In the examples $.5=\frac{5}{10}$ and $.25=\frac{25}{100}$, if we annex a cipher to .5 it becomes $.50=\frac{50}{100}$, having the same denominator with .25, therefore,—

Q. What are both methods of reading decimals? 23, 24. How are 6-tenths, 1-ten-thousandth and 4-millionths written in one line? 34. How are 5-hundredths or 5-millionths written, and why? 37. What is the general direction? 38. What is a mixed decimal? 48. How may the following numbers be read, viz 45.2 5.62, and 3.005? [See 49, 50.]

52. Whole or mixed numbers and pure decimals are easily reduced to decimals having the same denominators by simply annexing ciphers

53. Reduce 2.5, 8.1, and 7.05, each to hundredths.

A. 2.50; 8.10; 7.05.

54. Reduce $3\frac{5}{10}$, 17.8, and .212, to thousandths.

A. 3.500; 17.800; .212.

55. Reduce the following numbers to decimals having the same denominators. 8.5; $3\frac{21}{100}$; $\frac{5}{100}$; $\frac{8}{1000}$; 756.3; $\frac{9}{100000}$; $981\frac{1}{10}$.

56. Answers. 8.50000; 3.21000; .05000; .00800; 756.30000; .00009; 981.10000.

57. In decimals, no single expression, containing any number of figures whatever, can fully equal unity.

58. Thus, .9 is 1-tenth less than 10-tenths, which make one unit; so .999999 is .000001, or 1 millionth less than 1.

59. It is observable also, that of two decimal expressions, the greater one, (no matter of how many figures either consists) has the greater number of tenths, or if the tenths be equal, a greater number of hundredths, and so on.

60. Thus .4 is greater than .399999, or .3 with any number of 9s that can possibly be annexed.

61. For .4 is (by 52) =.4000000, or equal to .4 with any number of ciphers annexed; now, .400000 is obviously greater than .399999.

---

62. Federal Money, by assuming the dollar, as the money unit, is perfectly adapted in all its inferior denominations, to the decimal notation.

63. For, as 10 dimes make one dollar; 10 cents 1 dime; and 10 mills one cent; dimes are 10ths of dollars; cents, 10ths of dimes or 100ths of dollars, and mills 10ths of cents, or 1,000ths of dollars.

64. Thus $3, 2 dimes, 4 cents and 5 mills are written decimally $3.245, that is, $$3\frac{245}{1000}$.

REDUCTION OF DECIMALS.

LXIX. 1. Reduction of Decimals is the changing of their forms, without altering their value.

CASE I.

To reduce a decimal fraction to a vulgar one.

RULE.

1. *Write under the given decimal its proper denominator, and it*

Q. How may whole numbers, or decimals of different denominators, be reduced to a common denominator? 52. Reduce 2.5, 8, and 7.05, each to hundredths. 53. Is then any decimal expression fully equal to unity? What is the difference in value between unity and .9? Unity and .999999? Which is the greater decimal, .4 or .399999? 60. How do you ascertain it? 61. What similarity has Federal Money to decimals? 62.

becomes a vulgar fraction, which may generally be reduced to lower terms.

2. Reduce .5 to a vulgar fraction. .5 is $\frac{5}{10}=\frac{1}{2}$. *A.* $\frac{1}{2}$.
3. Reduce .75 and .125 to vulgar fractions. *A.* $\frac{3}{4}$; $\frac{1}{8}$.
4. Reduce .875 and .15 to common fractions. *A.* $\frac{7}{8}$; $\frac{3}{20}$
5. Reduce .05 and .1875 to common fractions. *A.* $\frac{1}{20}$; $\frac{3}{16}$
6. Reduce .005 and .0005 to vulgar fractions. *A.* $\frac{1}{200}$; $\frac{1}{2000}$.
7. Reduce .00125 and 6.25 to vulgar fractions. *A.* $\frac{1}{800}$; $6\frac{1}{4}$.
8. Reduce 6.015 and 5.50 to vulgar fractions. *A.* $6\frac{3}{200}$; $5\frac{1}{2}$.

CASE II.

To reduce a vulgar fraction to a decimal.

RULE.

1. *Annex a cipher to the numerator and divide by the denominator if there be a remainder, annex another cipher and divide as before, and so on to any extent required.*

2. *The quotient will contain as many decimal places as there are ciphers annexed; but if there be not as many places, supply the defect by prefixing ciphers to the quotient.*

3. For, annexing one cipher to the numerator multiplies it by 10, which brings it into tenths, (as $\frac{2}{5} \times 10 = \frac{20}{5}$ tenths.)

4. Then as many times as the denominator is contained in the numerator, so many 10ths are contained in the fraction, (as $\frac{20 \text{ tenths}}{5} = .4$.)

5. Annexing another cipher brings the numerator into hundredths, then dividing by the denominator will show the hundredths contained in the fraction, and so on, ($\frac{1}{4} \times 100 = \frac{100}{4}$ hundredths$=.25$.)

6. When there are no tenths, hundredths, &c., the vacant places in the quotient must be filled with ciphers, to keep the significant figures of the quotient in their proper places.

7. Reduce $\frac{1}{2}$, $\frac{3}{4}$, $\frac{7}{8}$, $\frac{1}{25}$, and $\frac{1}{200}$, to decimal fractions.

2)1.0	4)3.00	8)7.000	25)1.00	200)1.000
A. .5	*A.* .75	*A.* .875	*A.* .04	*A.* .005

8. Reduce $\frac{51}{200}$ and $\frac{13}{625}$ to decimal fractions. *A.* .255; .0208.
9. Reduce $\frac{1}{8}$ and $\frac{2}{25}$ to decimal fractions. *A.* .125; .08.
10. Reduce to decimals $\frac{9}{16}$, $\frac{1}{40}$, $\frac{3}{8}$, $\frac{5}{8000}$, $\frac{1}{20000}$, $\frac{1}{250}$, $\frac{3}{40}$.

Answers. .5625; .025; .375; .000625; .00005; .004; .075.

11. Reduce $14\frac{1}{8}$ to a decimal fraction. Reduce the fractional part separately, then annex it. *A.* 14.125.

LXIX. *Q.* What is Reduction of Decimals? 1.

CASE I. *Q.* What vulgar fraction is equal to .5?—to .75?—to .4?—to .25? What is the rule? 1.

CASE II. *Q.* How is a vulgar fraction reduced to a decimal one? 1. How many decimal places must there be in the quotient? 2. Why is the cipher annexed? 3, 4. Give an example. Why annex two or more ciphers? 5 Why are ciphers in some instances to be prefixed to the quotient? 6

12. Reduce to decimals $9\frac{1}{5}$, $5\frac{1}{8}$, $6\frac{7}{8}$, $5\frac{1}{16}$, $74520\frac{1}{200000}$ and $\frac{1}{3}$.

A. 9.2; 5.125; 6.875; 5.0625; 74520.00005; .333333+.

13. Reduce $\frac{11}{2}$ to a decimal. $\frac{11}{2}=5\frac{1}{2}=5.5$, or $11.0\div2=5.5$, recollecting that the quotient figure or figures, before ciphers are annexed, is, of course, a whole number. *A.* 5.5.

14. Reduce to decimals $\frac{1111}{200}$, $\frac{4251}{5}$, $\frac{121}{4}$, $\frac{14551}{25}$, and $\frac{10}{3}$.

A. 5.555; 850.2; 30.25; 582.04; 3.33333+.

15. In the last example, the decimal will repeat 33, &c., for ever if we continue the operation.

16. Decimals which repeat one or more figures are called REPEATING DECIMALS, or REPETENDS.

17. REPEATING DECIMALS are also called INFINITE DECIMALS; those that terminate, or come to an end, FINITE DECIMALS.*

Q. How is a mixed number reduced to a decimal? 11. How is an improper fraction reduced? 13. What decimal is equal to $5\frac{1}{2}$?—to $6\frac{3}{4}$?—to $\frac{11}{2}$? What are Repeating Decimals? 16. What other names have they, and when? 17.

---

* REPEATING DECIMALS are also called CIRCULATING DECIMALS. When only one figure repeats, it is called a single repetend; but if two or more figures repeat, it is called a compound repetend: thus, .333, &c. is a single repetend, .010101, &c. a compound repetend.

When other decimals come before circulating decimals, as .8 in .8333, the decimal is called a mixed repetend.

It is the common practice, instead of writing the repeating figures several times, to place a dot over the repeating figure in a single repetend; thus, .111, &c. is written $\dot{1}$; also over the first and last repeating figure of a compound repetend; thus, for .030303, &c. we write $.\dot{0}\dot{3}$.

The value of any repetend, notwithstanding it repeats one figure or more an infinite number of times, coming nearer and nearer to a unit each time, though never reaching it, may be easily determined by common fractions; as will appear from what follows.

By reducing $\frac{1}{9}$ to a decimal, we have a quotient consisting of .1111, &c., that is, the repetend $\dot{1}$; since $\frac{1}{9}$ is the value of the repetend $\dot{1}$, the value of .333, &c., that is, the repetend $\dot{3}$, must be three times as much, that is, $\frac{3}{9}$ and $\dot{4}=\frac{4}{9}$; $\dot{5}=\frac{5}{9}$; and $\dot{9}=\frac{9}{9}=1$ or the whole.

Hence we have the following RULE for changing a single repetend to its equal common fraction:—Make the given repetend a numerator, writing 9 underneath for a denominator, and it is done.

What is the value of $.\dot{1}$? Of $.\dot{2}$? Of $.\dot{4}$? Of $.\dot{7}$? Of $.\dot{8}$? Of $.\dot{6}$? *A.* $\frac{1}{9}$, $\frac{2}{9}$, $\frac{4}{9}$, $\frac{7}{9}$, $\frac{8}{9}$, $\frac{6}{9}$.

By changing $\frac{1}{99}$ to a decimal, we shall have .010101, that is, the repetend $.\dot{0}\dot{1}$. Then, the repetend $.\dot{0}\dot{4}$, being 4 times as much, must be $\frac{4}{99}$, and $\dot{3}\dot{6}$ must be $\frac{36}{99}$, also, $.\dot{4}\dot{5}=\frac{45}{99}$

If $\frac{1}{999}$ be reduced to a decimal, it produces $.\dot{0}0\dot{1}$. Then the decimal $.\dot{0}0\dot{4}$, being 4 times as much, is $\frac{4}{999}$, and $.\dot{0}3\dot{6}=\frac{36}{999}$. This principle will be true for any number of places.

Hence we derive the following RULE for reducing a circulating decimal to a common fraction:—Make the given repetend a numerator; and the denominator will be as many 9s as there are figures in the repetend.

Change $.\dot{1}\dot{8}$ to a common fraction. *A.* $\frac{18}{99}=\frac{2}{11}$.

Change $.\dot{7}\dot{2}$ to a common fraction. *A.* $\frac{72}{99}=\frac{8}{11}$.

Change $.\dot{0}0\dot{3}$ to a common fraction. *A.* $\frac{3}{999}=\frac{1}{333}$.

18. In general, whether the decimal be finite or infinite, three or four places are sufficient for most practical purposes.

19. Change $\frac{1}{9}$ to a decimal fraction. *A.* .1111+ or .1111$\frac{1}{9}$.
20. Change $\frac{2}{3}$ to a decimal fraction. *A.* .6666+ or .666$\frac{2}{3}$.
21. Change $\frac{3}{32}$ to a decimal fraction. *A.* .0937.+
22. Change $8\frac{1}{16}$ to a decimal fraction. *A.* 8.062$\frac{5}{10}$
23. Change \$$\frac{1}{8}$ to cents and mills. *A.* \$.125=.12$\frac{5}{10}$=12$\frac{1}{2}$.
24. Change \$8$\frac{5}{16}$ to dollars and cents. *A.* \$8.3125=\$8.31$\frac{1}{4}$.
25. Change £50$\frac{5}{8}$ to a decimal form. *A.* £50.625.
26. Change 3$\frac{1}{25}$ miles to a decimal form. *A.* 3.04m.
27. Reduce to single fractions first and then to decimals $\frac{1}{2}$ of $\frac{2}{3}$; $\frac{3}{4}$ of $\frac{5}{8}$; $\frac{3}{\frac{2}{3}}$ and $\frac{\frac{1}{2}}{3\frac{1}{4}}$. *A.* .3333$\frac{1}{3}$; .46875; 4.5; .153846$\frac{2}{13}$.

CASE III.

To reduce a simple number of a given denomination to a decimal of a higher denomination.

RULE.

1. *Divide as in Reduction of whole numbers, annexing ciphers and pointing off the places for decimals in each quotient, as in the last Case, and for the same reasons.*

```
4  ) 3 . 0 0 qr.
1 2 ) . 7 5 0 0 d.
2 0 ) . 0 6 2 5 0 0 s.
    £ . 0 0 3 1 2 5
```

2. Reduce 3 farthings to the decimal of a pound.

3. Here 3qr.÷4qr.=.75 of a penny÷12d.=.0625 of a shil.÷20s.=.003125 of a pound. *A.* £.003125.

Q. How many places of decimals are generally sufficient? 18. What are Circulating Decimals? [See reference from 17.] What are single, compound, and mixed repetends? Of the decimals .333, &c., .0101, &c., and .8333, &c., which are the repeating figures? What is the proper name for each of these decimals? How are repetends distinguished from other decimals? How is the value of the repetend $.\dot{3}$ expressed, and why? What is the rule for it? What is the value of $.\dot{2}\dot{7}$ and $.\dot{0}\dot{1}$? What is the rule for reducing a circulating decimal to a vulgar fraction? Reduce to a vulgar fraction $.\dot{1}\dot{8}$, $.\dot{7}\dot{2}$ and $.\dot{0}0\dot{3}$. Describe the process of finding the value of the mixed repetend $.8\dot{3}$. What is the rule?

In the following example, viz. Change $.8\dot{3}$ to a common fraction, the repeating figure is 3, that is, $\frac{3}{9}$, and .8 is $\frac{8}{10}$; then $\frac{3}{9}$, instead of being $\frac{3}{9}$ of a unit, is, by being in the second place, $\frac{3}{9}$ of $\frac{1}{10}=\frac{3}{90}$; then $\frac{8}{10}$ and $\frac{3}{90}$ added together, thus, $\frac{8}{10}+\frac{3}{90}=\frac{75}{90}=\frac{25}{30}$, *Ans.*

Hence, to find the value of a mixed repetend—First find the value of the repeating decimals, then of the other decimals, and add these results together.

Change $.91\dot{6}$ to a common fraction. *A.* $\frac{91}{100}+\frac{6}{900}=\frac{825}{900}=\frac{11}{12}$. *Proof,* 11÷12 $=.91\dot{6}$.

Change $.20\dot{3}$ to a common fraction. *A.* $\frac{61}{300}$.

To know if the result be right, change the common fraction to a decimal again. If it produces the same, the work is right

Repeating decimals may be easily multiplied, subtracted, &c. by first reducing them to their equal common fractions

4. Reduce 35 rods to the decimal of a mile. *A.* .109375m.
5. Reduce 9 pence to the decimal of a pound. *A.* £.0375.
6. Reduce 2 quarters to the decimal of a ton. *A.* .025T.
7. Reduce 8 drams to the decimal of a ton. *A.* .000015625T
8. Reduce 3 gills to the decimal of a hogshead.
A. .001488$\frac{2}{21}$hhd.
9. Reduce 2$\frac{3}{8}$ nails to the decimal of a yard. The 2$\frac{3}{8}$na.=2.375 a.÷4na.=.59375qr. &c. *A.* .1484375yd.
10. Reduce 4$\frac{7}{8}$ pence to the decimal of a shilling. *A.* .40625.
11. Reduce 9$\frac{1}{4}$ shillings to the decimal of a dollar, (6s.) that is, to dollars and cents. *A.* \$1.541$\frac{2}{3}$.
12. Reduce 15$\frac{3}{4}$ shillings to dollars and cents. *A.* \$2.625.
13. Reduce 26$\frac{2}{5}$ hours to the decimal of a day. *A.* 1.1 day.

CASE IV.

To reduce compound numbers to decimals of higher denominations.

RULE.

1. *Divide as in the last Case, annexing each decimal quotient to the integer of that denomination.*

```
   4 ) 3 . 0 0 qr.
  12 ) 6 . 7 5 0 0 d.
  20 ) 1 0 . 5 6 2 5 0 s.
A. £ 9 . 5 2 8 1 2 5 .
```

2. Reduce £9. 10s. 6d. 3qr. to the decimal of a pound.

Recollect to make the decimal places in the quotient equal to those in the dividend, according to Case III.

3. Reduce £5. 11s. 4$\frac{1}{2}$ pence to the decimal of a £.
A. £5.56875.
4. Reduce 7s. 6d. 3qr. to the decimal of a £. *A.* £.378125.
5. Reduce 8oz. 17dwt. to the decimal of a lb. *A.* .7375lb.
6. Reduce 3qr. 3 na. to the decimal of a yd. *A.* .9375yd.
7. Reduce 2qr. 2na. to the decimal of an E. e. *A.* 5. E. e.
8. Reduce £8. 17s. 6$\frac{3}{4}$d. to the decimal of a £. *A.* £8.878125.
9. Reduce 5T. 2cwt. 3qr. 7lb. to the decimal of a ton.
A. 5.141T.
10. Reduce 3.5 shillings to the decimal of a £. *A.* £.175.
11. Reduce 12h. 15.3m. to the decimal of a day. *A.* .510625d.
12. Reduce 2m. 4fur. 20rd. 4yd. 1ft. 2 in. 1$\frac{1}{3}$b. c. to the decimal of a league. *A.* .8551.

CASE V.

To reduce a decimal of a higher denomination to integers of lower denominations.

CASE III. *Q.* How are 3 farthings reduced to the decimal of a pound? 3. What is this case? [See Case III.] What is the rule? 1. What decimal of a pound is 15 shillings?—is 7 shillings?—is 3 shillings? What decimal of a mile is 32 rods?—is 16 rods?

CASE IV. *Q.* How do you divide farthings, pence, &c. to reduce them to the decimal of a pound? 2. What is this case? [See the Case.] What is the rule? 1

1. This Case is the reverse of the last and proves it.

RULE.

2. *Multiply as in Reduction of whole numbers, and make the decimal places in each product equal to those of the multiplicand; then the several excesses on the left will constitute the compound number required.*

3. What is the value of .26 of a shilling?

```
     .2 6 s.
      1 2 d.
Ans. 3 . 1 2 d.
```

For, .26s. $=\frac{26}{100}\times 12$d. $=\frac{312}{100}$d. $=3\frac{12}{100}=3$ 12, that is, as .26 has two decimal places, so must the product have two.

4. What is the value of .845 of a £?

```
   £ . 8 4 5
         2 0 s.
 1 6 . 9 0 0 s.
         1 2 d.
 1 0 . 8 0 0 d.
           4 qr.
     3 . 2 0 0 qr.
```

Here, £.845×20s.=16.900s., making three decimal places, because £.845 has three;—next multiply only the .900s.; because the 16s. is an integer, and therefore a part of the answer. Do the same with the .800d.

5. What is the value of .645 of a £? *A.* 12s. 10d. $3\frac{1}{5}$qr.
6. What is the value of .375 of a £? *A.* 7s. 6d.
7. What is the value of .7375 of a lb.? *A.* 8oz. 17dwt.
8. What is the value of .9375 of a yd.? *A.* 3qr. 3na.
9. What is the value of .025 of a mile? *A.* 8rd.
10. What is the value of .0125 of a mile? *A.* 4rd.
11. What is the value of .025 of a ton? *A.* 2qr.
12. What is the value of .0025 of a £? *A.* $2.\frac{4}{10}$qr. $=2\frac{2}{5}$qr
13. What is the value of .005 of an hour? *A.* 18sec.
14. What is the value of.005 of a year? *A.* 1d. 19h. 48m.
15. Change .00025T. of round timber to inches. *A.* 21.6 in.
16. How many quarters in .534575yd.? *A.* 2 qr. $\frac{5532}{10000}$na.
17. What compound number will express .042675 of a year? *A.* 15d. 13h. 49m. $58\frac{8}{10}$sec.

ADDITION OF DECIMALS.

LXX. 1. Since decimals increase towards the left like whole numbers, like them, therefore, they may be added and subtracted, multiplied and divided.

CASE V. *Q.* What is the value of .26 of a shilling? 3. Why point off two decimal figures in the quotient? 3. What is the rule? 2. What is the value of .5 of a pound?—.75 of a cwt.?—.75 of a shilling?—.15 of a furlong?

LXX. *Q.* How are decimal operations performed, and why? 1. What is the rule for Addition of Decimals? 3. What is the sum of .2 .5 and .25? Add together 4.5 yards and 5.5 yards.

2. The only difficulty attending these operations consists in ascertaining where the decimal point should stand. This will be noticed in its proper place.

RULE.

3. *Write tenths under tenths, hundredths under hundredths, &c., and add as in whole numbers. Then make the places for decimals in the answer equal to the greatest number of decimal places in any of the given numbers.*

4. Add into one sum .5; .875; and .25: and into another sum 62.25, 350.009 and .0036.

```
     .5
     .875
     .25
A. 1.625
```

Add the numbers as they stand; thus 5 is 5; 5 and 7 are 12, carrying 1, &c.

```
     62.25
    350.009
       .0036
A.  412.2626
```

5. If in adding .875, .25 and .5 we reduce them all to 1000ths, the proper denominator for .875, we have $\frac{875}{1000}$, $\frac{250}{1000}$, and $\frac{500}{1000}$, whose sum is $\frac{1625}{1000}=1\frac{625}{1000}=1.625$, hence the reason for pointing off as the rule directs.

6. Add together 2.35C.; 450.009C.; .0839C. *A.* 452.4429 cords.

7. Add together 3.5689T.; 245.003T.; 6.8T. *A.* 255.3719 tons.

8. Add together .0632; .08; .456; .81 and 15. *A.* 16.4092.

9. Add together $16.375; $81.065 and 25 cents. *A.* $97.69.

10. Add together 8.5; 834.6758; 8.35; 4236.2. *A.* 5087.7258.

11. Write decimally and add 2 dollars and 30 cents; 4 dollars, 9 dimes, and 8 cents; 7 dollars and 8 mills; 3 dimes, 7 cents; 9 dimes and 2 mills. *A.* 15.56.

12. Write decimally and add $45\frac{3}{10}$hhd., $68\frac{4}{100}$hhd., $96\frac{25}{1000}$hhd., and $8\frac{3}{100}$hhd. *A.* 217.395hhd.

13. Find the sum of 6 tenths, 35 hundredths, 6 thousandths, and 35 hundred-thousandths. *A.* .95635.

14. What is the amount of 555 and 5 thousandths; 5 and 505 millionths; 620 ten-millionths; 36 billionths; 428 and 15 ten-thousandths; 5 million and 5 millionths? *A.* 5000988.007072036.

15. Reduce first to a decimal fraction by Case II., then add the following numbers, viz.—$45\frac{3}{4}$ yards; $10\frac{3}{16}$ yards; $67\frac{1}{2}$ yards; and $9\frac{1}{25}$ yards. *A.* 132.4775.

16. Add into one sum $\frac{3}{4}$, $\frac{5}{16}$, $\frac{2}{5}$, and $\frac{3}{8}$. *A.* 1.8375

17. What is the sum of $\frac{35}{100}$, $\frac{5}{100}$, $\frac{25}{1000}$, $\frac{295}{10000}$, $\frac{2}{10}$, $\frac{3756}{1000000}$, an 000005? *A.* .692065.

18. Reduce the following numbers to a decimal of the highest de-

---

Q. What is the sum of .3, .25, .2, and .25? How are vulgar fractions, wher connected with whole numbers, added? 15. What mixed decimal will express he sum of $2\frac{1}{2}$ and $2\frac{1}{8}$? How are compound numbers added decimally? 18

nomination mentioned in either, then add them together, viz.—10s. 3d., £3. 15s., £9. 4s. 6d. *A.* £13.4875.

19. Add together 6yd. 1qr., 5qr. 3 na., $17\frac{5}{8}$yd., and 256yd. $2\frac{1}{5}$qr. *A.* 281.8625 yards.

20. Find the sum of $60\frac{5}{8}$ years, $196\frac{3}{4}$ years, $675\frac{3}{5}$ years, $45\frac{3}{16}$ years, and $20\frac{3}{8}$ years. *A.* 998.3375 years.

21. Add decimally 19 miles, 3 furlongs, 20 rods; 31 miles, 5 furlongs; and 98 miles, 2 furlongs, 15 rods. *A.* 149.359375 miles.

SUBTRACTION OF DECIMALS.

RULE.

LXXI. 1. *Write tenths under tenths, hundredths under hundredths, &c.; then subtract as in whole numbers and point off as in Addition.*

2. Suppose a merchant bought 14.25 barrels of flour, and sold 8,375 barrels; how many barrels has he on hand?

```
1 4 . 2 5
  8 , 3 7 5
  ---------
  5 , 8 7 5
```

3. Say, 5 from 10 leaves 5, 1 to carry to 7. For $14.25=14\frac{250}{1000}-8\frac{375}{1000}$ (by LXIV. 27.)= 5,875 barrels, Ans.

4. From \$90.025 take \$8.6285. *A.* \$81.3965.

5. From \$38.036 take \$4.0375. *A.* \$33.9985.

6. Bought 513.025 barrels of flour, and sold 95.0375 barrels; how much had I left? *A.* 417.9875 barrels.

7. From 891.037 take 89.0738. *A.* 801.9632.

8. From 376,683 take 47.0931. *A.* 329.5899.

9. From 83.12 take 5.3758. *A.* 77.7442.

10. From 835.2 take .1234567. *A.* 835.0765433.

11. Subtract \$53.008 from \$100. *A.* \$46.992.

12. From \$5 take 5 dimes and 5 mills. *A.* \$4.495.

13. From 8 take 9 thousandths. *A.* 7.991.

14. From 1 take 1 hundredth. *A.* .99

15. From 1 take 1 millionth. *A.* .999999.

16. How much greater is 4 than 3.99999999? *A.* .0000001.

17. How much smaller is .999999 than 1? *A.* 1 millionth.

18. How much less than 1 is 1 trillionth?

A. .999999999999

19. From $19145\frac{3}{4}$ take $7918\frac{3}{8}$. *A.* 11227.375.

20. From \$$89\frac{2}{5}$ take \$37.125. *A.* 52.275.

21. From $17\frac{5}{8}$ barrels take $9\frac{3}{20}$ barrels. *A.* 8.475bl.

22. From £$95\frac{1}{8}$ take $18\frac{4}{5}$s. [See Case IV.] *A.* £94.185.

23. From $408\frac{3}{40}$ acres take $329\frac{3}{5}$ rods. *A.* 406.015A.

LXXI. *Q.* What is the rule for subtracting decimals? 1. What is the difference between .3 and .25?—between 1 and .1?—between .5 and .18?—between 2.5 and 3?—between .9 and unity?

24. From 10 eagles and $\frac{6}{10}$ of a mill take \$99.9998. *A.* $\frac{8}{10}$m.
25. From an unit subtract 5 millionths. *A.* .999995.
26. From $357\frac{5}{32}$lb. take $22\frac{3}{125}$lb. *A.* 335.13225lb.
27. From £8. 17s. $6\frac{3}{5}$d. take £3. 19s. $6\frac{9}{25}$d. *A.* £4.901.

MULTIPLICATION OF DECIMALS.

LXXII. 1. Every decimal has as many places as its denominator has ciphers, as $\frac{5}{10}$ is .5; $\frac{5}{100}$ is .05, &c.

2. *The product of any two decimals, then, must have as many decimal places as the product of their denominators has ciphers.*

3. Thus, .5×.7=.35, for $\frac{5}{10}\times\frac{7}{10}=\frac{35}{100}$=.35.

4. The product of the denominators of any two decimals has as many ciphers as both its factors. (XVIII.)

5. Thus, $\frac{25}{100}\times\frac{7}{10}=\frac{175}{1000}$, and $\frac{257}{10000}\times\frac{5}{10}=\frac{1285}{100000}$.

6. *Hence the product of any two decimal expressions must have as many decimal places as both its factors.*

7. Thus .5×.87=.435; for $\frac{5}{10}\times\frac{87}{190}=\frac{435}{1000}$=.435.

8. When the product has not as many places of figures as its factors have places of decimals, we must supply the deficiency by prefixing ciphers.

9. Thus, .2×.4=.08 because $\frac{2}{10}\times\frac{4}{10}=\frac{8}{100}$=.08.

GENERAL RULE.

10. *Multiply as in whole numbers, and point off so many places for decimals in the product as are equal to the decimal places in both the factors; but if the product has not so many places, prefix ciphers to make up the number.*

1. Multiply 4 5 . 6 2 5
by 5
A. 2 2 8 . 1 2 5

Here only one factor is a decimal, and it has 3 places; therefore make 3 decimal places in the product.

12. Multiply 81.235 wine gallons by 35. *A.* 2843.225.
13. Multiply 90,325 puncheons by .45. *A.* 40,646.25.

14. Multiply 3 . 2 5 1
by . 7
A. 2 . 2 7 5 7

Here are 4 decimal places in both the factors, therefore make 4 decimal places in the product.

15. Multiply 2.345 ale gallons by.15. *A.* .35175 of an ale gallon.

LXXII. *Q.* How is $\frac{5}{100}$ written decimally? What is the rule for it? 1. Why does .7 multiplied by .5 make .35? 4. What is the rule for ascertaining the decimal places? 2. How many ciphers have all such products? 4. What is the inference in respect to pointing off the product in decimals? 6. When are ciphers to be prefixed? 8. Why then is .08 the product of .2 by .4? 9. General Rule? 10. Multiply .8 by .6, by .06, by .006,—.8 by .5, by 05. by 0005;—.08 by .08 by .008.

16. Multiply 75.06 beer gallons by .19. *A.* 14.2614 gallons.

17. Multiply . 1 1 3 5
by . 3
A. . 0 3 4 0 5

Prefix 1 cipher, since the two factors have 5 decimal places. See 8.

18. Multiply .085 of a dollar by .39. *A.* $.03315.

19. Multiply .009 of a gallon by .05. *A.* .00045 of a gallon.

20. What will 3719.25 needles cost at $.005 a-piece?
A. $18.59ct. $6\frac{25}{100}$m.

21. Multiply .618 of a hogshead by .312. *A.* .192816hhd.

22. Multiply .521 of a bushel by .48. *A.* .25008bu.

23. Multiply .235 of a century by .45. *A.* .10575C.

24. Multiply .375 of a square inch by .00027. *A.* .00010125sq.in.

25. Multiply 8.165 of a minute by .00089. *A.* .00726685min.

26. What will 800 trees cost at $.375 a-piece? *A.* $300.

27. Multiply 800 and .008 together. *A.* 6.4.

28. Multiply 5 and .0005 together. *A.* .0025.

29. Multiply .16 and 500 together. *A.* 80.

30. Multiply .003 by 8500000. *A.* 25500.

31. Since decimals increase from the right to the left in a tenfold proportion, therefore,—

32. To multiply by 10, or 100, or 1,000, &c.—*Merely remove the separatrix one place farther towards the right for every cipher, and it is done.*

33. Multiply .3621 by 10;—by 100;—by 1,000;—by 10,000;—by 100,000. *A.* 3.621; 36.21; 362.1; 3621; 36210.

34. What would one million of flax seeds cost at $.000001 for each seed? *A.* $1.

35. Multiply 25 millionths by 18 thousandths. *A.* .000000045.

36. Multiply .02562 into $12\frac{1}{3}$. *A.* .31598.

37. What will be the cost of $333\frac{1}{3}$ Rohan potatoes at $.0645 a-piece? *A.* $$21\frac{1}{2}$.

38. What will $415\frac{1}{5}$ barrels of sugar weigh, the average weight of each being 495.00025 pounds? *A.* 205,524.1038.

39. Reduce the fractional parts of the following numbers to decimals, then multiply them together, viz. $30\frac{1}{4}$ by $5\frac{1}{2}$. *A.* 166.375.

40. Multiply decimally $30\frac{1}{4}$ sq. yd. by 9 sq. ft. *A.* 272.25 sq. ft.

41. Multiply 5.5 by $5\frac{1}{2}$ as a common fraction. *A.* $30.25=30\frac{1}{4}$.

42. Multiply $5\frac{1}{2}$ by $5\frac{1}{2}$ decimally. *A.* 30.25.

43. What will $18\frac{3}{4}$ hogsheads of molasses cost at the rate of $15\frac{3}{20}$ dollars per hogshead? *A.* $284.0625.

44. What will $7\frac{1}{2}$ loads of hay cost at £3. 10s. 6d. per load? Reduce both to decimals by LXIX. Case IV. *A.* £26.4375.

Q. How are decimals multiplied by 10 or 100, &c. easily? 32. Why has the process this effect? 31. What is the product of .1234 multiplied by 10?—by 100?—by 1,000?

45. Multiply £2. 3s. 8¾d. by 5 decimally. *A.* £10.921875.

46. Multiply 2m. 6fur. 30⅓rd. by 5. *A.* 14.221875 miles.

47. What will 10 tons 15cwt. of Russia iron cost at £2. 3s. 6d. per ton? The product is £23.38125. (LXIX. Case v.)

A. £23. 7s. 7½d.

48. What will 14 hogsheads 18.9 gallons of molasses cost at £3. 15s. 9d. per hogshead? *A.* £54. 3s. 2½d.+

49. Suppose a certain farm to consist of 200A. 3R. 25rd.; what will be its value at $25.375 per acre. *A.* $5097.996.+

50. If a man travels 30m. 3fur. 15½rd. a day for 26½ days, how far will he have traveled during that time? *A.* 806m. 1 fur. 30¾rd.

51. What will 10¾ bales of cotton cost, each bale weighing 3cwt. 2qr., at $10.62½ per cwt.? *A.* $399.76½+.

52. How many solid feet in a stick of timber 40ft. 9in. long, 1ft. 3in. wide, and 1ft. 9in. deep? What will it cost at 25cts. a foot?

A. 89ft. 243in. nearly; $22.285+.

DIVISION OF DECIMALS.

LXXIII. 1. We have seen that the factors in multiplication become the divisor and quotient in division. XXI. 53.

2. In multiplication of decimals, the product has as many decimal places as both its factors.

3. *In division of decimals, therefore, the divisor and quotient must have as many decimal places as the dividend.*

4. Thus .7×.5=.35, product: then, .35÷.5=.7, quotient.

5. The same result will follow from considering the decimals as vulgar fractions.

6. Thus .5 is $\frac{5}{10}$ or $\frac{50}{100}$; then $\frac{35}{100}\div\frac{50}{100}$(by LXVI. 54)$=\frac{35}{50}$=.7, quotient.

7. From the above it follows, that when the divisor is a whole number, the quotient alone has as many decimal places as the dividend.

8. And when the divisor alone has as many decimal places as the dividend, the quotient is a whole number.

GENERAL RULE.

9. *Divide as in whole numbers, and point off figures enough to make the decimal places in the divisor and quotient just equal to those in the dividend, prefixing ciphers to the quotient when necessary to make out the number.*

LXXIII. *Q.* How many decimal places has any product? 2. How many have the divisor and quotient? 3. Why? 1. Why is .7 the quotient of .35 divided by .5 ? 6. When does the quotient have as many decimals as the dividend? 7. When is the quotient a whole number? 8. What is the general rule? 9, 10, 11.

10. *When the dividend has not so many decimals as the divisor, make them first equal by annexing ciphers, in which case the quotient will be a whole number.*

11. *When any dividend, either with or without ciphers annexed, as above, does not contain the divisor, or when there is a remainder, annex ciphers, and for every such cipher reckon another decimal place in the quotient.*

12. What is the quotient of .75 divided by 5?

5) . 7 5
A. . 1 5

When the divisor is an integer, the quotient has as many decimals as the dividend. (See above, 7.)

13. Divide 87.745 barrels among 7 persons. *A.* 12.535.

14. Divide 9.031455 into 9 equal parts. *A.* 1.003495.

15. What is the quotient of .75 divided by .25?

. 2 5) . 7 5
A. 3

When the divisor has as many decimals as the dividend, the quotient is a whole number. (See 8.)

15. How many times .108 in .972? *A.* 9 times.

16. How many times .00009 in .00045? *A.* 5 times.

17. When 8 barrels of flour cost $80.75, what does one barrel cost?

8) 8 0 . 7 5 0 0 0 See the rule (11) for annexing ciphers
$ 1 0 . 0 9 3 7 5=$10. 9c. $3\frac{75}{100}$m. Answer.

18. Divide $50.5 into 16 equal parts. *A.* $3.15c. $6\frac{25}{100}$m

19. Divide 150.15 barrels into 32 equal parts. *A.* 4.6921875

20. At $.5 a yard, how many yards will cost $25?

. 5) 2 5 . 0
A. 5 0 yd.

See the rule (10) for annexing ciphers; which makes the quotient a whole number

21. Divide 30 hogsheads by .15 of a hogshead. *A.* 200.

22. How many times does 60.15 exceed .000005?

A. 12,030,000.

23. At .5 of a dollar or 50 cents a yard, how much may be bought for .35 of a dollar or 35 cents?

. 5) . 3 5
A. . 7 yd.

Make the 7 a decimal, then the decimals in .7 and .5 will equal those in .35. (Rule 9.)

24. Divide .192816 by .312. *A.* .618.

25. Divide .25008 by .48. *A.* .521.

26. What is the quotient of .00025 divided by .25?

. 2 5) . 0 0 0 2 5
A. . 0 0 1

Prefixing 2 ciphers to 1 makes the decimal places in .001 and .25 equal to those in .00025. (Rule 9.)

Q. Why is .75÷5=.15? 12. Why is .75÷.25=3? 15. Why is 80.75÷8=10.09375? 17. Why is .25÷.5=50? 20. Why is .35÷.5=.7? 23. What is the quotient of 10.8 divided by 12?—10.8 by .12?—.108 by 1.2?—.108 by .12?—.108 by 12?—1.08 by 1 2?—1.08 by .12?—10.8 by 1.2?—108 by .12?—108 by 1.2?—1.08 by 12?

27. Divide 37.035 dollars into 12345 equal parts. *A.* .003

28. Divide 1.77975 into 25425 equal parts. *A.* .00007.

29. Since decimals decrease from the left towards the right in a tenfold proportion; therefore,—

30. To divide by 10, 100, 1,000, &c.—*Merely remove the separatrix one place further towards the left for every cipher in the divisor.*

31. Divide 3752.3 by 10;—by 100;—by 1,000;—by 10,000;—by 00,000;—by 1,000,000.

A. 375.23; 37.523; 3.7523; .37523; .037523; .0037523.

32. Divide 1561.275 by 24.3;—by 48.6;—by 12.15;—by 6.075.

A. 64.25; 32.125; 128.5; 257.

33. Divide 8358 by .6073. *A.* 13762.55557.+

34. Divide .03315 by .085. *A.* .39.

35. Divide .264 by .2;—by .4;—by .02;—by .04;—by .002;—by .004. *A.* 1.32; .66; 13.2; 6.6; 132; 66.

36. Suppose a bushel of corn to contain 15,000 grains, and to cost $.90, what is the cost of a single grain? *A.* $.00006.

37. Divide 80 by 8 tenths;—800 by 8 hundredths.

38. Multiply 100 by 8 tenths;—10,000 by 8 hundredths.

39. Divide 5,000 by 5 thousandths;—5,000,000 by 5 millionths

40. Multiply 1 million by .005;—1 trillion by .000005.

41. How many times greater is 8 than .16?

42. Multiply 50 by 16 hundredths of a unit.

43. How many times greater is $16 than 2 cents or $.02?

44. Multiply 800 by 2 hundredths of a dollar.

45. How many times are 5 cents contained in 5 dollars?

46. Multiply 100 by 5 hundredths of a dollar.

47. Divide $196.08 by $5.16. *A.* 38.

48. Divide $2.15565 by $1.05. *A.* 2.053.

49. When the quotient repeats one figure or more, continue the division for three decimal places; then write a 9 under the repeating figure in the fourth place, and the quotient will express the exact decimal. (LXIX. Case II. 17.)

50. Divide $7 by 12 cents. ($\frac{3}{9}=\frac{1}{3}$.) *A.* 58.333$\frac{1}{3}$.

51. Multiply 58.333$\frac{1}{3}$ by 12 hundredths of a dollar.

52. What is the price of cloth by the yard when 6 yards cost $20? when 9 yards cost $4? *A.* 3\frac{2}{6}$ or 3.333\frac{1}{3}$; $.444$\frac{4}{9}$.

53. Divide 272$\frac{1}{4}$ by 30$\frac{1}{4}$ decimally. *A.* 9.

54. Divide 272$\frac{1}{4}$ by 9 decimally. *A.* 30.25.

55. Divide 30$\frac{1}{4}$ by 5$\frac{1}{2}$ decimally. *A.* 5.5.

56. How many times greater is £6. 5s. than 12s. 6d.? Reduce both to the decimal of a pound first. *A.* 10 times.

57. Divide £200. 16$\frac{1}{5}$s. by 2s. 6d. decimally. *A.* £1606.48.

58. Suppose 200A. 3$\frac{1}{5}$rd. of land to cost £1,700. 3s. 4d. 3$\frac{1}{5}$qr.; what is the price per acre? *A.* £8.5=£8. 10s.

59. If a ship in sailing from New York to Canton occupy 207 days

16h. 26m. and 24sec., how many such trips would require 170Y. 255d. 12h.? *A*. 300.

60. A gentleman having returned from a voyage at sea, found by actual calculation that he had sailed just 5,475 miles 6fur. 24rd., being on an average 60 miles 6fur. 29rd. 3yd. 10in. and $2\frac{2}{3}$b.c. each day during the voyage; how many days was he in performing the voyage? *A*. 90 days.

REDUCTION OF CURRENCIES.

LXXIV. 1. Formerly, all accounts in the United States were kept in the currency of Great Britain, that is, in pounds, shillings, pence, and farthings, called sterling money, which was at first of uniform value in both countries. See VII. table 2.

2. But previous to the adoption of federal money, the pound, and consequently its divisions, had, under the same names, received different values in different sections of the union, as in the following

3. TABLES OF CURRENCIES.

NEW ENGLAND STATES, VIRGINIA, KENTUCKY AND TENNESSEE.			N. JERSEY, PENNSYLVANIA, DELAWARE, MARYLAND AND LOUISIANA.		
1 pound[1]	=$3.33\frac{1}{3}$	= \$$\frac{10}{3}$	1 pound[3]	=$2.66\frac{2}{3}$	= \$$\frac{8}{3}$
6 shillings	=\$1.00	=£$\frac{3}{10}$	7s. 6 pence	=\$1.00	=£$\frac{3}{8}$
9 pence	=\$.$12\frac{1}{2}$	= \$$\frac{1}{8}$	$11\frac{1}{4}$ pence	=\$.$12\frac{1}{2}$	= \$$\frac{1}{8}$
$4\frac{1}{2}$ pence	=\$.$06\frac{1}{4}$	= \$$\frac{1}{16}$	5d. $2\frac{1}{2}$qr.	=\$.$06\frac{1}{4}$	= \$$\frac{1}{16}$

N. YORK, N. CAROLINA AND OHIO.			SOUTH CAROLINA AND GEORGIA.		
1 pound[2]	=\$2.50	= \$$\frac{10}{4}$	1 pound[4]	=\$$4.28\frac{4}{7}$	=\$$\frac{30}{7}$
8 shillings	=\$1.00	=£$\frac{4}{10}$	4s. 8 pence	=\$1.00	=£$\frac{7}{30}$
1 shilling	=\$.$12\frac{1}{2}$	= \$$\frac{1}{8}$	7 pence	=\$.$12\frac{1}{2}$	=\$$\frac{1}{8}$
6 pence	=\$.$06\frac{1}{4}$	= \$$\frac{1}{16}$	$3\frac{1}{2}$ pence	=\$.$06\frac{1}{4}$	=\$$\frac{1}{16}$

IN CANADA AND NOVA SCOTIA CURRENCY.[5]

5 shillings= \$1.00= £$\frac{1}{4}$, and £1= \$4.

IN ENGLISH OR STERLING MONEY.[6]

4s. 6d.= \$1.00= £$\frac{9}{40}$, and £1= \$$4.44\frac{4}{9}$= \$$\frac{40}{9}$.

LXXIV. *Q*. In what currency are our accounts kept? 2. How were they formerly kept? 1. When was federal money established? XXXI. 1. What is said of the value of the pound in the different states? 1, 2. How many shillings make a dollar in the different states, Canada, and Great Britain? See 3. Where do $12\frac{1}{2}$ cents pass for 1 shilling?—for 9 pence?—for $11\frac{1}{4}$ pence?—for 7 pence? What does $6\frac{1}{4}$ cents pass for in the different states? What part of a pound is \$1 in Maine, N. Carolina, Delaware, Georgia, Canada, and sterling money? See tables, 3.

---

1. \$1 is £$\frac{6}{20}$= £$\frac{3}{10}$, or £1= \$$\frac{10}{3}$. 2. \$1 is £$\frac{8}{20}$= £$\frac{4}{10}$, or £1 is \$$\frac{10}{4}$. 3. \$1 is 90d. and £1 is 240d.; then £$\frac{90}{240}$= £$\frac{3}{8}$, or £1 is \$$\frac{8}{3}$. 4. \$1 is 56d. and £$\frac{56}{240}$= £$\frac{7}{30}$, or £1 is \$$\frac{30}{7}$. 5. \$1 is £$\frac{5}{20}$= £$\frac{1}{4}$. 6. \$1 is £$\frac{54}{240}$ =£$\frac{9}{40}$, or £1 is \$$\frac{40}{9}$.

GENERAL RULE FOR THE REDUCTION OF CURRENCIES.

4. *Multiplying the pounds, with the shillings, pence, &c. reduced to the decimal of a pound, by the fraction that expresses what part £1 is of $1, found in the tables, produces federal money; and reversing the process produces pounds again.*

5. Or, which is the same thing—*Annexing a cipher to any sum of ounds in New England or New York currency, and dividing by half he number of shillings in a dollar, produces federal money.*

6. Change £900 N. E. currency to dollars. (£1=$$\frac{10}{3}$×900.)
7. Change $3,000 to N. E. currency. ($1=£$\frac{3}{10}$×$3,000.)
8. Change £270 N. E. currency to federal money.
9. Change 900 dollars to N. E. currency.
10. Change £150 Pennsylvania currency to dollars.
11. Change 400 dollars to the currency of Pennsylvania.
12. Change £3. 8s. 3d. S. Carolina currency to federal money.
NOTE.—£3.8s.3d.=£3.4125×$\frac{30}{7}$=$14.6250, *A.* See LXIX. C. IV
13. Change $14.625 to S. Carolina currency.
NOTE.—$14.625×$\frac{7}{30}$=£3.4125=£3.8s.3d. *A.* See LXIX. case V.
14. Change £17. 15s. 6d. N. E. currency to dollars.
15. Change $59.25 to New England currency.
16. Change £8. 15s. 6d. New York currency to dollars.
17. Change $21.9375 to N. Carolina currency.
18. Change £12. 7s. 6d. Virginia currency to dollars.
19. Change $41.25 to Tennessee currency.
20. Change £5. 15s. 6d. Delaware currency to dollars.
21. Change $15.40 to Maryland currency.
22. Change £2. 2s. Georgia currency to dollars.
23. Change $9 to South Carolina currency.
24. Change £252 Canada currency to dollars.
25. Change $1,008 to Nova Scotia currency.
26. Change £252 English currency to federal money.
27. Change $1,120 to English or sterling money.
28. A merchant in Philadelphia purchased for Messrs. Robinson & Pratt, of Quebec, 300 bales of cotton, which averaged 275 pounds per bale, at a cost of 10¼ cents per pound. Now, how many pounds, in Canada currency, must Robinson & Pratt remit to their agent for the payment of said cotton? *A.* £2,114. 1s. 3d.

(29.) LONDON, Jan. 4. 1840.

Messrs. Rice & Donaldson, of Philadelphia, U. S.

Bought of James Wellington.

462 yds.	Blue Broadcloth,	*a* 17s. 6½d.
418 do.	Black do.	*a* 14s. 9¾d.
519 do.	Black Silk	*a* 9s. 3d.
		£954. 16s. 7½d.

What sum in federal money will settle the above bill? *A.* $4243.6944.+

Q. What is the rule for reduction of currencies? 4, 5

RATE PER CENT.

LXXV. 1. PER CENT., from the Latin *per*, by, and *centum*, one hundred, signifies, *by the hundred.*

2. RATE PER CENT., therefore, signifies, *the rate by the hundred*, that is, the number of *hundredths* of any sum which is considered as advanced, expended, gained, or lost.

3. Since hundredths are decimals of two places, all calculations by the rate per cent. fall very properly within the province of decimals.

4. Thus 5 per cent. of any sum is $\frac{5}{100}$ of that sum; 6 per cent. is $\frac{6}{100}$=.06; 15 per cent. is $\frac{15}{100}$=.15, &c.

5. Then 5 per cent. of $20 is .05=$\frac{5}{100}$ of $20=$1, and 6 per cent. of $20 is .06 or $\frac{6}{100}$ of $20=$1.20.

6. Hence, to express any given rate decimally—*When it consists of one figure only, write a cipher and the separatrix before it; when of two figures, only the separatrix; when of three or more figures, call only the two right hand ones decimals.*

7. Thus 1 per cent. is .01; 10 per cent. is .10=.1; 23 per cent. is .23; 315 per cent. is 3.15.

CASE I.

To find the value of the rate per cent.

RULE.

8. *Either multiply by the rate per cent. as a decimal, or as a whole number; but in the latter case point off two figures in the product, (for dividing by* 100.)

9. Suppose a trader has $1,500, and wishes to lay out 6 per cent. of it in silks, how many dollars will the silk cost?

$ 1 5 0 0	Or $ 1 5 0 0	For 6 per cent.=.06 or
. 0 6	6	$\frac{6}{100}$. To divide by 100, cut
A. $ 9 0 . 0 0	*A.* $ 9 0 . 0 0	off two figures on the right.

10. What is 5 per cent. of $8,000? *A.* $400.
11. What is 25 per cent. of $400? *A.* $100.
12. What is 50 per cent. of $300? *A.* $150.
13. What is 75 per cent. of $600? *A.* $450.
14. What is 100 per cent. of $40? *A.* $40.

15. Hence, *when the rate is* 100 *per cent. the given sum itself is the value of the given rate; when* 75 *per cent. the value is* $\frac{3}{4}$ *of the given sum; when* 50 *per cent. it is* $\frac{1}{2}$; *when* 25 *per cent.* $\frac{1}{4}$; *and so on.*

16. Suppose a man who has $5,000 pays away 25 per cent. of it,

---

LXXV. *Q.* What is the meaning of per cent.? 1. What of rate per cent.? 2. How is the rate per cent. determined? 3. What is meant by 5 or 6 per cent.? 4. What is 5 per cent. of $20?—of $30?—6 per cent. of $20?—of $30? How is any per cent. expressed decimally? 6. What is the decimal expression for 1 per cent.?—for 9 per cent.?—for 10 per cent.?—for 23 per cent.?—for 315 per cent.? What is the rule for finding the value of the rate per cent.? 8. What is 1 per cent. of $2,000?—2 per cent. of $4,000?—6 per cent. of $10,000?

how many dollars has he left? The whole is 100 per cent.; 25 per cent. deducted, leaves 75 per cent. or $\frac{3}{4}$ of the whole. *A.* \$3,750.

17. Suppose a man having an estate worth in cash \$1,000, spends 15 per cent. of it in one year, 35 per cent. the second year, and the remainder in the third year; what per cent. does he spend the last year, and how many dollars is it? *A.* 50 per cent.; \$500.

18. Suppose a flour merchant sells, in the course of the year, 8,000 barrels of flour, at an average price of \$9$\frac{3}{4}$ a barrel, and deposits 20$\frac{2}{3}$ per cent. of the money in the bank; how many dollars does he deposit? *A.* \$16,120.

19. What is 6 per cent. of \$40.27? *A.* \$2.4162.

20. What is 15 per cent. of \$62.50? *A.* \$9.37$\frac{1}{2}$.

21. What is 20$\frac{2}{3}$ per cent. of 25.56lb.? *A.* 5.2824lb.

22. What is 5$\frac{3}{4}$ per cent. of \$65.375? *A.* \$3.759.+

23. Suppose a planter grows 40,000 pounds of cotton in a year, and sells 30$\frac{1}{4}$ per cent. of it for 15 cents a pound; 44$\frac{3}{4}$ per cent. of it for 16$\frac{1}{8}$ cents per pound; what will his sales amount to, and how much of his crop remains on his hands? *A.* \$4,701.37$\frac{1}{2}$; 10,000lb.

24. What is $\frac{3}{8}$ per cent. of \$200? *A*\$.75.

25. What is $\frac{3}{50}$ per cent. of \$53.625? *A.* \$.321$\frac{3}{4}$.

26. When the per centage is very small, it is sometimes expressed in cents; thus 30 cents of \$100 is 100 times smaller than 30 per cent, which is \$30 of \$100; therefore,—

27. When the rate per cent. is expressed by cents—*Consider two more places in the product decimals on that account.*

28. What is the difference between 20 per cent. of \$500, and 20 cents for every hundred dollars of \$500? *A.* \$99.

29. When the rate per cent. is 10 cents of \$100, what is its value on \$1,000,000? \$1,000.

30. When the rate per cent. is 5 cents of \$100, what is its value on \$6,834.5625? *A.* \$3.417.+

CASE II.

When the given sum is a compound quantity.

RULE.

31. *Reduce it first to a decimal of the highest denomination mentioned, then multiply by the rate as before; after which find the value of the decimal in a compound quantity again.* [LXIX. cases IV. and V.]

Q. When the rate is 100 per cent. what will always be its value? 15. What when 75 per cent.? 15. When 50 per cent.? 15. When 25 per cent.? 15. What then is the value of 100 per cent. of \$300?—75 per cent. of 400 guineas?—50 per cent. of 144 eagles?—25 per cent. of 800 gallons? What is the value of 25 per cent., 10 per cent., and 20 per cent., of \$1,000? Suppose a man has \$400, and loses 10 per cent. of it, how many dollars has he left? What is 1$\frac{1}{2}$ per cent. of \$200?—of \$500? What is 5 per cent. of \$120? What is $\frac{1}{2}$ per cent. of \$120?—of \$240? When the per centage is very small, how is it frequently expressed? 26. What is the direction in such cases? 27. When the per centage is 30 cents for every \$100, what does it amount to on \$2,000?—on \$15,000? When the given sum consists of several denominations, how do you proceed? 31.

32. What is 5 per cent. of £75. 9s. 6d.? (=£75.475.)
A. £3. 15s. $5\frac{1}{2}$d.+

33. What is 5 per cent. of £400. 17s. 3d.? A. £20. 10d. $1\frac{3}{5}$qr.

34. What is $101\frac{1}{2}$ per cent. of 2cwt. 3qr. 7lb.?
A. 2cwt. 3qr. 11lb.+

35. What is 10 per cent. of 500l. 6fur. 30rd.? A. 50l. 27rd.

CASE III.

To find the rate from having its value given.

36. What per cent. is $30 of $500? Since 6 per cent. of $500 is found by case I, thus, $500×.06=$30: then $30÷$500 must reproduce the rate.

RULE.

37. *Of the two given sums, divide the one expressing the value of the rate by the other.*

38. What per cent. of $800 is $56? A. 7 per cent.

39. What per cent. of $700 is $63? A. 9 per cent.

40. What per cent. of $200 is $30? A. 15 per cent.

41. If a man has 300 bushels of rye, and sells 60 bushels, what per cent. of the whole does he sell? A. 20 per cent.

42. Recollect that the rate per cent. is always equal to so many hundredths; thus .06 is $\frac{6}{100}$; therefore, it is 6 per cent.; .09 is $\frac{9}{100}$ =9 per cent.; .12=$\frac{12}{100}$=12 per cent.

43. Then, as the rate per cent. is restricted to two decimal places, call the other figures on the right of hundredths in the quotient decimals of hundredths, or of the rate per cent.

44. Thus, .062 is $6\frac{2}{10}=6\frac{1}{5}$ per cent.; .1525 is $15\frac{25}{100}=15\frac{1}{4}$ per cent.; .27375=$27\frac{375}{1000}=27\frac{3}{8}$ per cent.

45. What per cent. of $20 is $1.70? A. .085=$8\frac{1}{2}$.

46. What per cent. of $200 is $6.40? A. .032=$3\frac{1}{5}$.

47. What per cent. of $300 is $13.20? A. .044=$4\frac{2}{5}$.

48. What per cent. of $60.25 is $4.6995? A. $7\frac{4}{5}$.

49. What per cent. of $200 is $1? $1÷200=.005, and as there are no hundredths, the 5 is $\frac{5}{10}=\frac{1}{2}$ per cent. A. $\frac{1}{2}$ per cent.

50. What per cent. is 45 cents of $60? A. .0075=$\frac{75}{100}=\frac{3}{4}$.

51. Suppose a lawyer charges $5 for collecting a debt of $4,000, what per cent. is it? A. .00125=$\frac{125}{1000}=\frac{1}{8}$.

52. Suppose a grocer finds that in retailing 2,500 gallons of molasses, it does not hold out by 10 gallons, what per cent. of the whole is the loss? A. $\frac{2}{5}$ per cent.

CASE IV

To find any sum from having its rate per cent. and its value given

Q. What is the rule for finding the rate? 37. Why? 36. What per cent. is $5 on $20?—$4 on $24?—$6 on $150? How is the per centage expressed decimally? 42. When it is expressed by three or more decimal figures, what per cent. do all the figures except the two left hand ones express? 43. What per cent. then is .062?—is .1525? 44. Of what sum is $5 only 2 per cent.? What is the rule for it? 54. Of what sum is 2 per cent. $10? Of what sum is 8 per cent. $10?

53. A man failing in business is able to pay $720, which is only 6 per cent. of what he owed; how much, then, does he owe?

.06)$720.00 Since $12,000×.06=720, therefore $720
A. $12,000 ÷.06 must reproduce the $12,000.

RULE.

54. *Divide the value of the rate per cent. by the rate itself expressed as a decimal.*

55. If a man owes a note at bank, and pays 12 per cent. of it, making $300, what was the face of the note? *A.* $2,500.

56. Of what sum is $50 only 4 per cent.? *A.* $1,250.

57. Of what sum is $50; $\frac{1}{2}$ per cent.? (=.005.) *A.* $10,000.

58. Suppose I paid $750, which was $\frac{3}{4}$ per cent., for the collection of a certain debt; how large was the debt? *A.* $100,000.

STOCKS.

LXXVI. 1. STOCKS is a general name for all funds[1] invested in banks and other corporate[2] bodies.

2. STOCKS consist of shares[3] of an equal amount, as $50, or $100 each, but usually of $100, and as such are bought and sold.

3. The NOMINAL VALUE of a share is what it costs at first, and when it sells for that in the market, it is said to be *at par.*[5] The *nominal* value, then, is its *par* value.

4. The REAL VALUE of a share is what it actually sells for, which often varies at different times.

5. When stock sells for more than its par value, it is said to be *above par*, or *at an advance;*[6] but when for less, *below par*, or *at a discount.*[7]

RULE.

6. *The various values of stock being estimated at so much per cent., they are calculated as in the preceding article.*

LXXVI. *Q.* What is understood by stocks? 1. Of what do they consist? 2. What is meant by their nominal value? 3. What by their real value? 4. What names are used to designate the market price of stocks? 5. What is the rule for ascertaining the value of stocks? 6. What is the value of $500 stock at 8 per cent. advance?—at 5 per cent. discount?—at 2 per cent. above par?—at 10 per cent. below par?—at par?

1 FUNDS. Capital; a sum of money appropriated for commercial or other operations. A stock or capital intended to furnish supplies of any kind. Money lent to government. Abundance; ample stock or store.

2 CORPORATE. United in one body or community, as a number of persons who are allowed by law to *sue* and be *sued, &c.*, as if they were but one person. United, collectively one.

3 SHARE. A part; a portion; a quantity; dividend; a part contributed. The broad iron or blade of a plough. *To go shares*, to be equally interested.

4 NOMINAL. Not real; existing only in name.

5 PAR. The Latin for *equal.*

6 ADVANCE. Movement forward, or a gradual progressing in any thing. Promotion, preferment; first step towards agreement. A furnishing of goods or money for others *In advance*, in front; before.

7 DISCOUNT. A sum deducted; deduction for prompt pay

7. What is the value of \$2,500 of stock at 8 per cent. advance? $100+8=108$ per cent. $\times$ \$2,500=\$2,700. *A.*

8. A gentleman purchased 35 shares, of \$100 each, in the United States Bank, at $8\frac{5}{8}$ per cent. advance; what did they cost? *A.* \$3801.875.

9. What is the value of \$4,500 stock at a discount of $5\frac{1}{4}$ per cent.? *A.* \$4,263.75.

10. A merchant subscribed for 25 shares, of \$50 each, in a certain railroad, which declined afterwards $15\frac{3}{4}$ per cent.; what was its real value then. *A.* \$1,053.125.

11. What is the difference between \$2,200 stock at par value, and at an advance of $7\frac{3}{11}$ per cent.? *A.* \$160.

12. Suppose I purchased stock, the par value of which is \$20,000, for 2 per cent. discount, and sold it for 2 per cent. advance; how much did I make on it? *A.* \$800.

13. Suppose an original subscriber for 100 shares, of \$50 each, in the Exchange Bank, Providence, receives a dividend of \$100; what per cent. is that on his stock? [See LXXV. case III.] *A.* 2 per cent.

14. What per cent. is equal to a dividend of \$1,000 on \$20,000 stock? *A.* 5 per cent.

15. Suppose you receive a dividend of \$1,000, being at the rate of 5 per cent., on your bank stock; what amount of stock have you? [See LXXV. case IV.] *A.* \$20,000.

16. Suppose a merchant to have been an original subscriber for 500 shares, of \$50 each, in the Bank of America, payable by installments, as follows:—$\frac{1}{8}$ in three months, which he sold for $5\frac{1}{4}$ per cent advance; $\frac{2}{3}$ in six months, which brought him $6\frac{3}{5}$ per cent. advance; and the balance in nine months, which he was compelled to sell at $8\frac{3}{4}$ per cent. discount; what did he gain by the whole transaction? *A.* \$808.333.+

COMMISSION.

LXXVII. 1. COMMISSION is an allowance of so much per cent. made to factors, brokers, and other agents, for their services in buying and selling for their employers.

2. A Factor is a person employed by another at a distance, to transact business on his account.

3. A Broker is a person who deals in stocks, goods, &c., or exchanges money, either on his own account or for others.

LXXVII. *Q.* What is Commission? 1. Factor? 2. Broker? 3. What is the rule? 4. How many dollars does the commission of 2 per cent. on \$300 amount to?—5 per cent. on \$400? How many dollars must I pay for changing \$5,000, when the brokerage is $\frac{1}{4}$ per cent.? How much for changing \$2,000 at the same rate?

RULE.

4. *The rule is the same as that for estimating the value of stocks*

5. If my agent sells goods amounting to $400, what is his commission at $2\frac{1}{2}$ per cent.? $400 $\times$ $2\frac{1}{2}$ per cent. = $10. *A.*

6. My correspondent writes me that he has purchased goods to the amount of $5,000; what will his commission amount to at 3 per cent.? *A.* $150.

7. What is the commission on $417 at 1 per cent.?—at $1\frac{1}{2}$ per cent.?—at 2 per cent.?—at $3\frac{1}{2}$ per cent.?—at $4\frac{3}{4}$ per cent.?
A. $4.17; 6.25\frac{1}{2}$; $8.34; 14.59\frac{1}{2}$; $19.8075.

8. Suppose a factor purchases 300 pounds of indigo for $2.50 a pound; what will his commission amount to at $6\frac{1}{2}$ per cent.?
A. $48.75.

9. What does a broker exact for changing $3,700 in bills on the Bank of America, New York, for the same sum on the Phœnix Bank, Connecticut, the rate of exchange being $\frac{3}{8}$ per cent.? *A.* 13.87\frac{1}{2}$.

10. Suppose a broker purchase on your account $20,000 stock in the "Great Western Canal," at $13\frac{1}{8}$ per cent. advance; what will the stock cost, and what will his brokerage amount to at the rate of $\frac{1}{8}$ per cent.? *A.* $22,625, cost; 28.28\frac{1}{8}$.

11. If a broker receive 80 cents for changing $400, at what per cent. does he reckon the exchange? *A.* $\frac{1}{5}$ per cent.

12. Suppose I remit to my factor in New York, $6,000, for the purchase of flour, and he writes me that the flour cost $10 a barrel, and that his commission is $2\frac{1}{2}$ per cent.; how many barrels of flour shall I receive, after deducting his commission? *A.* 585 barrels.

13. Suppose a commission merchant sells on my account $200\frac{1}{3}$ bales of cotton, each bale containing 315lb. 8oz., at 13 cents a pound, for a note at six months, and charges $2\frac{1}{2}$ per cent. for selling, and 3 per cent. more for guaranteeing the paper; what will be the balance due me? *A.* $7759.587.+

14. A commission merchant sold 200hhd. 7gal. 3qt. 1.504pt. of molasses for $30 a hogshead, for which he charged 3 per cent. commission; what will be the balance due his employer?
A. $5823.666.+

INSURANCE.

LXXVIII. 1. **Insurance** is a security, by paying a stipulated sum, called a premium, to indemnify the party insured against such losses on ships, houses, goods, &c., as may happen from storms, fire, or other accidents.

LXXVIII. What is Insurance? 1. A policy? 2. How is the cost of insuring estimated? 2. How many dollars must you pay for procuring an insurance on your house, valued in the policy at $10,000, the rate being $\frac{1}{4}$ per cent.?

2. The CONTRACT OF INDEMNITY[1] is called a POLICY, and the premium paid for it is usually stated at so much per cent. as in the foregoing articles.

3. What is the premium for insuring an East India ship valued at $25,000, at $15\frac{1}{2}$ per cent.? *A.* 3,875.

4. What is the premium for insuring $2,000 at $2\frac{1}{2}$ per cent.?—3 per cent.?—$4\frac{3}{8}$ per cent.—$5\frac{1}{4}$ per cent.?—$7\frac{3}{20}$ per cent.—$8\frac{4}{15}$ per cent.? *A.* $50; $60; $87.50; $105; $143; 165.33\frac{1}{3}$.

5. If you effect an insurance on your house for $5,000 at $\frac{3}{8}$ per cent. per annum, what would it amount to in 5 years? *A.* $93.75.

6. A manufacturer effected an insurance on his factory building and machinery, both valued in his policy at $15,600, paying a premium of 2 per cent. per annum. In the second year the establishment suffered by fire, as was estimated, $1,200; how much did he save by the insurance? *A.* $576.

7. Suppose you pay a premium of 30 cents on $100 for insuring $10,000 on your house, $2,000 on your furniture, $450 on your books, $600 on your span of horses, and $350 on your harnesses, buffalo robes and saddles; how many dollars does the insurance on all these articles cost? *A.* $40.20.

LOSS AND GAIN.

CASE I.

LXXIX. To find the sum gained or lost.

RULE

1. *Find the value of the given rate per cent. as before.*

2. If I buy goods amounting to $1,675, and sell them for 15 per cent. gain, what are my profits? *A.* $251.25.

3. If I sell goods for 10 per cent. loss, which cost me $500, how many dollars do I lose by them? *A.* $50.

4. Suppose I buy $400\frac{1}{5}$ barrels of flour for $16\frac{3}{4}$ a barrel, and sell it for $\frac{3}{8}$ per cent. advance; how many dollars do I gain by it? *A.* $25,13$\frac{3}{4}$.+

CASE II.

To find what price must be demanded to gain or lose a certain per cent.

RULE.

5. *If the given rate is gain per cent. add it to* 100 *and multiply the*

LXXIX. *Q.* When goods worth $500 are sold for 10 per cent. loss, what do they bring? What is the rule? 1. When calico costs 30 cents per yard, at what price must it be sold to gain 5 per cent.?—to lose 6 per cent.?—to gain 10 per cent.? What is the rule? 5. How is the gain or loss per cent. ascertained? 13.

---

1 INDEMNITY. Security given to save from harm or loss; security against punishment.

cost by that sum; but if the rate is loss per cent. deduct it from 100 *and multiply by the remainder.*

6. If I buy a quantity of wheat for $200 and wish to gain 10 per cent. by the sale of it, what must I ask for it? *A.* $220.

7. A man paid $50 for goods which he purchased at auction, and which he was glad to sell at a loss of 10 per cent.; what did he receive for them? *A.* $45.

8. A man bought a cow for $44 and sold it for 15 per cent. advance; what did he get for it? *A.* $50.60.

9. A merchant bought a hogshead of molasses for $44, and by accident, 9 gallons leaked out; what must he sell the remainder for per gallon, so as not to lose but 10 per cent.? *A.* $73\frac{1}{3}$ cents.

10. A merchant bought 108 barrels of flour for $$10\frac{1}{4}$ a barrel; paid for carting, 6 cents a barrel, and for assistance in storing it $1,50; now how much must he ask a barrel for it to gain 20 per cent.?
A. $12.388+.

11. A wholesale dealer in flour bought in one week the following lots, viz.—118 barrels for $$9\frac{1}{4}$ per barrel, 212 barrels for $$9\frac{1}{2}$ per barrel, 315 barrels for $$9\frac{1}{3}$ per barrel, and 400 barrels for $10 per barrel. His store rent was $12.50 a week; clerk hire, $17 per week, and insurance $\frac{1}{30}$ per cent.; what price per barrel will cover all the expenses and afford him 15 per cent. profit? *A.* $11.018.+

CASE III.

To find the gain or loss per cent.

RULE.

12. *First find what the gain or loss is by subtraction, then divide the remainder by the actual cost.*

13. A merchant bought molasses for 24 cents a gallon, which he sold for 30 cents; what was his gain per cent.?
A. .25=25 per cent.

14. A grocer bought a hogshead of wine for $75, from which several gallons leaked out; the remainder he sold for $60. What did he lose per cent.? [.02 is 2 per cent., but .2 is 20 per cent.]
A. 20 per cent.

15. A man bought a piece of cloth for $20, and sold it for $25; what did he gain per cent? *A.* 25 per cent.

16. A grocer bought a barrel of flour for $8, and sold it for $9; what was the gain per cent.? *A.* $12\frac{1}{2}$ per cent.

17. A merchant bought a quantity of goods for $318.50, and sold them for $299.39; what was his loss per cent.? *A.* 6 per cent.

18. If I buy a horse for $150, and a chaise for $250, and sell the chaise for $350 and the horse for 100, what is my gain per cent.?
A. $12\frac{1}{2}$ per cent.

19 Bought 20 barrels of rice for $20 a barrel, and paid for transportation of it 50 cents a barrel; what will be my gain per cent. in selling it for $$25.62\frac{5}{10}$ a barrel? *A.* 25 per per cent.

INTEREST.

LXXX. 1. INTEREST[1] is the premium[2] given by the borrower to the lender for the use of money.

2. Interest is usually reckoned at so much per cent. like stocks; it however differs from them in limiting the per cent. to definite[3] periods[4] of time.

3. Thus, 6 per cent. per annum[5] means $6, and 8 per cent. per annum, $8; each for the use of $100 for one year, and in the same proportion for a longer or shorter period of time.

4. The rate per cent. in most countries is established by law, and if a higher rate be exacted it is called usury.[6]

5. The legal[7] rate varies in different countries; thus, it is 5 per cent. in England, and 6 per cent. in our country with the following exceptions.

6. In New York, S. Carolina, Michigan and Wisconsin, it is 7 per cent.; in Georgia, Alabama and Mississippi, 8 per cent.; and in Louisiana, 5 per cent.*

7. When no mention is made of the *rate* of interest, the lawful one is always understood.

8. If A borrows of B $1,000, and agrees to pay him 6 per cent. interest, that is, $6 for every $100, making $60 interest, A evidently owes B at the year's end, $1000 and $60, making $1060.

9. The sum lent is called the PRINCIPAL; the per cent. agreed on, the RATE; and the principal and interest added together the Amount.

CASE I.

To find the interest for one year. [See LXXV. 8.]

RULE.

1. *Multiply the principal by the given rate, and cut off two more*

LXXX. Q. What is Interest? 1. How is it reckoned? 2. In what respect does it differ from per cent.? 2. What does 6 per cent. per annum mean? 3. What is usury? 4. What legal rates are mentioned? See 5, 6. How is the rate known when it is not mentioned? 7. How many dollars are due for the use of $1,000 for 1 year, when the rate is 6 per cent.? 7. What is the proper name for each of these terms? 9.

CASE I. Q. What is the rule for finding the interest for 1 year? 1.

---

1 INTEREST. Concern; advantage; good; share; portion. Regard to private property; any surplus advantage.

2 PREMIUM. Award; compensation; prize; bounty; sometimes synonymous with interest.

3 DEFINITE. Limited; determinate; certain; fixed; exact.

4 PERIOD. Circuit; epoch; a definite portion of time; end; stop; time between one occurrence and another.

5 PER ANNUM, from the Latin, *per*, by, and *annum*, a *year*; hence the origin of *annually*, meaning yearly.

6 USURY. *Formerly*, interest; *now*, illegal interest; a premium greater than that allowed by law.

7 LEGAL. Lawful; according to law.

* In some of the Southern States, it is not considered usury to receive a much higher rate; thus, in Louisiana 10 per cent. is the usual rate.

figures in the product, for cents or decimals, than there are cents or decimals in the given sum.

2. What is the interest of $5000 at 6 per cent. for one year What is the amount?

$ 5 0 0 0	Principal.	$ 5 0 0 0	Principal.
6	Rate.	3 0 0	Interest.
A. $ 3 0 0 , 0 0	Interest.	*A.* $ 5 3 0 0	Amount.

3. What is the interest of $8000 for 1 year at 6 per cent.? Wha is the amount? *A.* Interest $480; Amount $8,480.

4. What is the interest of $1200 for 1 year at 5 per cent.?—at 4 per cent.?—at 3 per cent.? *A.* $60; $48; $36.

5. What is the amount of $500 for 1 year at 7 per cent.?—at 8 per cent.?—at $5\frac{1}{2}$ per cent.? *A.* 35; $40; $27.50.

6. What is the interest of $650.62, and of $453,625, for 1 year at 7 per cent.?

$ 6 5 0 . 6 2	$ 4 5 3 , 6 2 5	*A.* $45.54c. $3\frac{4}{10}$m.
7	7	*A.* $31.75c. $3\frac{75}{100}$m.
$ 4 5 . 5 4 3 4	$ 3 1 , 7 5 3 7 5	See rule for pointing off.

7. What is the interest of $3019.20 for 1 year at 6 per cent.?—at 5 per cent.? *A.* 181.152; $150.96.

8. What is the amount of $1250.375 for 1 year at $3\frac{1}{2}$ per cent.?—at $5\frac{3}{4}$ per cent.? *A.* $1294.138+; $1322.271+

CASE II.

To find the interest for more years than one.

RULE.

1. *Multiply the interest of one year, found as before, by the number of years.*

2. What is the interest of $375.875 for 3 years at 6 per cent.? The interest for 3 years must be 3 times as much as for 1 year; thus, $375.875 × 6 per cent. × 3 years = $67.6575, Ans.

What is the interest and amount—

3. Of $200 for 2 years at 5 per cent.? *A.* $20; $220.
4. Of $1,700 for 5 years at 8 per cent.? *A.* $680; $2,380.
5. Of $7.25 for $2\frac{1}{2}$ years at 4 per cent.? *A.* $.725; $$7.97\frac{1}{2}$.
6. Of 44 cents for 15 years at 10 per cent.? *A.* $.66; $1.10.
7. Of 62ct. 5 mills for 12 years at 5 per cent.? *A.* $$.37\frac{1}{2}$; $1.
8. Of £400 for 3 years at 5 per cent.? *A.* £60; £460.

CASE III.

To find the interest for years and months when the rate is 6 per cent.

Q. What is the interest of $400 for 1 year at 2 per cent.?—at 5 per cent.?—7 per cent.?—10 per cent.?—$5\frac{1}{2}$ per cent.? At 8 per cent. what will be the amount of $200 for 1 year?—of $300?—of $500?—of $600?

CASE II. What is the rule for finding the interest for two or more years? 1. What is the amount of $500 at 6 per cent. for 1 year?—for 2 years?—for 3 years?—for 4 years?—for 5 years?

RULE.

1. *Multiply by half the number of months, and cut off two figures, as before.*

2. For at 6 per cent. for every 12 months, the rate for any period of time is just half the months of that time.

3. What is the interest of $400 at 6 per cent. for 4mo.?—for 1Y. 6mo.?—for 2Y. 3mo.?

$400	$400	$400	For 4 mo.÷2=2, rate; 1 Y. 6 mo.=18 mo.÷2=9, rate; 2 Y. 3 mo.=27 mo.÷2=13½, rate.
2	9	13½	
A. $8.00	*A.* $36.00	*A.* $54.00	

4. Recollect, when no mention is made of the rate of interest, 6 per cent. is understood.

What is the interest and amount—

5. Of $600 for 10 months? *A.* $30; $630
6. Of $1,600 for 8 months? *A.* $64; $1,664
7. Of $34.675 for 6 months? *A.* $1.04; $35.715¼
8. Of $13,000 for 1Y. 3 months? *A.* $975; $13,975
9. Of $375.50 for 2Y. 3 months? *A.* $50.69¼; $426.19¼
10. Of $689.30 for 12Y. 3mo.? *A.* $506.635+; $1195.935+
11. Of $313.06 for 1Y. 11 months? *A.* $36+; $349.06.
12. Of £500 for 2Y. 1 month? *A.* £62. 10s.; £562. 10s
13. Of 17,000 eagles for 3Y. 5 months? *A.* 3,485E.; 20,485E
14. Of $595.38 for 1 month? *A.* $2.976+; $598.356+.

CASE IV.

To find the interest for days when the rate is 6 per cent.

1. Since any number of days is a certain part of a month, half this part will of course express the rate per cent. for the days.

RULE.

2. *First reduce the days to the fractional part of a month* (=30 *days*); *next half this fraction, for the rate per cent. for the days, with which multiply, and point off as before.*

3. Recollect that a fraction is *halved* either by dividing its numerator, or by multiplying its denominator, by 2.

4. What is the interest of $60 for 15 days?

15 days=½mo.÷2=¼, the rate or multiplier.

$60
¼
A. $.15

5. What is the amount of $200 for 15 days? *A.* $200.50.

6. What is the amount of $800.40 for 15 days? *A.* $802.401.

CASE III. When the time consists of years and months, and the rate is 6 per cent., how do you proceed? 1. What is the interest of $200 for 1 year and 8 months?—for 2 years 6 months?—for 3 years 4 months? When no per cent. is mentioned in this work, what per cent. is understood? 4.

CASE IV. *Q.* What is the rule for days when the rate is 6 per cent.? 1. Why? 2. How is a fraction *halved?* 3. What is the interest of $60 for 15 days?—for 10 days?—for 5 days?—for 6 days?—for 1 day?

7. What is the interest of \$240 for 10 days? \$ 2 4 0
10 days=$\frac{1}{3}$mo.÷2=$\frac{1}{6}$, the rate and multiplier. $\frac{1}{6}$
A. \$. 4 0

8. What is the amount of \$3,000 for 10 days? *A.* \$3005.
9. What is the amount of \$480.60 for 10 days? *A.* \$481.401.
10. What is the interest of \$120.60 for 20 days? \$ 1 2 0 . 6 0
20 days=$\frac{20}{30}$mo.=$\frac{2}{3}$mo.÷2=$\frac{1}{3}$, the rate. $\frac{1}{3}$
A. \$. 4 0 2 0

11. What is the amount of \$40.80 for 20 days? *A.* \$40.936.
12. What is the amount of \$3678.90 for 20 days? *A.* \$.3691.163.
13. What is the interest of \$360.60 for 19 days? \$ 3 6 0 . 6 0
19 days=$\frac{19}{30}$mo.÷2=$\frac{19}{60}$, the rate. $\frac{19}{60}$
A. \$1 . 1 4 1 9

14. What is the amount of \$420.30 for 19 days? *A.* \$421.63+.

15. When the days are an *even* number, it will be more convenient to find what part of a month half their number is, for the rate; thus, for 20 days say 10 days are $\frac{1}{3}$, which is the rate for 20 days.

16. What is the interest of \$24.42$\frac{1}{2}$ for 6 days? Half the days is 3; 3d.=$\frac{3}{30}$ or $\frac{1}{10}$mo.; then $\frac{1}{10}$ is the rate. *A.* \$.024425=2ct. 4m.+

17. What is the interest of \$2.442$\frac{5}{10}$ for 6 days? *A.* 2$\frac{4}{10}$m.+

18. What is the interest of \$.24425 for 6 days? *A.* \$.000$\frac{2}{10}$+.

19. What is the amount of \$600 for 20 days? *A.* \$602.

20. What is the interest of \$60 for 29 days?—for 28 days?—for 27 days?—for 26 days?—for 25 days?—for 24 days?—for 23 days?—for 22 days?—for 21 days?
A. \$.29; \$.28; \$.27; \$.26; \$.25; \$.24; \$.23; \$.22; \$.21.

21. What is the interest of \$1.20 for 20d.?—for 19d.?—for 18d.?—for 17d.?—for 16d.?—for 15d.?—for 14d.?—for 13d.?—for 11d.?
A. 4 mills; 3$\frac{8}{10}$m.; 3$\frac{6}{10}$m.; 3$\frac{4}{10}$m.; 3$\frac{2}{10}$m.; 3m.; 2$\frac{8}{10}$m.; 2$\frac{6}{10}$m. 2$\frac{2}{10}$m.

22. What is the interest of \$960 for 10d.?—for 9d.?—for 8d.?—for 7d.?—for 6d.?—for 5d.?—for 4d.?—for 3d.?—for 2d.?—for 1d.?
A. \$1.60; \$1.44; \$1.28; \$1.12; \$.96; \$.80; \$.64; \$.48; \$.32; \$.16.

CASE V.

To compute the interest for years, months, and days.

RULE.

1. *Calculate the interest for the days as above, and the interest for the years and months by case* III., *then add both together.*

Q. When the days are an even number, what different method is suggested? 15. What is the interest of \$120 for 10 days? Of \$1.20 for 10 days? Of \$2.40 for 15 days?—for 20 days?—for 25 days?

Case V. *Q.* To find the interest of \$600 for 1 year 6 months 15 days, how do you obtain the multiplier? 2. What is the rule for such cases? 1.

2. What is the interest of $600 for 1 year, 6 months, and 15 days?

Say 1Y. 6mo.=18mo.÷2=9; and 15d.= ½mo.÷2=¼; the whole rate then is 9¼.

$$\begin{array}{r} \$\,6\,0\,0 \\ 9\tfrac{1}{4} \\ \hline 1\,5\,0 \\ 5\,4\,0\,0 \\ \hline A.\ \$\ 5\,5.5\,0 \end{array}$$

What is the amount—

3. Of $1,800 for 2 years 6 months and 10 days? *A.* $2,073
4. Of $15,000 for 1 year 4 months and 25 days? *A.* $16,262.50.
5. Of $6,000 for 2 years 3 months and 12 days? *A.* $6,822.
6. Of $9,030 for 3 years 8 months and 2 days? *A.* $11,019.61.
7. Of $1,260 for 5 years and 25 days? *A.* $1,643.25.
8. Of $1,320 for 3 months and 11 days? *A.* $1,342.22.
9. Of $60 for 2 months and 1 day? *A.* $60.61.
10. Of $30 for 1 month and 1 day? *A.* $30.155.
11. Of $3 for 7 years 7 months and 7 days? *A.* 4.368\frac{5}{10}$.
12. Of 60 cents for 10 years 10 months and 10 days? *A.* $.991.
13. Of 30 cents for 3 years 3 months and 3 days? *A.* $.358+.
14. Of $1,625 for 1 day? *A.* 1.625\frac{2}{10}$+.
15. Of $300 for 5 years 5 months and 5 days? *A.* $397.75.

16. What is the interest of $600 for 1 year 7 months and 15 days? 1Y. 7m. 15d.=19½mo.÷2=9¾ per cent. *A.* $58.50.

17. Hence, we may take half the months and days together, or separately, as is most convenient.

18. What is the interest of $2,400 for 3 years 9 months and 10 days? 45⅓mo.÷2=22⅔, the multiplier. *A.* $544.

19. What is the amount of $40.80 for 5 years 3 months and 7 days? 63$\frac{7}{30}$÷2=31$\frac{37}{60}$ per cent. *A.* 53.699$\frac{6}{10}$.

20. What is the amount of $300 for 2 years 8 months and 5 days? 32⅙mo.÷2=16$\frac{1}{12}$ per cent. *A.* $348.25.

21. What is the amount of $1,200 for 12 years 11 months and 29 days? *A.* $2,135.80.

NOTE.—Few examples occur more difficult than the last, and it is solved with one-third the usual number of figures required by other methods.

CASE VI.

To compute the interest between different dates.

RULE.

1. *Write down the later year, and on the right, the months and days of that year that have elapsed; do the same with the other date.*

2. *Subtract the earlier from the later date, as in Compound Subtraction, always reckoning* 30 *days to the month, and* 12 *months to*

Q. What is the multiplier for 2 years 6 months 10 days?—for 1Y. 4m. 25d.?—for 2Y. 3m. 12d.?—for 2Y. 6mo. 5d.? What is the interest of $30 for 1Y. 4mo. 15d.?—for 1Y. 6mo. 10d.?—for 2Y. 6mo. and 5d.?

CASE VI. *Q.* Describe the whole process in casting the interest between different dates 1. 2.

the year; the remainder will be the required time, with which proceed as before.

3. What is the interest of $180 from September 20th, 1836, to April 5th, 1839?

Years.	mo.	d.
* 1 8 3 9 .	3 .	5
1 8 3 6	8	20
Time 2 .	6 .	15

In 1839, 3 months [viz. Jan. Feb. March] and 5 days have elapsed; and there have elapsed in 1836, 8 months and 20 days. Subtract, saying 20d. from 30d. leaves 10d., and 5d.=15d. &c. Next find the interest of the $180 for the 2Y. 6mo. and 15d. as before. *A.* $27.45.

4. What is the interest of $320.40 from March 27, 1827, to February 12, 1828?

5. Note.—According to the table at the bottom of the page, if the date were February 20th, for instance, it would be expressed 1828, 1mo. 20d.; but as it is the 12th instead of 20th, call the later date 1828, 1m. 12d. Against March in the same table we find 2m. 20d.; then call the earlier date 1827, 2m. 27d. *A.* $16.821.

6. What is the interest of $1,640 from July 5th, 1826, to June 20th, 1828? *A.* $192.70.

7. What is the interest of $415.30 from April 1st, 1828, to September 11th, 1830? *A.* $60.9106.

8. What is the interest of $840 from June 11th, 1820, to April 20th, 1822? *A.* $93.66.

9. What is the interest of $60.50 from March 1st, 1819, to November 21st, 1825? *A.* $24.4016.+

10. What is the interest of $240 from October 31st, 1832, to July 5th, 1834?

Y.	m.	d.
1 8 3 4 .	6 .	5
1 8 3 2 .	9 .	31
Time 1 .	8 .	4

Since 30d. in reckoning interest=1m., then 31d.=1m. 1d.; therefore add 1m. to 9m.=10m., and subtract 10m. 1d. instead of 9m. 31d. *A.* $24.16.

11. What is the interest of $840.60 from May 31st, 1828, to June 25, 1830? *A.* 104.234\frac{4}{10}$.

12. What is the interest of $120.50 from December 1st, 1815, to May 31, 1820? The required time is 4Y. 5m. 30d., but as 30d.=1 month, call the whole 4Y. 6 months. *A.* $32.535.

13. What is the interest of $75.80 from August 31, 1827, to July 1st, 1830? *A.* Time, 2Y. 10mo.; Int. 12.88\frac{6}{10}$.

Q. How is April 5th, 1839, and September 20th, 1836, written down? 3
In taking 1832Y. 9m. 31d. from 1834Y. 6m. 5d. how is the 31d. subtracted? 10

* Suppose the date of each month in the year were the 20th, for instance, the time that has elapsed would be expressed as follows, viz.

January	0m. 20d.	May	4m. 20d.	September	8m. 20d.
February	1m. 20d.	June	5m. 20d.	October	9m. 20d
March	2m. 20d.	July	6m. 20d.	November	10m. 20d.
April	3m. 20d.	August	7m. 20d.	December	11m. 20d

14. What is the interest of $75.80 from March 1st, 1836, to May 31, 1840? *A.* $19.329.

15. What is the amount of $14.30 from December 25th, 1837, to January 19th, 1840? *A.* 16.073\frac{2}{10}$.

16 What is the amount of $715.20 from November 10th, 1827, to February 20th, 1828? *A.* $727.12.

17. What is the amount of $50.60 from February 14th, 1829, to August 4th, 1831. *A.* 58.105\frac{6}{10}$.

18. What is the amount of $500 from March 31st, 1830, to July 1st, 1835? *A.* $657.50.

19. Suppose that A, on the 11th of October, 1835, borrows of B $1700.50, for which he is to pay interest; what does A owe B on the 11th of January, 1837? *A.* $1,828.03$\frac{3}{4}$.

20. Suppose a note of $300 which is on interest from July 5th, 1836, is paid June 17th, 1838; what sum of money will cancel the debt? *A.* $335.10.

21. If a merchant borrows $10,000, April 1st, 1828, and October 16th, 1830, pays $11,000, how much will he then owe, including the interest? *A.* $525.

22. A and B, on settling their accounts, found a balance due A of $320, for which B gave his note, dated Boston, July 5th, 1838, with interest. On the 11th of December, 1840, B paid A $400; how did the balance stand then? *A.* A owes B $33.28.

CASE VII.

To compute the interest more accurately for days.

1. By reckoning 30 days to the month and 12 months to the year, we get for 360 days the interest for 365, being a gain of 5 days interest in a year.*

2. Since 5 days are $\frac{1}{73}$ of 1Y., or 13 days, for instance, $\frac{13}{365}$ of 1Y., therefore—

3. *Multiplying the interest for* 1 *year, found as before, by the days, and dividing by* 365, [*but in leap year by* 366,] *or deducting* $\frac{1}{73}$ *from the interest computed by case* V., *gives the true interest.*

4. What is the interest of $14,600 for 90 days at 7 per cent.? 14,600×.07×90÷365. *A.* $252.

5. What is the interest of $438,000 for 300 days? What the interest for [300d.÷30d.=] 10 months, computed by case V.

A. $21,600; $21,900.

6. When our national debt fell a little short of $123,735,000

CASE VII. *Q.* What error is there in the last rule? 1. What is the general direction? 3. Why deduct $\frac{1}{73}$, or why divide by 73? 2. What are the multiplying and dividing terms in casting the interest on $4,000 accurately for 125 days?—for 150 days in the year 1840?

---

* The error in small sums, however, is a mere trifle; but in large sums it becomes too important to be neglected.

what would be the difference between computing its interest by the present method, and computing it by case v., reckoning the rate at 6 per cent. and the time 200 days? *A*. $56,500.

7. What is the interest of $274,500 at 4 per cent. for 150 days of a leap year? *A*. $4,500.

8. Suppose the bank of England, through which the national revenue, being about $210,000,000, is annually collected, should receive the interest on that sum for 292 days, how much would be gained by computing the interest by months instead of by days? *A*. $140,000.

CASE VIII.

To find the interest at any rate per cent.

RULE.

1. *Find the interest for the given time, as if it were* 6 *per cent., then multiply it by the given rate and always divide by* 6.

2. For 1 per cent. is $\frac{1}{6}$ of 6 per cent., 2 per cent. $\frac{2}{6}$ or $\frac{1}{3}$, 3 per cent. $\frac{3}{6}$ or $\frac{1}{2}$, &c.*

3. What is the interest for $600 for 1 year 2 months and 15 days at 5 per cent.?

```
      $ 6 0 0
          7 1/4
      -------
        1.5 0
      4 2.0 0
      -------
    $ 4 3.5 0 at 6 per cent.
            5 given rate.
  -----------
6 ) 2 1 7.5 0
A. $ 3 6.2 5 at 5 per cent.
```

Or, since 5 per cent. is $\frac{1}{6}$ less than 6 per cent., we may simply deduct $\frac{1}{6}$ of the interest of 6 per cent. from itself, as follows,—

```
Thus, 6)$43.50 int. at 6 per cent.
          7.25
A.     $36.25 int. at 5 per cent.
```

4. What is the interest of $240 for 2 years 6 months at 3 per cent.? *A*. $18.

CASE VIII. *Q*. What is the rule for finding the interest for any period of time at any rate per cent.? 1. Why? 2. In computing the interest of $600 for 1Y. 2m. 15d. at 5 per cent., what would you multiply and divide by? 3. Consult the reference from 2 at the bottom and tell how the process may be abbreviated when the rate is either 2, 3, 4, 5, 7, 8, 9, or 12 per cent. When the principle is 60 dollars and the time 10 days, what is the interest at 2 per cent.?—at 3 per cent.?—at 4 per cent.?—at 8 per cent.?—at 12 per cent.?

---

* Hence when the given rate is not 6 per cent. we may find the interest at 6 per cent. as before, then proceed with that interest according to the directions in the adjacent Table.

For 1 per cent.	$= \frac{1}{6} =$		—divide	by 6.
For 2 per cent.	$= \frac{2}{6} = \frac{1}{3}$		—divide	by 3.
For 3 per cent.	$= \frac{3}{6} = \frac{1}{2}$		—divide	by 2.
For 4 per cent.	$= \frac{4}{6} = \frac{2}{3}$		—deduct	$\frac{1}{3}$.
For 5 per cent.	$= \frac{5}{6} =$		—deduct	$\frac{1}{6}$.
For 7 per cent.	$= \frac{7}{6} = 1\frac{1}{6}$		—add	$\frac{1}{6}$.
For 8 per cent.	$= \frac{8}{6} = 1\frac{1}{3}$		—add	$\frac{1}{3}$.
For 9 per cent.	$= \frac{9}{6} = 1\frac{1}{2}$		—add	$\frac{1}{2}$.
For 10 per cent.	$= \frac{10}{6} = 1\frac{2}{3}$		—add	$\frac{2}{3}$.
For 11 per cent.	$= \frac{11}{6} = 1\frac{5}{6}$		—add	$\frac{5}{6}$
For 12 per cent.	$= \frac{12}{6} = 2$		—multiply	by 2

5. Of \$360 for 1Y. 8m. at 5 per cent.? *A.* \$30.
6. Of \$480 for 5Y. 4m. at 1 per cent.? *A.* \$25.60.
7. Of \$6.60 for 1Y. 8m. at 7 per cent.? *A.* \$0.77.
8. Of \$3.20 for 3Y. 4m. at 11 per cent.? *A.* \$1.173.+
9. Of \$120 for 8Y. 4.m. at 12 per cent.? *A.* \$120
10. Of \$620 for 2Y. 6m. at 9$\frac{3}{4}$ per cent.? *A.* \$151.12$\frac{1}{2}$.
11. Of \$1.20 for 9Y. 2m. at 11$\frac{5}{8}$ per cent.? *A.* \$1.278$\frac{75}{100}$.
12. Of \$720 for 1Y. 4m. at 13$\frac{2}{3}$ per cent.? *A.* \$131.20.

13. What is the interest of \$1,200.60 from January 1st to the 18th of April following, at 11$\frac{1}{2}$ per cent.? *A.* \$41.037$\frac{1}{10}$.+

14. What is the amount of \$500 from July 13th, 1833, to October 27th, 1835, at 5$\frac{3}{4}$ per cent.? *A.* \$565.804$\frac{9}{10}$.+

CASE IX.

A concise method of finding the interest at 7 per cent., being the legal rate in the STATE OF NEW YORK.

RULE.

1. *Compute the interest as before for 6 per cent., then add $\frac{1}{6}$ of this interest to itself.* [*See case* VIII. 2. *reference.*]

2. What is the interest of \$360 for 1 year 6 months and 15 days at 7 per cent.?

The time is 18$\frac{1}{2}$ months÷2=9$\frac{1}{4}$, the rate.

```
        $ 3 6 0
            9 ¼
        -------
            9 0
        3 2 4 0
 ⅙ )    3 3 . 3 0
          5 . 5 5
 A. $   3 8 . 8 5
```

3. What is the interest of \$60 for 2 years 4 months at 7 per cent.? *A.* \$9.80.

4. What is the amount of \$120.60 for 1 year 6 months 10 days at 7 per cent.? *A.* \$133.497$\frac{5}{10}$.

5. What is the amount of \$241.20 at 7 per cent. for 6 months 20 days?—for 1m. 1d.?—for 1Y. 4m. 5d.?—for 2Y. 6m. 25d.?
A. \$250.58; \$242.653; \$263.946; \$284.582.

6. What is the difference between the legal interest of \$10,000 in New York, and the legal interest of the same sum in New England, from April 1st, 1836, to October 1st, 1840? *A.* \$450

CASE X.

To find the Principal at interest.

1. What is the interest of \$1 for 1 year 8 months? *A.* 10 cents.

2. If the interest of \$1 for 1Y. 8m. is 10 cents, what must be the interest of \$2 for the same time? What of \$3?—of \$5?—of \$75?—of \$819.75? *A.* \$.20; \$.30; \$.50; \$7.50; \$81.97$\frac{1}{2}$.

3. If the interest of \$1 is 10 cents, what sum will be required to

CASE IX. *Q.* What is the rule for the State of New York in which the rate is 7 per cent.? 1.

draw 20 cents interest? What to draw 30 cents?—50 cents?—$7.50?—$81.97½? *A.* $2; $3; $5; $75; $819.75.

RULE.

4. *Divide the given interest by the interest of $1, for the given rate and time, the quotient will be the required principal.*

5. What sum of money put at interest 1 year and 8 months will gain $20.60?

NOTE.—Divide $20.60 by the interest of $1 for 1Y. 8m. *A.* $206.

7. What principal will gain $200 in 4Y. 2m? *A.* $800.

8. Suppose a gentleman's income from his property is $4,800 per annum, how much is he worth? *A.* $80,000.

9. Suppose the interest of a certain sum for 2Y. 6m. is $300; what is that sum? *A.* $2,000.

10. What is that sum the interest of which at 7 per cent. for 5Y. 4m. is $728. *A.* $1,950.

11. What sum is that whose interest is $200.104 $\frac{8}{10}$ for 10 years 9 months, at 4 per cent.? *A.* $465.36.

12. Suppose a note which was dated February 15th, 1828, had accumulated $60 interest on the 15th of August, 1830; what was the principal of the note? *A.* $400.

13. A merchant on a note receivable dated May 5th, 1820, calculated the interest up to September 20th, 1823, the time of its maturity, and made it $3,645, noting it down. When the debtor called to pay the note, it could not be found; can you tell what was the face of the note, and what sum ought to be considered full payment of the same.

A. $18,000 principal+$3,645=$21,645 full payment.

CASE XI.

To find the rate per cent.

RULE.

1. *Divide the given interest by the interest of the principal at 1 per cent. for the given time, the quotient will be the rate per cent.*

2. For the required rate is of course as many times greater than 1 per cent. as the given interest is greater than the interest at 1 per cent.

3. What is the interest of $2,000 for 1 year at 1 per cent.?—at 6 per cent.? *A.* $20; $120.

4. If $20 interest is gained in one year on $2,000; what per cent will gain six times as much—that is $120 in the same time?

$120÷20=6 per cent. *A.*

CASE X. What is the rule for finding the principal? 4. When the interest is $20.60 and the time 1Y. 8m., how is the principal found? 5. What is a gentleman worth when his interest money is $2,400 per annum? When it is $3,600 per annum?

CASE XI. What is the rule for finding the rate per cent.? 1. What is the reason of the rule? 2. When $60 dollars are paid as the interest of $600 for 1 year and 8 months, what must be the rate of interest?

5. What per cent. is $3 645 interest on $18 000 for 3 years 4 months 15 days? *A.* 6 per cent.

6. Suppose a note of $2 918 has acquired from February 1st, 1825, to December 16th, 1827, 587.24\frac{3}{4}$, what must have been the rate per cent.? *A.* 7 per cent.

CASE XII.

To find the time in interest.

1. What is the interest of $200 for one year? *A.* $12.

2. If the interest of $200 for one year is $12, how many years will be required to gain $24? *A.* 24÷12=2 years.

3. Therefore, as many times as the interest of 1 year is contained in the given interest, so many years will be required to gain the true interest.

RULE.

4. *Divide the given interest by the interest of the principal for one year at the given rate.*

5. How long will $600 be in gaining $72? *A.* 2 years.

6. How long will $1,000 be in gaining $120?—$180?—$240?—$300?—$360? *A.* 2Y. 3Y. 4Y. 5Y. 6Y.

7. What time will be required for $356.76 to gain $53.514? *A.* 2 years 6 months.

8. How long will $1,800 be in gaining $702? *A.* 6Y. 6m.

9. What time will be required for $2,500 to gain $470 at 8 per cent.? *A.* 2Y. 4m. 6d.

10. Suppose I receive $157 as the lawful interest of $1,200 for a certain time, what is that period of time? *A.* 2Y. 2m. 5d.

CASE XIII.

To find the date from which the interest begins or ends, that is, to find the later or earlier date.

1. When will $5,000 on interest from October 6th, 1830, gain $762.50.

2. The time found by the last case is 2Y. 6m. 15d.

1 8 3 0 Y.	9 m.	6 d.
2 Y.	6 m.	15 d
1 8 3 3 Y.	3 m.	21 d

Since 3m. 21d. of the year 1833 have elapsed, it brings the year, by the reference in case VI. 3. to April 21st. *A.* April 21st, 1833.

3. To find the earlier date, we may subtract the time between the two dates, from the later date, thus,

April 21st, 1833=1833Y. 3m. 21d.=the later date.
2Y. 6m. 15d.=the time between the dates
A. 1830 October 6=1830Y. 9m. 6d.=the earlier date.

RULE.

4. *Having found the intermediate time by the last case add it to*

CASE XII. *Q.* What is the rule for finding the time? 4. Why? 3. What time will be required for $500 to draw $30 interest?—to draw $45?—$60?

the earlier date for the later date, and subtract it from the later date for the earlier date.

5. Paid, July 5th, 1838, $48 interest on a note of $600; from what date did the interest commence? *A.* March 5th, 1837.

6. Suppose a note of $600 is dated March 5th, 1837, when will the interest amount to $48? *A.* July 5th, 1838.

7. Suppose I pay $42, August 23d, 1833, it being the lawful interest on a note of $250, what is the date of the note? *A.* June 11th, 1836.

8. On the 5th of June, 1820, a merchant paid $260, the interest on a note of $1000; when did the interest commence? *A.* February 5th, 1816.

9. If a man tells you he holds in his hand a note of $3,600, the interest of which from its date to Oct. 29, 1830, at 7 per cent., amounts to $1,438$\frac{1}{2}$, and requests you, without seeing the note, to name its date, could you do it? *A.* February 14th, 1825.

INTEREST ON NOTES.

LXXXI. 1. A Note is a written promise to pay a certain sum of money, or its value in goods, on demand, that is, when demanded; or at some future day, and hence, all notes are called promissory notes.

2. A Negotiable Note is one which is made payable to A. B., *or order.*

3. By *indorsing* a note is understood that the person to whom it is made payable writes his name on the back of it, and thereby becomes responsible for its payment.

4. When, however, a note is not paid at maturity, the responsibility of the indorser ceases unless he be notified of the fact within he time prescribed by law.

5. A person holding a negotiable note may transfer it by indorsing it, and whoever buys it may lawfully demand payment of the signer of the note, and if the signer refuses, from inability or otherwise, to pay it, the purchaser may lawfully demand payment of the indorser.

6. If a note be made payable to A. B., *or bearer*, then the signer only is responsible to any one who may purchase it.

7. Unless a note be written payable on some specific future time, it should be written "on demand," but should the words "on demand," be omitted, the note is supposed to be recoverable by law.

8. When the words "with interest," are omitted, a note is not sup-

---

Case XIII. How is the dates from which the interest begins or ends obtained? 4. How far will the 3 first months and 21 days carry the year?

LXXXI. *Q.* What is a note? 1. When is a note negotiable? 2. What does indorsing a note imply? 3. Is the indorser holden after it is due? 4. How is a negotiable note transferred? 5. When is the signer only responsible? 6. When should a note be written "on demand?" 7.

posed to be on interest. Except when a note payable at a future day becomes due; it is then considered on interest from that time till paid, though no mention be made of interest.

9. No mention need be made of the *rate* of interest; that particular is settled by law, and will be collected according to the laws of the State where the note is dated.

10. If two persons jointly and severally sign a note, it may be collected by law of either. A note is not valid, unless the words "for value received," be expressed.

11. When a note is given payable in any article of merchandise or property other than money, deliverable on a specified time, such articles should be tendered in payment at said time; otherwise, the holder of the note may lawfully demand the value in money.

12. The person who gives a note is called the SIGNER OR DRAWER, and when the note is indorsed by a third person, the PRINCIPAL, because the holder of the note looks *first* to him for payment.

13. The sum or debt for which a note is given, is called the PRINCIPAL OR FACE of the note: the person indorsing it the INDORSER, and the person to whom it is indorsed when sold, the INDORSEE or ASSIGNEE.

14. When a partial payment of a note is made, the creditor specifies in writing on the *back* of the note, the sum paid, and the time when it is paid, acknowledging it by subscribing his name, which is then called an *Indorsement.* Interest when paid, is indorsed as such in like manner.

CASE I.

When there are no indorsements.

RULE.

15. *Find the interest of the note from its date, or the time when the interest commenced up to the time it is due or paid; then adding this interest to the face of the note, will give the sum due,* (*as in case* V.)

(16.) $500. BOSTON, January 1, 1836.

For value received, I promise to pay William Marshall five hundred dollars in six months, with interest.

A. $515 due July 1, 1836. PETER STIVER.

(17.) $800. NEW YORK, April 1st, 1828.

For value received, I promise to pay Peter Parley, Esq eight hundred dollars, in two years with interest. [See LXXX. 6.]

A. $912 due April 1, 1830. S. G. GOODRICH.

Q. Are all notes on interest? 8. Need the rate of interest be specified? 9. What is said of a note given by two persons jointly and severally? 10. What of the words "for value received?" 10. What of articles specified to be delivered? 11. What do you understand by the Signer or Drawer of a note? 12. By the Principal, and why? 12. By the Indorser? 13. By the Indorsee or Assignee? 13. By the face or principal of the note? 13. How is an indorsement made? 14.

(18.) PROVIDENCE, June 10, 1826.

For value received, I promise to deliver unto John Northam two bales of good cotton, each bale to contain four hundred pounds, [valued at 12 cents per pound] on or before the twenty-fifth day of October, eighteen hundred and twenty-eight. STEPHEN TRADER.

The above cotton was not tendered in payment till October 25th, 1829, when it had declined 25 per cent. and was therefore refused, [see 11.] but when the note came to maturity it had advanced 15 per cent.; now what would the debtor have lost if he had furnished the cotton according to agreement, and how much does the creditor gain by refusing it when tendered? [See 8 and 11.] *

A. Debtor's loss $14.40; Creditor's gain $29.76.

CASE II.

When there is only one indorsement or payment.

19. When a settlement is made within a short time from the date or commencement of interest, it is generally the custom to proceed according to the rules in the two following cases.

RULE.

20. *Find the amount of the principal for the whole time, also the amount of the payment from the time it was paid to the time of settlement; then deduct the amount of the payment from the amount of the principal, the remainder will be the balance due.*

(21.) $200. BOSTON, July 1, 1838.

For value received, I promise to pay William Grey, or order, two hundred dollars on demand, with interest.

JOSHUA HUCKINS.

On this note there was the following indorsement. Received, December 16th, 1838, seventy-five dollars. What was the balance due March 16th, 1839?

Principal, - - - - - - - - - - - - - -		$200.00
Interest from July 1st, 1838, to March 16th, 1839, - - - -		8.50
Amount of principal for $8\frac{1}{2}$ months - - - - - -		$208,50
Payment, - - - - - - - - - - - -	$75.00	
Interest from Dec. 16, 1838, to March 16, 1839, - -	$1.12\frac{1}{2}$	
The amount of payment for 3mo. deducted, - - - - -		$ $76.12\frac{1}{2}$
Balance due March 16, 1839, - - - - - -		Answer, $$132.37\frac{1}{2}$.

(22.) $1,800. BOSTON, September 10th, 1830.

For value received, I promise to pay John Frothing, or order, eighteen hundred dollars in one year with interest.

JEHIEL JOHNSON.

The following payment was made. January 25th, 1831, received six hundred dollars. What was the balance due July 15th, 1831? *A.* $1,274.50.

CASE II. *Q.* When there is only one indorsement how do you proceed? 20.

---

* The note, of course, does not draw interest till after its maturity, and the debtor's loss, had he delivered the goods at the time specified, would have been 15 per cent. on the face of the note=$14.40. The creditor's gain will be the difference between the face of the note with the interest accruing for one year ($101.76,) and the value of the cotton at the time it was tendered ($72.00.)

(23.) $600. NEW YORK, August 8th, 1835.

For value received, I promise to pay Messrs. Brown and Catlin, six hundred dollars, on demand, with interest.

THOMAS TIPPLETON.

On this note was the following indorsement. Received, May 1st, 1836, four hundred and fifty dollars. What is the sum due July 15th, 1836? [See LXXX. 6.] *A.* $182.841.

CASE III.

When there are two or more indorsements

RULE.

24. *Find the amount of* EACH *payment and of the principal as before; then deduct the total amount of the payments from the amount of principal.*

(25.) $300. HARTFORD, April 1, 1825.

For value received, I promise to pay Rufus Stanly, or order, three hundred dollars, on demand, with interest.

SIMEON THOMPSON.

On this note were the following indorsements. Oct. 1st, 1825, received one hundred dollars; April 16th, 1826, received fifty dollars; Dec. 1st, 1827, received one hundred and twenty dollars. What was the balance due April 1st, 1828?

Principal,		$300.00
Interest of the principal to April 1st, 1828,		54.00
Amount of principal for 36 months,		$354.00
First payment Oct. 1st, 1825,	$100.00	
Interest to April 1st, 1828, (30mo.)	15.00	
Second payment, April 16, 1826,	50.00	
Interest to April 1st, 1828,	5.87	
Third payment, Dec. 1st, 1827	120.00	
Interest to April 1st, 1828, (4mo.)	2.40	
Amount of payments deducted,		$293.27
Balance due April 1st, 1828	Answer,	$60.73

(26.) $500. NEW HAVEN, July 1st, 1825.

For value received, I promise to pay Peter Trusty, or bearer, five hundred dollars, on demand, with interest.

JAMES CARELESS.

Indorsements.—Received, July 16th, 1826, two hundred dollars. Received, January 1st, 1827, forty dollars. Received, March 16th, 1827, two hundred and thirty dollars. What sum remained due July 16th, 1828?

Results, 591.25; 516.10. *A.* $75.15.

(27.) $1,000. PORTLAND, January 16th, 1820.

For value received, we jointly and severally promise to pay to Stimpson and Ripley, or order, one thousand dollars, on demand, with interest.

WILLIAM BIRD.
JAMES BEMENT.

On the back of this note were the following indorsements, viz.—Received, March 16th, 1821, six hundred dollars. Received, May 1st, 1822, one hundred and twenty dollars. Received, July 16th, 1822, one hundred and eighty dollars. What was the balance due January 16th, 1823?

Results. 1180; 97650. *A.* $203.50.

CASE III. *Q.* When there are several indorsements what is the rule? 24.

(28.) $5,000. NEW YORK, June 1st, 1835.

Four months from date, we jointly and severally promise to pay Timothy Dickens and Co. five thousand dollars, with interest, value received. JAMES ROVER.

JOHN TOWNSEND.

Indorsements.—Received, October 1st, 1835, one thousand dollars. Received, January 15th, 1836, one thousand dollars. Received, March 26th, 1836, one thousand dollars. Received, April 1st, 1836, one thousand dollars. What was the balance due July 16th, 1836? [See LXXX. 6.]

Results. 539375; 4132413. *A.*$1261.337.

---

29. The foregoing rule should be restricted to cases in which settlement is made within a year from the commencement of interest; for, beyond that period, its error, in comparison with the legal rule below, becomes too important to be neglected.

30. The rule below is essentially the same as that established by the United States' court, and adopted by most of the federal courts in the Union.

GENERAL RULE.

31. *Find the amount of the principal to the time of the first payment; subtract the payment from this amount, and then find the amount of the remainder to the time of the second payment; deduct the payment as before; and so on to the time of settlement.*

32. *But if any payment is less than the interest then due, find the amount of the sum due to the time when the payments added together shall be equal, at least, to the interest already due; then find the balance, and proceed as before.*

EXAMPLES,

In which every payment exceeds the interest then due.

(33.) $2,400. BOSTON, Oct. 1st, 1830.

For value received, I promise to pay Joseph Careless, or bearer, twenty-four hundred dollars, on demand, with interest.

JOHN SLACK.

Indorsement, April 1st, 1831, of $200.
Indorsement, Dec. 16th, 1831, of $300.
Indorsement, April 1st, 1832, of $900.
} Time. 6mo., 8mo. 15d., 3mo. 15d. 1Y. 3mo. 20d.

Settlement, July 21st, 1833. What was the balance due?

Principal of the note, - - - - - - - - - -	$2,400.00
Interest, April 1st, 1831, - - - - - - - - -	72.00
Amount, April 1st, 1831, - - - - - - - - -	$2,472.00
Payment, April 1st, 1831, - - - - - - -	200.00
Balance, April 1st, 1831, - - - - - - -	$2,272.00
Interest, Dec. 16th, 1831, - - - - - - - - -	96.56
Amount, Dec. 16th, 1831, - - - - - - - - -	$2,368.56
Payment, Dec. 16th, 1831, - - - - - - - - -	300.00

Q. What is the General Rule when each payment exceeds the interest then due? 31. What, when any payment is less than the interest due? 32. From what source is the rule derived? 30. When may the former rule be employed? 29.

Balance, Dec. 16th, 1831, - - - - - - - -	$2,068.56
Interest, April 1st, 1832, - - - - - - - -	36.20
Amount, April 1st, 1832, - - - - - - - -	$2,104.76
Payment, April 1st, 1832, - - - - - - - -	900.00
Balance, April 1st, 1832, - - - - - - - -	$1,204.76
Interest, July 21st, 1833,	94.37*
Balance, July 21st, 1833, - - - - - - - -	*A.* $1,299.13

34. On a note of hand for $1,000, payable July 1st, 1835, were received the following indorsements, viz.

Received, April 21st, 1836, $200.
Received, Aug. 6th, 1836, $150.
Received, April 26th, 1837, $300.
Received, Sept. 11th, 1837, $240.
Received, April 1st, 1838, $210.
Settlement, July 1st, 1839.

Time. 9, 20, 3, 15, 8, 20, 4, 15, 6, 20, 1, 3.
Results. 84833; 71318; 44408; 21407; 11 21.
Answer, $12.05.

EXAMPLES

In which every payment does not exceed the interest then due.

35. On a note given for $600, dated March 1st, 1822, with interest, there were indorsed the following sums.

May 1st, 1823, received $200.
June 16th, 1824, received $ 80.
Sept. 17th, 1825, received $ 12.
Dec. 19th, 1825, received $ 15.
March 1st, 1826, received $100.
Oct. 16th, 1827, received $150.

Time—1, 2, 1, 1, 15, 1, 8, 15, 1, 7, 15, 10, 15.

Settlement, August 31st, 1828. What was the balance due?

Principal, - - - - - - - - - - -		$600.00
Interest, May 1, 1823, - - - - - - - - -		42.00
Amount, May 1, 1823, - - - - - - - - -		$642.00
Payment, May 1, 1823, - - - - - - - - -		200.00
Balance, May 1, 1823, - - - - - - - - -		$442.00
Interest, June 16, 1824, - - - - - - - - -		29.83
Amount, June 16, 1824, - - - - - - - - -		$471.83
Payment, June 16, 1824, - - - - - - - - -		80.00
Balance, June 16, 1824, - - - - - - - - -		$391.83
Interest, March 1, 1826, - - - - - - - - -		40.16
Amount, March 1, 1826, - - - - - - - - -		$431.99
Payment, Sept. 17, 1825, - - - - - - - -	$ 12.00	
Payment, Dec. 19, 1825, - - - - -	$ 15.00	
Payment, March 1, 1826, - - - - - - - -	$100.00	$127.00
Balance, March 1, 1826, - - - - - - - - -		$304.99
Interest, Oct. 16, 1827, - - - - - - - - -		29.74
Amount, Oct. 16, 1827, - - - - - - - - -		$334.73
Payment, Oct. 16, 1827, - - - - - - - - -		150.00
Balance, Oct. 16, 1827, - - - - - - - - -		$184.73
Interest, Aug. 31, 1828, - - - - - - - -		9.70
Balance, Aug. 31, 1828, - - - - - - -		Answer, $194.43

36. On a note dated June 16th, 1820, given for $900, with interest, were indorsed the following payments:

* When the mills are 5 or more, add another cent; but when less than 5, reject them

Received, July 1st, 1821, $150.
Received, Sept. 16th, 1822, $ 90.
Received, Dec. 10th, 1824, $ 10.
Received, June 1st, 1825, $ 20.
Received, Aug. 16th, 1825, $200.
Received, March 1st, 1827, $300.

Time. 1, 15, 1, 2, 15, 2, 11, 1, 6 15, 1, 6.
Results. 80,625; 77,470; 68,027 44,319. *A.* $483.08.

Settlement, Sept. 1st, 1828. What was the balance due?

(37.) $1,600. For value received, I promise to pay Rufus Stanly, r order, sixteen hundred dollars, with interest.

Albany, July 1st, 1830. JONATHAN OVERTON.

Indorsements.—Received, Oct. 16th, 1830, $200. Jan. 1st, 1831, $200. May 26th, 1831, $500. November 1st, 1831, $15. February 11th, 1832, $25 June 6th, 1832, $11. November 26th, 1832, $11. December 1st, 1832, $5. January 11th, 1833, $24, and the balance November 26th, 1835. What was the balance? Time. 3, 15, 2, 15, 4, 25, 1, 7, 15, 2, 10, 15.
Results. 143,266; 125,355; 78,889; 78,763. *A.* $946.14.

CONNECTICUT RULE.

Established by the Supreme Court of the State of Connecticut in 1804.

38. "*Compute the interest to the time of the first payment; if that be one year or more from the time the interest commenced, add it to the principal, and deduct the payment from the sum total. If there be after payments made, compute the interest on the balance due to the next payment, and then deduct the payment as above; and, in like manner, from one payment to another, till all the payments are absorbed; provided the time between one payment and another be one year or more. But if any payments be made before one year's interest hath accrued, then compute the interest on the principal sum due on the obligation, for one year, add it to the principal, and compute the interest on the sum paid, from the time it was paid up to the end of the year; add it to the sum paid, and deduct that sum from the principal and interest, added as above.**"

"*If any payments be made of a less sum than the interest arisen at the time of such payment, no interest is to be computed, but only on the principal sum for any period.*"

39. For value received, I promise to pay Peter Trusty, or order, one thousand dollars, with interest. June 16th, 1824.

$1,000. JAMES PAYWELL.

INDORSEMENTS.

July 1st, 1825, received $250.
Aug. 16th, 1826, received $157.
Dec. 1st, 1826, received $ 87.
Feb. 16th, 1828, received $218.

Time. 1, 15, 1, 1, 15, 1, 8, 15, 6, 6, 10

Settlement, Aug. 26th, 1828. What was the balance?

Q. How do you dispose of the first payment by the Connecticut rule? 38 What is to be done with the other payments? 38. What exceptions are mentioned? 38.

* If a year does not extend beyond the time of payment; but if it does, then find the amount of the principal remaining unpaid, up to the time of settlement, likewise the amount of the payment or payments from the time they were paid to the time of settlement, and deduct the sum of these several amounts from the amount of the principal.

Principal of the note,		$1,000.00
Interest, July 1, 1825,		62.50
Amount, July 1, 1825,		1,062.50
Payment, July 1, 1825,		250.00
Balance, July 1, 1825,		812.50
Interest, Aug. 16, 1826,		54.84
Amount, Aug. 16, 1826,		867.34
Payment, Aug. 16, 1826,		157.00
Balance, Aug. 16, 1826,		710.34
Interest for one year,		42.62
Amount, Aug. 16, 1827,		752.96
Payment, Dec. 1, 1826,	$87.00	
Interest, Aug. 16, 1827,	3.69	90.69
Balance, Aug. 16, 1827,		662.27
Interest, Feb. 16, 1828,		19.87
Amount, Feb. 16, 1828,		682.14
Payment, Feb. 16, 1828,		218.00
Balance, Feb. 16, 1828,		464.14
Interest, Aug. 26, 1828,		14.70
Balance, Aug. 26, 1828,	Ans.	$478.84.

(40.) $875. For value received, I promise to pay Daniel Burgess, or order, eight hundred and seventy-five dollars, with interest
Hartford, January 10th, 1821. HENRY FROTHING.

Indorsements.—Received $260, August 10th, 1824. $300, December 16th, 1825. $50, March 1st, 1826. $150, July 1st, 1827. What was there due September 1st, 1828? Time—3, 7, 1, 4, 6, 1, $9\frac{1}{2}$, 6, 15, 1, 2. Results—80,313; 56,818; 54,990; 41,777. *A.* $447.01

COMPOUND INTEREST.

LXXXII. 1. Compound Interest is the premium given for the use of both the principal and its interest when the latter becomes due and remains unpaid. This is sometimes called interest upon interest.

2. Simple interest implies, as we have seen, (LXXX. 3,) that the interest is payable annually; hence, to find the compound interest, we may proceed as follows: *

RULE.

3. *Find the amount of the principal for one year, (unless a different time be named,) then of this amount as before, and so on to the time of settlement.*

4. *Subtract the given sum from the last amount, and the remainder will be the compound interest required.*

LXXXII. *Q.* What is Compound Interest? 1. What is the rule for finding the amount? 3. What, for finding the compound interest? 4. Why should the interest be compounded annually? 2.

* Compound interest, though just, is not legal

5. What is the compound interest of $156, for 2 years, and what is the amount?

```
            $ 1 5 6=given sum or first principal.
                  6=rate per cent. understood.
              9.3 6=interest for the first year.
          1 5 6    =principal for the first year.
          1 6 5.3 6=amount, principal the second year.
                  6=rate per cent. understood.
          9.9 2 1 6=interest for the second year.
      1 6 5.3 6    =principal for the second year.
A.  $ 1 7 5.2 8 1 6=amount for two years.
      1 5 6        =given sum deducted.
  A.  $ 1 9.2 8 1 6=compound interest for two years.
```

6. What is the compound interest of $500 for 4 years? *A.* 131.238\frac{4}{10}$+.

7. What is the compound interest of $15,000 for 5 years at 7 per cent.? *A.* 6038.275\frac{8}{10}$.

8. What is the amount of $13,000 for 3 years at compound interest, the rate being 4½ per cent.? *A.* $14835.159.

9. What will $600 amount to at compound interest in 4 years at 7 per cent., the interest being payable semi-annually? Find the amount of $600 for 6 months; then of this amount for another 6 months; and so on for the whole time. *A.* $790.079.

10. What will be the compound interest of $140 for 3 years, it being payable semi-annually? *A.* $27.16.

11. What is the compound interest of $240, payable quarterly, for 2 years, at 7 per cent.? *A.* $35.728.

12. What is the compound interest of $1,000 for 2 years at 3½ per per cent. payable quarterly? *A.* $72.18.

13. What is the compound interest of $750 for 5 years and 6 months, payable annually? Find the amount for 5 years, then for 6 months. *A.* $283.78.

14. What is the amount at compound interest of $300 at 7 per cent. for 3 years 4 months and 15 days? *A.* $377.15.

15. If a note of $60.60, dated October 25th, 1836, with the interest payable annually, be paid October 25th, 1840, what will it amount to at compound interest? *A.* $76.51.

16. Find the balance due on the following note, (by LXXXI. 31, 32,) compounding the interest annually.

$1,000. On demand, for value received, I promise to pay John Stearns, or order, one thousand dollars, with interest.

JOSEPH DISCOUNT.

Hartford, August 1st, 1830.

This note has $500 indorsed on the back of it January 16th, 1836, and was paid in full February 1st, 1840. *A.* $1107.46

17. If the number of colored persons in the United States at the present time (1840) be, as is supposed by some, three millions, and their rate of increase 25 per cent. in ten years, what will be their number in 1860?—in 1900? *A.* 4,687,500; 11,444,090.

18. As $2 at compound interest amounts to 2 times as much as $1; $3, 3 times as much, and so on, we may make a table containing the amount of £1 or $1 for several years, by which the amount of any sum may be easily found by simply multiplying once.

TABLE,

Showing the amount of £1 or $1, for 30 years at 5, 6, and 7 per cent. compound interest.

Years.	5 per cent.	6 per cent.	7 per cent.
1	1.050000	1.060000	1.070000
2	1.102500	1.123600	1.144900
3	1.157625	1.191016	1.225043
4	1.215506	1.262477	1.310795
5	1.276281	1.338225	1.402552
6	1.340095	1.418519	1.500730
7	1.407100	1.503630	1.605781
8	1.477455	1.593848	1.718186
9	1.551328	1.689479	1.838459
10	1.628894	1.790848	1.967151
11	1.710339	1.898299	2.104852
12	1.795856	2.012197	2.252191
13	1.885649	2.132928	2.409845
14	1.979931	2.260904	2.578534
15	2.078928	2.396558	2.759032
16	2.182875	2.540352	2.952164
17	2.292018	2.692773	3.158815
18	2.406619	2.854339	3.379932
19	2.526950	3.025600	3.616528
20	2.653297	3.207136	3.869685
21	2.785963	3.399564	4.140563
22	2.925260	3.603539	4.430403
23	3.071524	3.819750	4.740530
24	3.225100	4.048935	5.072367
25	3.386355	4.291870	5.427434
26	3.555673	4.549383	5.807352
27	3.733457	4.822347	6.213868
28	3.920130	5.111688	6.648838
29	4.116136	5.418389	7.114257
30	4.321943	5.743493	7.612255

19. What is the compound interest of $20.15 for 4 years at 6 pe cent.? By the Table the amount of $1 for 4 years is $1.262477×$20.15=$25.438+from which subtracting $20.15 leaves $5.288\frac{7}{10}$.+ *A.* $$5.288\frac{7}{10}$.+

20. What is the compound interest of $2,000 for ten years at 7 per cent.? *A.* $1,934.30. At 6 per cent.? *A.* $1,581.696. At 5 per cent.? *A.* $1,257.79.

21. What is the compound interest of $300 for 20 years at 7 per cent.? *A.* $860.90. At 6 per cent.? *A.* $662.14.

22. What is the amount of $600 for 30 years at 6 per cent. compound interest? *A.* $3,446.10. For 15Y. 6mo.? [See 13.] *A.* $1,481.07.

23. To what sum will $500 amount in 17 years 4 months and 15 days at compound interest? *A.* $1,376.68.

24. What is the amount of $200 for 45 years, at 7 per cent. compound interest? As 45 years extend beyond the Table, find the amount for any number of years in it at first, say 20 years, then of this amount for 20 more; finally for the remaining 5 years. *A.* $4,200.49.

25. What is the amount of $6,000 for 60 years, the compound interest being at the rate of 7 per cent.? *A.* $347,678.56.

26. What is the amount of $600 for 11 years 10 months and 23 days, at 6 per cent.? the interest compounded annually? *A.* $1,200.294$\frac{4}{10}$.

27. What is the amount of $600 for 16 years 8 months at 6 per cent. simple interest? *A.* $1,200.

NOTE.—By the last two examples, it appears that any sum at 6 per cent. compound interest, will double in 11 years 10 months and from 22 to 23 days, while at simple interest it would require 16 years and 8 months.*

DISCOUNT.

LXXXIII. 1. DISCOUNT is that deduction which is made for paying money before it is due.

2. PRESENT WORTH of any sum implies that it is payable at a future day without interest.

3. The PRESENT WORTH, then, is and ought to be such a sum as would at interest amount to the debt when due.

4. Thus the present worth of $106, due 1 year hence, is $100, and the discount $6; for $100 at interest for that time amounts to $106.

5. The discount of any sum is less than its interest; thus the dis

---

LXXXIII. *Q.* What is Discount? 1. Present worth? 2. What does it imply? 3. What is the present worth and what the discount of $106, due 1 year hence? 4. What is the interest of $106 for 1 year? 5. Which then is the most, the interest or the discount? 5.

---

* It seems there is considerable difference between simple and compound interest even for a short time, and when the latter is permitted to accumulate for ages it amounts to a sum almost incredible. For example, suppose a cent had been put at interest at the commencement of the Christian era, it would have amounted at the end of the year 1827, to only $1,106$\frac{2}{10}$. But the compound interest of the same sum for the same time would have amounted to a sum greater than can be contained in 6,000,000 of globes, each equal to our earth in magnitude and all of solid gold; or to $172,616,474,047,552, 529,470,760,914,974,711,959,976,620,354$\frac{56}{100}$ nearly

count of $106 for a year is $6, but the interest of $106 for that time is $6.36.

6. The *debt* then may be regarded as the *amount*, the *present worth* as the *principal*, and the *discount* as the *interest* of this principal but not of the debt.

7. Hence finding the present worth is the same process in effect as that for finding the principal in Interest, Case XI., which may be expressed thus,—

RULE.

8. *Divide the given sum or debt by the amount of* $1 *for the given time.*

9. *The quotient will represent the present worth, which taken from the debt will leave the discount.*

10. What is the present worth of $133.20, payable 1 year and 10 months hence, and what the discount?

NOTE.—The amount of $1 for 1Y. 10m. is $1.11, then $133.20÷$1.11=$120=for the present worth, and $120 from $133.20 leaves $13.20 for the discount. *A.* $120; $13.20.

11. For the proof, find the interest of $120 for 1 year 10mo. then its amount; and if it make $133.20 the work is right.

12. What is the present worth of $660 due 1 year and 8 months hence? What its discount? A. $600; $60.

13. What sum of ready money is equivalent to $460 due 2 years and 6 months hence. What sum is equal to the discount?
A. $400; $60.

14. If I pay a debt of $1,350, 5 years and 10 months before it is due, what sum ought I to pay and what discount ought to be made me? *A.* $1000; $350.

15. Suppose you have owing to you $3065.62½ payable in 2 years 8 months 15 days, and money is worth no more than 5 per cent.; what sum of ready money can you afford to take, and what will the discount amount to? *A.* $2,700; $365.62½.

16. What is the difference in value between $699.25 cash, and $751.116, due 1Y. 6m. hence, when money is worth only 4 per cent.?
A. $9.35.

17. If I am offered goods for $2,500 cash, or for $2,821.50 on "9 months;" which is the best offer, and by how much?
A. Cash by $200.

18. Suppose a merchant contracts a debt of $24,000, to be paid in four installments, as follows, viz: one fifth in 4 months; one quarter in 9 months; one sixth in one year and 2 months, and the rest in 1 year and 7 months; what is the present worth of the whole sum?
A. $22,587.651.

Q. What terms in Discount resemble those in Interest? 6. Which operation in the one is the same in effect as in the other? 7. Rule? 8. What is the discount of $104 for 4 months?—of $208 for 8 months?—of $109 for 1 year 6 months?

19. Suppose I contract to receive flour at different times, from New-York, on 9 months' credit, and receive as follows, viz:

Jan. 16, 1830, 180 barrels at $10 per barrel. $———
Feb. 20, 1830, 900 barrels at 9\frac{1}{2}$ per barrel. $———
April 16, 1830, 850 barrels at 10\frac{1}{4}$ per barrel. $———
June 21, 1830, 600 barrels at $11 per barrel. $———
Oct. 10, 1830, 950 barrels at 10\frac{1}{2}$ per barrel. $———

Now suppose I remit the cash in payment as often as I receive a lot of flour, what ought to be the sum total of all my remittances, when money (being "tight") is worth at least 10 per cent.?

A. $33,151.162.

DISCOUNT BY COMPOUND INTEREST

RULE.

LXXXIV. 1. *Divide the given sum by the amount of $1 at compound interest for the stated time; the quotient will be the present worth, which, subtracted from the given sum, will leave the discount.*

2. For the quotient, which is the present worth; multiplied by the divisor, which is the amount of $1 for the whole time; must reproduce the dividend, which is the given sum or amount.

3. What is the present worth of $561.80, due 2 years hence, reckoning 6 per cent. per annum, compound interest? *A.* $500.

4. What sum in cash is equivalent to $687,512$\frac{45}{100}$, payable 2 years hence, deducting 7 per cent., compound interest? *A.* $600.50.

5. How much discount for the cash ought to be made on $2,127.778$\frac{6}{10}$, due 6 years hence, reckoning compound interest yearly? *A.* 627.778\frac{6}{10}$.

6. Suppose I propose to sell you the following note at a discount of 7 per cent. per annum, compound interest: what sum do I ask for it?

$11,792.47. New York, April 1st, 1828.

For value received, I promise to pay on the first day of September, eighteen hundred and thirty, unto Peter Hunks, or order, eleven thousand seven hundred and ninety-two $\frac{47}{100}$ dollars.

William Neverfail.

A. $10,008.09.

7. Suppose a father's estate was so divided between two sons, one 20 years old and the other only one year old, that each on arriving "at age" should receive an equal portion. Suppose, also, that when the younger brother was 21, the older brother's portion, by means of annual loans at compound interest amounted to $3,207.136, how many dollars was each to receive when 21 years old? *A.* $1,060.

LXXXIV. *Q.* What is the rule for finding the discount when the interest has been compounded? 1. What is the reason for it? 2.

BANKING

LXXXV. 1. A BANK is an incorporated institution, that deals in money. Its capital, which is limited by law, is usually owned in shares by persons called *Stockholders*.

2. The proper business of a bank is to make and lend notes called "*bank bills*," which circulate as money, because the bank is obliged to redeem them with specie.

3. When the banks loan money, it is their custom to take the interest in advance; that is, to deduct it from the face of the note at the time the money is lent. The note is thence said to be *discounted*.

4. The face of every note, therefore, should exceed the sum received or wanted, as much as will just equal the interest of the note to the time when it is payable

5. Hence the sum discounted is called the AMOUNT; the interest deducted the DISCOUNT, and the remainder the *proceeds*, or more correctly, the PRESENT WORTH.

6. A note to be discounted or bankable, must be made payable at a future day, and to the order of some person who indorses it.

7. The indorser, however, is not responsible for its payment unless notified that the note is due and demanded, but not paid.*

8. The banks take interest for 3 days more than the time specified in the note, because the debtor is not obliged by law to make payment till the same 3 days have elapsed, which are thence called *days of grace*.

RULE.

9. *Cast the interest on the note for 3 days more than the time specified; then deduct the interest from the face of the note, and the remainder will be the sum loaned.*

10. What is the bank discount on $600, payable in 60 days? The interest of $600 for [60d.+3d.=] 63d.=$6.30. *A.* $6.30.

11. What is the bank discount on $1,200, payable 90 days hence, and what would be its present worth? *A.* $18.60; $1,181.40.

12. $1,800. Sixty days after date, for value received, I promise

LXXXV. Q. What is a Bank? 1. What is said of its capital? 1. What, of its business? 2. When is a note said to be discounted? 3. When a particular sum is wanted at bank, what sum should be named in the note? 4. What are meant by bank Discount, Amount, and Present Worth? 5. What particulars must be observed in writing a bankable note? 6. What is meant by a *protest?* 7. What by 3 *days of grace?* 8. What is the rule for ascertaining the sum loaned or received? 9. What sum would be received on a note of $60 for 2 months?—for 4 months?—for 6 months?—for 8 months?—for 10 months?

* A legal notice in writing is called a *protest*.

to pay Peter Parley, or order, at the Etna Bank (N. York,) eighteen hundred dollars. PETER PAYWELL.

Suppose "old Mr. Peter Parley" indorses the above note, and it is discounted, what sum would Mr. Paywell receive? *A.* $1,777.95.

13. What sum would be the present worth of $1,200 discounted at bank and payable in 60 days, at 7 per cent.? *A.* $1,185.30.

14. A merchant sold 250 bales of cotton, each weighing 300 pounds, for $12\frac{1}{2}$ cents per lb. which cost him the same day 10 cents per lb.; he received in payment good paper for 4 months time. Now supposing he gets this note discounted at bank, what will be his profits? *A.* $1,682.81$\frac{25}{100}$.

15. To find what sum or amount must be named in a note in order to obtain a particular loan at bank.

RULE.

16. *Deduct the bank discount on* $1 *for the given time from* $1, *and divide the desired loan by the remainder, the quotient will be the sum or amount required.*

17. For if the quotient, which is the required amount, be multiplied by the divisor which is the present worth of $1, for the given time, the process must re-produce the dividend, which is the given loan.

18. Most paper at our banks is discounted either for 95 days or 4 months. The interest of $1 for 3 days (grace) is $.0005, and for 95d.+3d.=$.01633 nearly; for 4mo.+3d.=$.0205; then $1−$.01633=$.98367; and $1−$.0205=$.9795, therefore:

19. *The divisor under this rule for any note payable in* 95 *days, is* $.98367, *and for* 4 *months*, $.9795.

20. Suppose I want a loan at bank of $14,842.50 for 60 days: what sum must be named in the note to obtain that amount of money? The interest of $1 for 63 days (=.0105) deducted from $1 leaves $.9895, for a divisor. *A.* $15,000.

21. Suppose your note for 6 months is discounted at bank, and $484.75 passed to your credit; what must have been the face of the note? *A.* $500.

22. If I want from a bank at Rochester, New York, $5,786.50 for my note at 6 months, what must be the face of the note? *A.* $6,000.

23. Suppose "old Mr. Peter Parley" wants a loan himself at bank of $994.50 for 30 days, at which time he expects to be able to refund it from the profits of his story books, and that Mr. Paywell reciprocates the favor shown to him above, by indorsing it; what sum must be specified in the note to obtain that loan? *A.* $1000.

EQUATION[1] OF PAYMENTS.

LXXXVI. 1. In how many months will 1 dollar gain as much interest as 2 dollars will gain in 6 months? *A.* 6×2=12 months.

2. In how long time will 1 dollar gain as much interest as 5 dollars will gain in 12 months? *A.* 60 months.

3. How many months is the use of 1 dollar equivalent[2] to the use of 10 dollars for 20 months? *A.* 200 months.

4. How long ought you to lend B 1 dollar to repay him for his kindness in lending you 100 dollars for 4 months?

A. 400 months=$33\frac{1}{3}$ years.

5. In what time will the use of 100 dollars be equivalent to the use of 300 dollars 6 months? 300 dollars for 6 months is the same as 1 dollar for [300×6=] 1800 months, and 100 dollars is the same of course as $\frac{1}{100}$ of 1800 months; that is, 1800÷100=18mo.

A. 1Y. 6mo.

6. A having lent B 200 dollars for 9 months, wishes a like favor of B, but needs only 50 dollars; how long may A keep the 50 dollars without doing any injustice to B? *A.* 3 years.

7. Suppose *A* lends B 8 dollars to be paid in 2 months, and 12 dollars to B. paid in 7 months, making in all 20 dollars lent B. Now how long ought B to lend A 1 dollar to repay him for his kindness? How long ought B to lend A 20 dollars?

2× 8=16; therefore $8 for 2 mo.=$1 for 16 months.
7×12=84; therefore $12 for 7 mo.=$1 for 84 months.
$20)100(5mo. *A*

8. Then B ought to lend A 1 dollar 100 months, but 20 dollars only $\frac{1}{20}$ as long; that is, 100÷20=5 months. Therefore 20 dollars payable in 5 months is the same as if 8 dollars of the $20 were payable in 2 months, and the remaining 12 dollars in 7 months.

9. PROOF—The interest of $8 for 2 months is . . 8 cents.
The interest of $12 for 7 months is . . 42 cents.
The interest of $20 for 5 months is . . 50 cents.

RULE.

10. *Multiply each payment by the time, and divide the sum of these*

LXXXVI. Q. In how many months will 1 dollar gain as much interest as 6 dollars in 3 months?—as 6 dollars in 4 months?—in 8 months? How long ought A to lend me 12 dollars to reciprocate[1] my favor in lending him 6 dollars for 2 months? How long ought I to lend you 20 dollars to recompense you for lending me one time 8 dollars for 2 months, and at another time 12 dollars for 7 months? [See 7.] Why divide by 20 dollars? 8. What is the rule? 10.

---

1 EQUATION, [L. *æquatio.*] Literally, a making equal, or an equal division.
2. EQUIVALENT. Equal in value or worth; equal in force, power, or effect. Of the same import or meaning.
1. RECIPROCATE, [L. *recipricus.*] To exchange; to interchange; to give and return mutually.

*several products by the sum of the payments; the quotient will be the mean or equitable time for the payment of the whole.**

11. A owes B $200 to be paid in 6 months, $300 in 12 months $500 in 3 months; what is the equated time for the payment of the whole? *A.* $6\frac{3}{10}$.

12. What is the equated time for paying $2,000, of which $500 is due in 3 months, $360 in 5 months, and $600 in 8 months, and the balance in 9 months? *A.* $6\frac{960}{2000}=6\frac{12}{25}$ mo.

13. A merchant bought goods amounting to $1,200, $\frac{1}{4}$ of which he was to pay in cash, $\frac{1}{3}$ in 6 months, and the balance in 10 months what was the equitable time for the payment of the whole?

A. $6\frac{1}{6}$ months.

14. A merchant proposed to sell goods amounting to $4,000 on 8 months credit; but the purchaser preferred to pay $\frac{1}{2}$ in cash and $\frac{1}{4}$ in 3 months; what time should be allowed him for the payment of the remainder? *A.* 2Y. 5m.

15. A having sold B a bill of goods amounting to $1,200, left it optional with him either to take them on 8 months'credit, or to pay $\frac{1}{2}$ in cash, $\frac{1}{5}$ in two months, $\frac{1}{6}$ in 4 months, and the remainder at an equated time for paying the balance on the terms first named. What was that time? *A.* 4Y. 4m.

* This rule proceeds on the supposition, that what is gained by keeping the money after it is due is equal to what is lost by paying it before it is due. But this is not exactly true, for the gain is equal to the interest, while the loss is equal only to the discount, which is always less than the interest. However, the error is so trifling, in most cases which occur in business, as not to make any material difference in the result

SIMPLE PROPORTION,[1]

OR

THE RULE OF THREE,

SOMETIMES CALLED THE GOLDEN RULE.

BY ANALYSIS.[2]

LXXXVII. 1. If 1 hat costs \$5, what will 4 hats cost? *A.* \$20.

2. If 1 quarter of a yard of blue satin costs $37\frac{1}{2}$ cents, what will 1 yard cost? What will 315yd. 3qr. cost? *A.* \$1.50; \$473.62$\frac{1}{2}$.

3. If 1 pound of sugar will cost $9\frac{3}{4}$ cents, what will be the cost of 1cwt.?—of 1cwt. 3qr.? *A.* \$9.75. \$17.062$\frac{1}{2}$.

4. If \$1.125 will buy 1 gallon of wine, how many hogsheads may be bought for \$70.875? *A.* 1 hogshead.

5. If 1cwt. 3qr. of sugar cost \$17.50, what will 1 quarter cost? *A.* \$2.50. What will 1 pound cost? *A.* 10 cents.

6. If 6 bushels of wheat cost \$12, what will 1 bushel cost?—5 bushels cost?—115 bushels cost? *A.* \$2; \$10; \$230.

7. If 400 barrels of flour cost \$4000, what will 89 barrels cost? Find the price of 1 barrel first. *A.* \$890.

8. If a farm consisting of 300 acres sells for \$6,150, what would a small farm of 50 acres sell for at that rate? *A.* \$1,025.

9. When 10 yards of cotton cloth clost \$1.50, what will be the cost of 10 pieces, each containing $52\frac{1}{2}$ yards? *A.* \$78.75.

10. When tea is £5. 16s. by the cwt., what will 1qr. of a cwt. cost? What will 20 chests, each weighing 10cwt. 1qr. cost?
A. £1. 9s.; £1,189.

11. If 6 ounces of silver will make 15 spoons, how many spoons can be made from 8 silver tankards; each weighing 2lbs. 6oz.
A. 600=50doz.

12. If 50 dozen silver spoons are made from 8 silver tankards, each weighing 2lb. 6oz., how much silver will be required to make $1\frac{1}{4}$ dozen, or 15 spoons? *A.* 6 ounces.

13. If it require \$300 to gain \$15 interest in a year, how much will be required to gain \$100. *A.* \$2000.

14. If 10 men will mow a certain meadow in 13 days, how long a time will be required for 25 men to do the same? 1 man will be 10 times longer than 10 men. *A.* $5\frac{1}{5}$ days.

LXXXVII. Q. When 5 pounds of cheese cost 60 cents, what will 11 pound cost? What will $5\frac{1}{2}$ pounds of sugar cost at 30 cents for 3 pounds? When 4 gallons of wine cost \$5, what is the price of a single quart?

---

1. **Proportion,** [L. *proportio.*] The comparative relation of any one thing to another; symmetry; equal or just share; form; size.

2. **Analysis,** [G. *analusis.*] The separation of a compound into the parts that compose it; a resolving or consideration of any thing in its separate parts; it is opposed to *synthesis*, [G. *sunthesis*,] which means *the putting of two or more things together.* Analysis in *Arithmetic* is finding the whole by first finding the value of unity.

15 If 100 men can complete a job of work in 25 days, in how many days will 7 men do the same? *A.* $357\frac{1}{7}$.

16. If 100 men can do a job of work in 25 days, how many men will be required to do the same in 10 days? The less days the more men will be required. *A.* 250 men.

17. If the interest of a certain sum is $10 for 3 years, what is the interest of the same sum for 12 years? *A.* $40.

18. If 2,400 men can do a job of work in 6 months, how many men, working at the same rate, would do the same job in 4 months? *A.* 3,600.

19. If 461 bottles will hold 5hhd. 30gal. 3qt. of cider, how many hogsheads will 1151 such bottles hold? *A.* 3453qt.=13hhd. 44gal. 1qt.

20. If $\frac{2}{3}$ of a barrel of flour cost $4.80, what will $\frac{1}{3}$ of a barrel cost? *A.* $2.40. What will 1 barrel cost? *A.* $7.20.

21. If $\frac{5}{6}$ of a load of hay cost $10, what will $513\frac{1}{4}$ loads cost? *A.* $6,159.

22. If $\frac{3}{5}$ of a yard of cloth cost $6, what is it a yard? If $\frac{3}{5}$ be $6, then $\frac{1}{5}$ is $\frac{1}{3}$ of $6 (=2) and $\frac{5}{5}$ are $\frac{5}{3}$ of $6=$10.—[See LII. 10.] *A.* $10.

23. If $\frac{3}{7}$ of a barrel of cider is sufficient for a family 9 weeks, how long will 1 barrel last them?—1 barrel will last them $\frac{7}{3}$ of 9 weeks. How long would 50 barrels last the same family? *A.* 50 times 21= 1050w.÷52w.=$20\frac{5}{26}$ years.

24. If $\frac{3}{25}$ of a hogshead of molasses cost $2, what will $120\frac{2}{3}$hhd cost? 1hhd. cost $\frac{25}{3}$ of $2. *A.* $$2011\frac{1}{9}$.

25. If $\frac{2}{5}$ of a yard of broadcloth cost $2.40, what will $\frac{3}{8}$ cost? 1yd. cost $\frac{5}{2}$ of $2.40=$6, then $\frac{3}{8}$yd. is $\frac{3}{8}$ of $6. *A.* $2.25.

26. When $\frac{5}{11}$ of a ship is valued at $20,000, what is $\frac{3}{4}$ of it worth? *A.* $33,000.

27. When $\frac{6}{7}$ of a gallon of oil costs $\frac{3}{4}$ of a dollar, how much will 1 gallon cost? How much is $\frac{7}{6}$ of $\frac{3}{4}$? *A.* $$\frac{7}{8}$. What will 40 gallons cost? *A.* $35.

28. When $\frac{7}{8}$ of a dollar will buy $\frac{3}{5}$ of a bushel of corn, how much will $200 buy? *A.* $137\frac{1}{7}$bu.

29. If $\frac{5}{8}$ of a pound of cassia cost $\frac{3}{4}$ of a dollar, what will $\frac{2}{3}$ of a pound cost? 1lb. cost $\frac{8}{5}$ of $$\frac{3}{4}$; and $\frac{2}{3}$lb cost $\frac{2}{3}$ of $\frac{8}{5}$ of $$\frac{3}{4}$. *A.* $$\frac{4}{5}$=80 cents.

30. If $\frac{4}{5}$ of a yard of cloth cost $\frac{3}{8}$ of a dollar, what will $\frac{6}{11}$ of a yard cost? *A.* $.255. What will $40\frac{2}{3}$ yards cost? *A.* $$19.06\frac{1}{4}$.

31. If $\frac{3}{5}$ of $\frac{2}{3}$ of a cask of lime cost $\frac{7}{8}$ of a dollar, what will $\frac{3}{4}$ of $\frac{4}{5}$ of a cask cost? 1 cask costs $\frac{5}{2}$ of $$\frac{7}{8}$. *A.* $$1.31\frac{25}{100}$.

Q. If 10 men can perform a job of work in 5 days, how many men would be required to do the same in 10 days?—in 20 days? [See 16.] When 4 bushels of rye cost $3, what will $\frac{5}{6}$ of that quantity cost? If $\frac{3}{4}$ of a dollar will buy 5 yards of calico, what ought $\frac{2}{5}$ of a yard to cost? When $\frac{3}{5}$ of a barrel of flour sells for $6, why is $\frac{5}{3}$ of $6 the price of 1 yard? 22. If $\frac{5}{6}$ of a yard of cloth will make 10 stocks, how many stocks may be made with $3\frac{1}{4}$ yards?

32. A man traveled $\frac{3}{4}$ of $\frac{2}{3}$ of a mile in $\frac{7}{80}$ of an hour; how far would he go at that rate in $\frac{2}{3}$ of $\frac{5}{8}$ of an hour? *A.* $1\frac{1}{14}$m. How far in $\frac{3}{4}$ of 24 hours? *A.* $77\frac{1}{7}$ miles.

33. If $\frac{2}{3}$ of a barrel of flour costs $2\frac{4}{5}$ times $\frac{3}{4}$ of 2 dollars, what will $5\frac{3}{5}$ barrels cost? *A.* $35.28.

BY RATIO[1].

34. If 6 yards of cloth cost $8, what will 1 yard cost? Since 3 yards are $\frac{3}{6}$ or $\frac{1}{2}$ of 6 yards, then 3 yards will cost $\frac{1}{2}$ of the price of 6 yards, that is $\frac{1}{2}$ of $8, which is $4. *A.* $4.

35. If 5 hats cost $41 what will 30 cost? What part of 5 is 30? *A.* $\frac{30}{5}=6$. How many are 6 times $41? *A.* $246.

36. If 12 cows cost 432 dollars, what will 8 cows cost? What part of 12 is 8? How much is $\frac{2}{3}$ of $432? *A.* $288.

37. When 112 bushels of wheat cost $168, what will 80 bushels cost? What part of 112 is 80? How much is $\frac{5}{7}$ of $168? *A.* $120.

38. At the rate of $50 for 400 dozen of eggs, what will 1000 dozen cost? What is the ratio of 400 to 1000? [See LXII. case XI.] How much is $2\frac{1}{2}$ times $50. *A.* $125.

39. When 5 bushels of wheat cost $8, what will be the cost of 300 bushels? What is the ratio of 5 to 300? How many are 60 times $8? *A.* $480.

40. If 4 gallons of molasses cost $9\frac{5}{8}$ shillings, how many dollars will 40 gallons cost? Ratio 10. *A.* $$16.04\frac{1}{6}$.

41. Suppose a man travels $187\frac{1}{2}$ miles in 5 days, how far will he travel in 25 days? Ratio 5. *A.* $937\frac{1}{2}$ miles.

42. If 1 bag of salt cost $5, what will $500 purchase? What part of 5 is 500? *A.* 100 bags.

43. If 15 gallons of oil cost $$26\frac{3}{8}$, what will 4 gallons cost? How much is $\frac{4}{15}$ of $$26\frac{3}{8}$? *A.* $7\frac{1}{30}$ or $7.033+.

44. If 2cwt. 2qr. of sugar cost $15.625, what will 50cwt. cost? [See LXII. case XII.] What part of 10 qr. is 200 qr.? *A.* $312.50.

45. Suppose a stage runs at the rate of 7 miles and 4 furlongs in 45 minutes; how long will it be in running 8m. 6fur.? *A.* $52\frac{1}{2}$min.

46. If £1. 17s. 6d. will purchase 20 gallons of wine, how many gallons will 18s. 9d. purchase? *A.* 10 gallons.

47. If $\frac{3}{4}$ of a barrel of rice cost $$17\frac{3}{5}$, what will 750 barrels cost? The ratio is $\frac{750}{\frac{3}{4}}$ =(by LXII. case X.) 1000. *A.* $17,600.

48. If $\frac{3}{4}$ of a barrel of wine costs $30, what will $\frac{2}{3}$ barrel cost? What part of $\frac{3}{4}$ is $\frac{2}{3}$? [See LXII. case XI. 16.] How much is $\frac{8}{9}$ of $30? *A.* $$26\frac{2}{3}$.

49. If $\frac{3}{5}$ of a yard of silk cost $\frac{5}{8}$ of a dollar, what will $\frac{19}{20}$ of a yard cost? *A.* $$\frac{95}{96}$ or $0.9895.+

Q. When 6 loaves of bread are bought for 48 cents, why do 3 loaves, at that rate, cost 24 cents? When 20 papers of pins cost 120 cents, what will 10 papers cost, and why? When $\frac{2}{3}$ of $\frac{3}{4}$ of a dollar buys 8 skeins of silk, how many skeins may be bought for $\frac{2}{5}$ of a dollar?

---

1 For the meaning of Ratio, and the rule for finding it, see LXII. case XI.

50. When $3\frac{2}{3}$ pounds of butter cost 75 cents, what will $2\frac{3}{4}$ pounds cost? What part of $3\frac{2}{3}$ is $2\frac{3}{4}$? How much is $\frac{3}{4}$ of 75? *A.* $56\frac{1}{4}$ cents.

51. If $\frac{5}{8}$ of a ship cost $20,000, what will $\frac{3}{4}$ of her be worth at that rate? What part of $\frac{5}{8}$ is $\frac{3}{4}$? *A.* $24,000.

52. How many yards of cloth which is $\frac{3}{4}$yd. wide, are equal to 5 yards which is $\frac{7}{8}$yd. wide? Ratio $\frac{\frac{7}{8}}{\frac{3}{4}}$ or $\frac{7}{6}$. *A.* $5\frac{5}{6}$ yards.

53. When $\frac{1}{2}$ of $\frac{3}{4}$ of a pound of butter costs $12\frac{1}{2}$ cents, what will 40 firkins, each containing 25 pounds, cost? *A.* 333.33\frac{1}{3}$.

BY STATEMENT.

54. If 8 yards of cloth cost 63 cents, what will 24 yards cost?

By Ratio.—24 yards will cost $\frac{24}{8}$ of 63 cents.=$1.89. *A.*

By Analysis.—1yd. cost $\frac{1}{8}$ of 63cts=$7\frac{7}{8}$ct.×24yds.=$1.89 A.: or to avoid the fraction, multiply by the 24 first, and divide by the 8 afterwards; but before doing this a statement is often made of the terms employed, as follows:—

```
    1st. 2nd.   3d.
    Yds. Yds.   cts.
     8 : 2 4 :: 6 3
         6 3
        -----
         7 2
     1 4 4
    -------
 8 ) 1 5 1 2
    ---------
   $ 1 . 8 9 A.
```

Observe that the 1st and 2nd terms are of the same kind, and the 3rd term of the same kind with the answer; also that the 2nd term is the multiplying number, and the 1st term the dividing number: all of which must be observed in every statement.

Notice also the colons between the different terms. These colons are in common use, to show that the ratio of 8 to 24, which is ($\frac{24}{8}$=) 3, is the same as that of 63 to the answer 89, which is also ($\frac{189}{63}$=) 3.

55. If 419 books cost $1,257, what will 750 cost? *A.* $2,250.

56. If 750 books cost $2250, what will 419 cost? *A.* $1,257.

57. If 750 books cost $2,250, how many will $1,257 buy? *A.* 419.

58. You doubtless have noticed that the greater the multiplying term in comparison with the dividing term, the greater is the answer, and the reverse.

59. Hence we have the following direction, which will greatly assist you in arranging the first and second terms:

60. *Take the greater of these terms for the second term if the answer ought to be greater than the third term, otherwise take the smaller for the second term.*

61. If 1cwt. of iron costs $8.25, what will 27cwt. cost? 1cwt. 27cwt.:: $8.25. *A.* $222.75.

62. If 27cwt. of iron cost $225, what will 1cwt. cost? *A.* $8.25.

63. If 35 cows cost $700, what will 89 cows cost? *A.* $1,780.

Q. When 8 yards of cloth cost 63 cents, how do you find by *ratio* what 24 yards will cost? 54. How is the same question performed by *analysis?* 54. How is it done by *statement?* 54. Which terms must be of the same kind? 54. What must the third term be like? 54. Which term is the multiplying number? 54. Which the dividing number? 54. What is meant by the colons between the different numbers? 54. What simple direction is given in respect to the arrangement of the first and second terms? 60. How is this ascertained? 58

64. If 91 horses cost $4,788.875, what will 75 cost?

A. $3,946.875.

65. If 1824 barrels of flour will cost $15,750.64, what will 2736 barrels cost?

1824 bar.: 2736 bar.::$15750.64. But since 1824 and 2736 have 912 for a common divisor, we may use in their stead the quotients 2 and 3, for multiplying the third term by 912 and dividing the result by 912 cannot alter that term: thus,—

2 : 3::$15,750.64, then $15,750.64 × 3 ÷ 2 = $23,625.96. *A.*

66. *When then the first and second terms have common divisors, divide by the greatest divisor and substitute the quotients for those terms.*

67. When 183,945 yards of cloth costs $674,465.872, what will 147156 yards cost? (Gr. com. div. 36789.) *A.* $539,572.697$\frac{6}{10}$.

68. If 63 yards of tape cost 45 cents and 3 mills, what will 21 yards cost? *A.* 15c. 1m.

69. If 415 bales of cotton sell in London for £5,260.2s. 6d., what are 2536 bales worth at that rate?

£5,260. 2s. 6d. = 1262430d. Then 415 bales : 2536 bales::1262430 pence. *A.* 7714512d. = £32143.16s.

It is more convenient to reduce the 3d term to pence. The answer will of course be pence.

70. Hence when the third term is a compound number:—*Reduce it first to the lowest denomination in it; then proceed as before, recollecting that the answer will appear in the same denomination to which it was reduced, which may then be brought into any other denomination required.*

71. If a merchant pay in London £56. 11s. 3d. for 25 yards of broadcloth, what would 35$\frac{3}{5}$ yards cost? *A.* £80. 13s. 11d.

72. Suppose a merchant buys 2cwt. 3qr. 10lb. of sugar for $35.625, how many hundred weight at that rate may be bought for $468.75? *A.* 3,750lb. = 37cwt. 2qr.

73. If it costs $369.625 to make a fence over a distance of 2 miles 6fur. 16rd. 4yd. 1ft., what length of fence may be made at that rate for $1,108.875? *A.* 8m. 3fur. 10rd. 2yd

74. If 2hhd. 42gal. 2qt. of wine cost $193.20, what will 5hhd. cost?

Here 2hhd. 42gal. 2qt. = 674qt. : 5hhd. = 1260qt. Then 674qt.: 1260qt.::$193.20. *A.* $361.175.

75. *Hence when the first or second term is of a different denomination, it must be brought to the same by Reduction.*

76. If 4 yards of cloth cost $17.35, what will 101yd. 3qr. cost? *A.* $441.34.+

77. If 2cwt. 3qr. 10lb. of hay cost 4.12\frac{1}{2}$, what will 5T. 15cwt. 2qr. 20lb. cost? *A.* $167.46+

Q. When the third term is a compound number, what is to be done with it? 70. In what denomination is the answer? 70. What reduction is often required in reference to the other terms? 75

78. Suppose 108yd. 2qr. 1na. of cotton cloth cost in Manchester, (England) £4.13s. $6\frac{1}{2}$d., what would 500yd. 2qr. cost? *A* £21.11s. 3d.

79. A gentleman invested $2,000 in coal at the rate of $8.50 for 19cwt., how many tons did he buy? *A.* 223T. 10cwt. 2qr. $8\frac{14}{17}$lb.

80. Suppose it costs $49 to move a certain building 38rd. 3yd.; how many miles at that rate may the same building be moved for $5,000? *A.* 12m. 2fur. 13rd. $\frac{1}{2}$yd. 1ft. $11\frac{25}{49}$in.

RECAPITULATION.

81. THE RULE OF THREE is so called, because it has *three terms* given to find a fourth (the answer); which shall have the same ratio to the third, as the second has to the first.

82. The FIRST and SECOND terms are always of the same kind, and the THIRD of the same kind with the FOURTH or ANSWER.

GENERAL RULE.

83. *State the question by making the third term of the same kind with the answer; then consider whether the answer ought to be greater or less than the third term; if greater, make the second term greater than the first, but if less, make the second term less than the first.*

84. *Reduce the first and second terms to the same denomination, and the third term to the lowest denomination in it, then multiply the second and third terms together, and divide their product by the first, the quotient will be the fourth term or answer, in the same denomination with the third term.*

CONTRACTIONS OF THE RULE.

85. Reduce the fractional ratio of the first and second terms to its simplest form, then multiply the third term by it.

86. Or divide the first and second terms by their greatest common divisor, then substitute the quotients for the terms themselves, and proceed as before.

87. Or proceed analytically to find the whole by first finding the value of unity.

88. If 17 yards of satinet cost $12.75, what will 51 yards cost?

BY STATEMENT. 17yd. : 51yd. :: $12.75. ($12.75×51÷17=) $38.25. *A.* $38.25.

BY RATIO. 17 : 51=$\frac{51}{17}$=3 : then $12.75×3=$38.25. *A.* $38.25.

BY ANALYSIS. $12.75÷17=75ct. for 1 yard; then 75ct.×51yd. =$38.25. *A.* $38.25.

89. If 51 yards of satinet cost $38.25, what will 17 yards cost? Ratio $\frac{1}{3}$: 1 yd =75cts. *A.* $12.75.

90. If $38.25 will buy 51 yards of satinet, what will $12.75 buy? Ratio $\frac{1}{3}$. 75 cents buys 1yd. *A.* 17 yards.

91. If $12.75 will buy 17 yards of satinet, what will $38.25 buy? Ratio 3 : 75 cents for 1yd. *A.* 51 yards.

Q. What is the Rule of Three? 81. What similarity is there between the terms? 82. General rule? 83, 84. What are the three methods by which this rule is abbreviated? 85, 86, 87

92. When 108 barrels of flour cost $837, what will $43\frac{1}{5}$ barrels cost? Ratio $\frac{2}{5}$: 1bl. =$$7\frac{3}{4}$. *A.* $334.80.

93. Suppose $600 bushels of wheat cost $1,200; how many bushels may be bought for $7,200? *A.* 3,600 bushels.

94. If $7,200 dollars will purchase 3,600 bushels of wheat, what will 600 bushels cost? *A.* $1,200.

95. When you pay $13.50 per month (=4 weeks) for board, how much will pay your bill for 22 weeks? *A.* $74.25.

96. Suppose you give 30 bushels of rye for 120 bushels of potatoes, how much rye must you give at that rate for 600 bushels of potatoes? *A.* 150 bushels.

97. If 4cwt. 1qr. of sugar cost $45.20, what will 21cwt. 1qr. cost? [See ex. 75.] *A.* $226. Ratio 5.

98. Suppose you pay $120 for 60 yards of cloth; what does it cost by the ell English? *A.* $2.50.

99. When 4 tuns of wine cost $322.56, what will 1 tierce cost? *A.* $13.44. What will 1 barrel cost? *A.* $10.08. What will 1 pint cost? *A.* 4 cents.

100. When a merchant compounds with his creditors for 40 cents on a dollar, how much is A's part, to whom he owes $2,500? How much is B's part, to whom he owes $1,600? *A.* A's $1,000; B's $640.

101. When the velocity of a locomotive on a railroad is 35 miles an hour, how far does it move in 30sec.? *A.* $\frac{7}{24}$m. or 93rd. $5\frac{1}{2}$ft.

102. If a steamboat cross the Atlantic (3000 miles) in 12 days, what is her average velocity per hour? *A.* $10\frac{5}{12}$miles.

103. The surface of the planet Jupiter contains 24,884,000,000 square miles. How many inhabitants would it accommodate, if 1,120 occupy 4 square miles? *A.* 6,967,520,000,000.

104. How many minutes would there be in 16 weeks, provided there were 2,160 in 3 days? *A.* 80,640 minutes.

105. If a man's family expenses are $2.50 per day, and his salary $1,537.40, with perquisites amounting to $\frac{3}{8}$ of a dollar per day, how much can he save annually? *A.* $$761.77\frac{1}{2}$.

106. Jupiter moves in its annual course 90,000 miles every 3 hours. How far does it move in 7 weeks? *A.* 35,280,000 miles.

107. How much flour will a family consume in 4 years, if 275 pounds supply them 37 days? *A.* 55 barrels 71lbs. $5\frac{23}{37}$oz.

108. A man pays $25 for a load of corn containing 30 bushels, how many loads can he buy for $92.50? *A.* 3 loads and 21 bushels.

109. The breadth of the dark space between the two rings of Saturn is 2,839 miles. How long would sound be in passing through it at the rate of 1,142 feet in a second? *A.* 3h. 38m. $46\frac{14}{571}$s.

110. The planet Uranus is 1,705,000,000 miles distant when nearest us. How long would a cannon ball be in reaching it, moving 12,000 miles in 24 hours? *A.* 389Y. $98\frac{1}{3}$.d.

111. If five times four were thirty-three,
What would the fourth of twenty be? *A.* $8\frac{1}{4}$.

112. If a steeple 150 feet high cast a shade 375 feet in length, how long is that staff whose shadow at that time is 8 feet? *A.* $3\frac{1}{5}$

113. Divide $240 in the proportion of 3 to 2. *A.* **$160.**

114. Suppose 3cwt. 2qr. 10lb. of sugar cost $52.625, what will 21cwt. 2qr. 10lb. cost? *A.* $315.75.

115. If you receive $89 interest on $1,780 for one year, what is the rate per cent.; that is, what is it on $100. *A.* 5 per cent.

116. If 12 men build a wall in 20 days, how many men can do the same in 5 days? *A.* 48 men.

117. Suppose a wall is built by 48 men in 5 days; what number of men could do the same, if they were allowed to be four times as long about it? *A.* 12 men.

118. If 4 men dig a trench in 48 days, how many men could do it in the sixth part of that time? *A.* 24 men.

119. Suppose a man, by traveling 10 hours a day, performs a journey in 4 weeks, without desecrating the Sabbath; how many weeks would it take him to perform the same journey, provided he travels only 8 hours per day, and pays no regard to the Sabbath?
A. 4 weeks and 2 days.

120. Suppose a certain pasture, in which are 20 cows, is sufficient to keep them 6 weeks, how many must be turned out, that the same pasture may keep the rest 6 months? *A.* 15 cows.

121. If a certain garrison is manned with 1,000 men, and with provisions enough for 18 months, how many must leave the garrison, that the rest may be able to hold out against a siege of 2 years?
A. 250 men.

122. Suppose a man of war that has 1,800 marines on board, and provisions enough for 18 months, should lose a fourth part of her men, how long would their provisions serve the rest? *A.* 2 years.

123. Suppose that 50 yards of carpeting 1 ell English wide, will carpet a room; how many yards of carpeting that is only 3 quarters wide, will do the same? *A.* $83\frac{1}{3}$ yards.

124. If 7s. 6d. in New Jersey currency is equal to 8s. in New York currency, what sum of the former is equal to £720 of the latter?
A. £675.

125. A mason was engaged in building a wall, when another came up and asked him how many feet he had laid; he replied that the part he had finished bore the same proportion to 1 league which $\frac{3}{17}$ does to 87. How many feet had he laid? *A.* $32\frac{142}{1479}$ feet.

126. A, standing on the bank of a river, discharges a cannon, and B, upon the opposite bank, counts six pulsations at his wrist between the flash and the report; now if sound flies 1,142 feet per second, and the pulse of a person in health beats 75 strokes in a minute, what is the breadth of the river? *A.* $5,481\frac{3}{5}$ft. or 1m. $201\frac{3}{5}$ ft.

RULE OF THREE IN FRACTIONS

GENERAL RULE.

LXXXVIII. 1. *State the question and perform the operation as before, except you are to multiply and divide the terms according to the rules to which the numbers respectively belong.*

2. If $\frac{7}{8}$ of a barrel of flour costs $8.40, what will $13\frac{1}{8}$bl. cost? $\frac{7}{8}$bl. : $13\frac{1}{8}$bl. : : $8.40. For dividing by $\frac{7}{8}$ see LXVI. 18. *A.* $126.

Or 1 barrel costs $\frac{8}{7}$ of $8.40=$9.60 × $13\frac{1}{8}$bl. *A.* $126.

Or the ratio of $\frac{7}{8}$ to $13\frac{1}{8} = \frac{13\frac{1}{8}}{\frac{7}{8}} = 15 \times \$8.40$. *A.* $126.

3. If $\frac{3}{16}$ of a dollar will buy 480 pins, how many dozen pins will $2 purchase? Ratio $10\frac{2}{3}$. *A.* $426\frac{2}{3}$ dozen.

4. Suppose you pay 5s. 3d. for $\frac{3}{4}$ of a gallon of oil; what will 19 barrels 3 gallons cost? Ratio 802 : 1gal.=7s. *A.* $701.75.

5. If $5 will purchase 30 yards of calico, how many yards will $\frac{2}{3}$ of a dollar purchase? Ratio $\frac{2}{15}$. *A.* 4 yards.

6. When $\frac{5}{7}$ of a hogshead of wine is bought for $8.50, what does $\frac{2}{5}$ of a pint cost? Either reduce by LXII. case XIV. $\frac{2}{5}$pt. to the fraction of a hogshead, $(=\frac{2}{2520}=\frac{1}{1260})$; or $\frac{5}{7}$hhd. to the fraction of a pint; then $\frac{5}{7}$hhd. : $\frac{1}{1260}$hhd. : $8.50. *A.* $9\frac{44}{100}$ mills.

7. When £$\frac{3}{5}$ will purchase in London 24 dozen steel pens, how many pens will $\frac{5}{6}$ of a penny purchase? *A.* $1\frac{2}{3}$ pens.

8. If $\frac{3}{5}$ of a bushel of wheat cost $\frac{15}{16}$ of a dollar, what will $\frac{2}{3}$ of a dollar purchase? [See LXVI. 18.] *A.* $\frac{32}{75}$ bushel.

9. What will $5\frac{1}{2}$ yards of broadcloth cost in London, if $\frac{5}{8}$ of a yard cost £$\frac{1}{5}$? *A.* £$1\frac{19}{25}$.

10. If $52\frac{2}{5}$ yards of cloth cost $$75\frac{1}{2}$, what will $3,676\frac{7}{10}$ yards cost? 52.4 yards : 3676.7 yards : : $75.50. *A.* $5297.535+.

11. If $37\frac{1}{2}$lb. of sugar cost $$5\frac{1}{4}$, what will $205\frac{3}{5}$lb. cost? *A.* $28.784.

12. When $40\frac{3}{4}$ acres of land are purchased for $$219\frac{1}{6}$, what will 5 farms cost at the same rate, each farm containing $195\frac{5}{8}$ acres? First reduce all the terms to decimals. *A.* $5260.672.

13. If 6lb. 3oz. of silver will make 2 silver tankards, how many such tankards would 513lb. 3oz. 15dwt. make? Reduce the first and second terms to decimal expressions of the same denomination. See LXIX. case IV. *A.* $164\frac{26}{100}$ tankards.

14. What will $5\frac{1}{2}$ yards of satin cost, if $\frac{5}{8}$ of a yard costs 7s. 6d.?

Since $\frac{5}{8}$yd.=.625yd.:$5\frac{1}{2}$yd.	$\frac{5}{8}$yd. : $5\frac{1}{2}$yd. : : 90d.	*A.* £3. 6s.
=$\frac{11}{2}$yd.=5.5yd., and 7s. 6d.	$\frac{5}{8}$yd. : $\frac{11}{2}$yd. : : 90d.	*A.* £3. 6s.
=90d. or £.375; the num-	$\frac{5}{8}$yd. : $\frac{11}{2}$yd. : : $\frac{90}{1}$d.	*A.* £3. 6s.
bers are susceptible of the	$\frac{5}{8}$yd. : 5.5yd. : : 90d.	*A.* £3. 6s.
adjoining statements, each of	$\frac{5}{8}$yd. : 5.5yd. : : £.375.	*A.* £3. 6s.
which is to be performed	.625yd. : 5.5yd. : : £.375.	*A.* £3. 6s.
without any further reduction.		

LXXXVIII. *Q.* When questions in the Rule of Three contain fractions now are they performed? 1.

15. What will $563\frac{1}{4}$ bushels of early apples cost, when $\frac{3}{4}$ of bushel costs 60 cents? *A.* \$450.60

16. If $\frac{3}{4}$ of a yard of muslin costs $\frac{5}{16}$ of a dollar, what will $\frac{3}{20}$ of a yard cost? *A.* \$$\frac{1}{16}$=$6\frac{1}{4}$ cents.

17. When $\frac{1}{2}$ of $\frac{3}{4}$ of a gallon of wine costs \$$\frac{5}{8}$, what will $5\frac{1}{2}$ gallons cost? *A.* \$$9.16\frac{2}{3}$.

18. If 3 yds. of cloth that is $2\frac{1}{2}$ yds. wide will make a cloak, how many yds. $\frac{3}{4}$ of a yd. wide will make the same cloak? *A.* 10yds.

19. If 12 men do a piece of work in $12\frac{1}{4}$ days, how many men will do the same in $6\frac{1}{8}$ days? *A.* 24 men.

20. A merchant owning $\frac{2}{3}$ of a vessel, sells $\frac{2}{3}$ of his share for \$500; what was the whole vessel worth? *A.* \$1,125.

21. If $1\frac{1}{2}$lb. of indigo cost \$3.84, what will 49.2lb. cost?
A. \$125.952.

22. If \$$29\frac{3}{4}$ buy $59\frac{1}{2}$yds. of cloth, what will \$60 buy? *A.* 120yds.

23. How many yards of cloth can you buy for \$$75\frac{1}{2}$ at the rate of $267\frac{3}{4}$ yards for \$$37\frac{3}{4}$? *A.* $535\frac{1}{2}$ yards.

24. If 7 times $\frac{3}{9}$ of $\frac{7}{8}$ of an estate be worth \$15,000, what is $\frac{2}{9}$ of $\frac{3}{8}$ of it worth? *A.* \$612.244+.

25. If I pay 29 cents for $\frac{1}{16}$ of a yard of broadcloth, for what can I buy $3\frac{1}{4}$ yards? *A.* \$15.08.

26. When you can buy $\frac{2}{3}$ of $\frac{3}{5}$ of a barrel of flour for $3\frac{1}{4}$ dollars, what must you pay for $19\frac{1}{2}$ barrels? *A.* \$158.4375.

27. When the price of cotton cloth is $\frac{3}{5}$ of a shilling, sterling money, for 5 nails, for what can 30 yards be purchased? *A.* £2. 17s. $7\frac{1}{5}$d.

28. When $\frac{1}{8}$ of a dollar will buy $\frac{1}{7}$ of a pennyweight of gold, how much can be bought for \$8,000. *A.* 38lb. 1oz. 2dwt. $20\frac{4}{7}$gr.

29. A man paid $\frac{1}{8}$ of a dollar apiece for 24 apple trees, $\frac{1}{6}$ of a dollar apiece for 30 pear trees, $\frac{3}{5}$ of a dollar apiece for 15 plum trees; how many of an equal number of each kind could he have bought for \$10.70? *A.* 12 trees.

30. Suppose a merchant who has contracted for the transportation of 3cwt. 2qr. 15lb. for \$2.40 concludes to forward a greater quantity viz. 5T. 12cwt. 1qr. 10lb.; what charge for the latter would be in proportion to what he agreed to pay for the former? See Ex. 13
A. \$73.87+

THE DOCTRINE[1] OF PROPORTION.

DEDUCED[2] MAINLY FROM THE TWO PRECEDING CHAPTERS.

LXXXIX. 1. From the preceding exercises it appears that—*Ratio is the mutual relation of one quantity to another of the same kind, and is indicated by a colon written between the quantities, as* 5 : 20, *the ratio of which is* $\frac{20}{5}=4$.

LXXXIX. What is Ratio? 1.

1 DOCTRINE, [L. *doctrina.*] Whatever is taught; a principle; learning; knowledge; the truths of the gospel.
2 DEDUCED, [L. *deduco.*] Drawn from; inferred.

2. The first term of a ratio is called the ANTECEDENT, and the second, the CONSEQUENT, and both together form a COUPLET, as 12 : 3. Here 12 is the Antecedent, and 3 the Consequent.

3. GEOMETRICAL RATIO expresses the quotient arising from dividing the consequent by the antecedent. This quotient or its equivalent whole number is sometimes called the *Index* or *Exponent* of the ratio, as 5 : 20, whose index would be $\frac{20}{5}=4$.

4. Quantities between which there exists ratio must be of the same kind, or else we cannot form any judgment of their equality or inequality. Thus, 2 hours and 6 barrels have no ratio one to the other, for neither can be said to be either greater or less than the other.

5. What is the ratio of 5 to 405? *A.* 5 : 405, or $\frac{405}{5}=81$.

6. Which has the greater ratio, 6 : 72 or 5 : 60? *A.* Each=12.

7. Which couplet is the greater one, 4 : 83 or 8 : 166?
A. Each=$20\frac{3}{4}$.

8. What is the difference between the ratio of 3 yards to 15 yards and that of 5 yards to 75 yards? *A.* 10.

9. How much greater is the couplet 4 : 3 than the couplet 5 : 3?
A. $\frac{3}{20}$.

10. What is the difference between 2 couplets which have each 40 for its antecedent, but 120 and 1,800 for their consequents. *A.* 42.

11. What is the difference between the couplet 2 yards : 1 yard and the couplet 1,800,000 yards : 900,000? *A.* 0.

12. Hence small quantities may have the same ratio, and even a greater one, than quantities many times larger; from which it is clear, that the ratios of any two quantities can be predicated only on their relative magnitude.

13. For example, the sun may have a less ratio to the moon than a bullet to a small particle of matter, though the former quantities are, either of them, immensely greater than the latter.

14. When two couplets have the same ratio, they are said to form a proportion, and the terms, to be proportionals.

15. HENCE, PROPORTION MAY BE DENOMINATED AN EQUALITY OF RATIOS, AND IS DENOTED USUALLY BY TWO COLONS PLACED BETWEEN TWO COUPLETS.

16. For example, 2 : 10 : : 8 : 40, which is read thus:—

The ratio of 2 to 10 is equal to the ratio of 8 to 40.

Or the ratio of 2 to 10 is the same as that of 8 to 40.

Or 2 has the same relation to 10 that 8 has to 40.

Or 2 is contained in 10 as often as 8 is in 40; for 5 times 2 are 10, and 5 times 8 are 40.

17. The proportional terms take their names from their location;

Q. What names have the terms? 2. Give an example. How is geometrical ratio expressed? 3. What is essential to the existence of ratio? 4. What is the ratio of 5 to 20?—of 22 to 11?—of 8 to 30?—of $\frac{1}{2}$ to $\frac{3}{4}$?—of $\frac{2}{10}$ to $\frac{2}{5}$? When are four numbers proportional? 14. What then is Proportion? 15 What are the different methods of reading examples? 16. How are the proportional terms distinguished? 17. Give an example. [See 18.]

thus, the first and last are called the *extreme terms*, or the Extremes, and the second and third, the *middle terms*, the *mean terms*, or the Means.

18. Thus in 2 : 6 : 4 : 12, the 2 and 12 are the *extremes*, and the 6 and 4 the *means*.

19. Proportion generally consists of four terms, but it may exist with only three, when the first number has to the second the same atio as the second has to the third, which is thence called continued proportion; as, 2 : : 6 : 18; for 3 times 2 are 6, and 3 times 6 are 18.

20. Hence, we need not hesitate to pronounce any set of numbers proportional, when we can prove that an equality of ratios exists between them.

21. Arithmetical Ratio is the difference between two quantities, as 6—3, and is denoted by the sign between the couplets.

22. For example, 7—5=8—6 is a proportion; for the ratio of each couplet is 2, and the extremes 7 and 6 added together are equal to the means 5 and 8 added together, and universally—

23. In Arithmetical Proportion the sum of the extremes is equal to the sum of the means.

24. Geometrical[1] Proportion is an equality of Geometrical Ratios, and Arithmetical Proportion an equality of Arithmetical Ratios.

25. The terms *Geometrical* and *Arithmetical* are generally used as above, because they are employed in the same sense in *Geometry.*[2]

26. In the proportion 4 : 3 : : 8 : 6, the ratios $\frac{3}{4}$ and $\frac{6}{8}$ are equal; for $\frac{6}{8}$ reduced is equal to $\frac{3}{4}$.

27. Again, if the ratios $\frac{3}{4}$ and $\frac{6}{8}$ are equal, it follows, that, by reducing them to a common denominator, their numerators will become equal, as before. This is actually the case, for they make $\frac{24}{32}$ and $\frac{24}{32}$, but the first numerator, 24, is in reality the product of the two extremes, 4 and 6, and the second numerator the product of the two means, 3 and 8.

28. Hence we deduce an important principle, viz. That in every geometrical proportion the product of the two extreme terms is equal to the product of the two mean terms.

29. For example, 4 : 5 : : 8 : 10. Here the product of the two extremes, 4 and 10, is 40, and the product of the two means, 5 and 8, is also 40. The numbers are therefore proportional.

Q. Of how many terms does Proportion consist? 19. What is an Arithmetical Ratio? 21. Give an example. See 21. What is said of the sum of the extremes? What is Geometrical Proportion? 24. Arithmetical Proportion? 24. Whence the origin of these terms? 25 How is the proportion 4 : 3 : : 8 : 6 proved to be such? 26. What other proof is adduced? 27. In the last process, which terms are multiplied together? 27. What important principle is deduced from the operation? 28. Give an example? 29.

---

1 Geometrical. According to the rules or principles of geometry.[2]

2 Geometry, [*Geometria.*] The science of magnitude [3] in general, comprehending the doctrine and relations of whatever is susceptible of being increased or diminished

3 Magnitude, [L. *magnitudo.*] Extent of dimensions of parts; extent; bulk; grandeur; importance.

30. From the principle just deduced it follows, that the order of the terms of a proportion may be changed, provided they be so placed that the product of the extremes shall equal the product of the means; thus:

10 : 40 :: 20 : 80 for 10 × 80 = 40 × 20. Each ratio 4.
10 : 20 :: 40 : 80 for 40 × 20 = 80 × 10. Each ratio 2.
20 : 10 :: 80 : 40 for 20 × 40 = 10 × 80. Each ratio 2.
20 : 80 :: 10 : 40 for 20 × 40 = 80 × 10. Each ratio 4.
40 : 80 :: 10 : 20 for 40 × 20 = 80 × 10. Each ratio 2
40 : 10 :: 80 : 20 for 40 × 20 = 10 × 80. Each ratio 4.
80 : 40 :: 20 : 10 for 80 × 10 = 40 × 20. Each ratio 2.
80 : 20 :: 40 : 10 for 80 × 10 = 20 × 40. Each ratio 4.

31. Hence proportionals in changing places may vary their ratios, but observe that their equality is maintained, which is the true characteristic of every proportion.

32. Both antecedents, or both consequents, or even all the terms of a proportion, may be multiplied or divided by the same number without disturbing the equality of the ratios; consequently the terms will still be proportional.

33. For the principle of equality cannot be affected by the process, because in each instance of multiplying the same factor is contained in each product, and in each instance of dividing, the same factor is cancelled from each dividend, as—

4 :	8 ::	6 :	12	the given proportion.
5		5		Multiplying antecedents by 5;
20 :	8 ::	30 :	12	for 20×12=8×30.
	4		4	Multiplying consequents by 4;
20 :	32 ::	30 :	48	for 20×48=32×30.
10		10		Dividing antecedents by 10;
2 :	32 ::	3 :	48	for 2×48=32×3.
	8		8	Dividing consequents by 8;
2 :	4 ::	3 :	6	for 2×6=4×3.
9 :	9 ::	9 :	9	Multiplying all the terms by 9;
18 :	36 ::	27 :	54	for 18×54=36×27.
3	3	3	3	Dividing all the terms by 3
6 :	12 ::	9 :	18	for 6×18=12×9.

34. Two geometrical proportions may be multiplied or divided one by the other, term by term, with results still proportional:

35. For the fractional ratios of each couplet being equal, the process is the same in effect as multiplying or dividing equal fractions by equal fractions, the results of which will of course be equal; as—

1st Prop.	3 :	6 ::	5 :	10	Multiplying the 1st
2nd Prop.	2 :	8 ::	3 :	12	and 2nd, term by term.
3d Prop.	6 :	48 ::	15 :	120	for 6×120=48×15.
1st Prop.	3 :	6 ::	5 :	10	Dividing by the 1st,
2nd Prop.	2 :	8 ::	3 :	12	for 2×12=8×3.

36. The terms of one proportion may be added to or subtracted

from the corresponding terms of another proportion, on the principle, that if equal fractions be either taken from or added to other equal fractions, the results must be equal.

1st Prop.	5 :	7 : :	10 :	14
2nd Prop.	3	4	6	8
3d Prop.	2 :	3 : :	4 :	6
	10 :	14 : :	20 :	28

Subtract the 2nd from the **1st**, then add the three together; for $10\times 28 = 14\times 20$.

37. Note.—We have seen that a number is squared by multiplying it by itself, [VIII. 22,] and cubed by multiplying its square by the same number again. [VIII. 27.] These terms have a general name, called powers; thus 5 before it is multiplied is called the first power; 5×5 or 25, the square or 2nd power of 5; $5\times 5\times 5$ or 125, the cube or 3d power of 5; and $5\times 5\times 5\times 5$ or 625, the 4th power of 5, and so on, every power being named from the number of times the given number is used as a factor.

38. Proportionals, we have seen, may be multiplied by themselves term by term; that is, they may be squared, or cubed, or raised to any power whatever without affecting the proportion; as—

1st proportion,	4 :	3 : :	8 :	6
Squared,	16 :	9 : :	64 :	36
Cubed,	64 :	27 : :	512 :	216
4th power,	256 :	81 : :	4096 :	1296
5th power,	1024 :	241 : :	32768 :	7776

39. Proportion is susceptible of other useful changes; but those already noticed are sufficient for our present purpose.

APPLICATION OF GEOMETRICAL PROPORTION:

ILLUSTRATING MORE FULLY THE RULE OF THREE.

XC. 1. Since the product of the extremes is equal to the product of the means, one product may be taken for the other; consequently,*

2. *In dividing either product by one extreme, the quotient will be the other extreme, and dividing by one mean, the quotient will be the other mean.*

3. Find the value of x, that is, find the fourth term in the proportion 3 : 30 : : 7 : x. Thus, $30\times 7=210\div 3=70$. *A.* $x=70$.

4. Find the third term or the value of x in the proportion 4 : 20 : : x : 25. *A.* $x=5$.

5. Find the second term or the value of x in the proportion $2\frac{1}{2}$: x : : 8 : 20. *A.* $x=6\frac{1}{4}$.

Q. What changes in the order of the terms are admissible? 30. What variation is there in the ratios? 31. Give an example. What operations may be performed with those terms? 32. Why is not the proportion destroyed? 33. On what principle can one proportion be multiplied by another? 35. How may one proportion be added to or subtracted from another? 36. Why so? 36 What is meant by a square, cube, &c.? 37. How may a proportion be squared, cubed, &c.? 37

* From Lacroix

6 What is the value of x, or the first term in the proportion x : $75\frac{1}{2}$: : 42 : 151? A. $x=21$.

7. When one couplet of a proportion is 6 : 24, and another $563\frac{1}{2}$: x, what is the value of x, or what is the numeral consequent for the latter couplet? A. $x=2,254$.

8. When 6 is the consequent and 18 the antecedent of one couplet in a proportion, what will be the antecedent of the other couplet if its consequent be 1200? A. 3,600.

9. When the first term in a proportion is 20, the second term 849, the third term 6,750, and the fourth term x, what is the value of x? A. $x=286,537\frac{1}{2}$.

10. The operation by which, when any three terms of a geometrical proportion are given, we find the fourth, is called the Rule of Three, or the Rule of Proportion.

11. The Rule of Three, then, is based on the principle, *that the product of the extremes is equal to the product of the means:* consequently, dividing the product of the second and third terms by the first, as directed by the rule, is the same thing as dividing the product of the means by one of the extremes to find the other extreme.

12. If 30 barrels of flour cost $240, what will 90 barrels cost?

13. In this example not only the ratio of 30 barrels to 90 barrels, which is 30 : 90, is given, but also the antecedent of the next couplet, for the ratio of the price must correspond with the ratio of the quantities; that is, if one quantity is 3 times greater than the other, it will cost 3 times as much; if 4 times greater, 4 times as much; the question then may be resolved into the following proportion—

30 barrels : 90 barrels : : 240 dollars : x. Here the product of the means is 21,600, which, being divided by one of the extremes, gives a quotient of 720. *That is, multiply the second and third terms together, and divide the product by the first; the quotient will be the fourth term or answer.* A. $720.

14. If 20 pounds of butter cost $5, what will 80 pounds cost? 20 pounds : 80 pounds : : 5 dollars. Here the first consequent is 80, because 80 pounds will cost more than 20 pounds. *Multiply and divide as before, or multiply the third term by the ratio of the first and second.* A. $20.

15. If 20 men mow a meadow, consisting of 30 acres, in a day, how many acres would 40 men mow, at that rate, in the same time? Here the more men, the more acres will be mowed, which is called *Direct Proportion*, or a case in which *more requires more.* A. 60A.

16. If 40 men mow 60 acres in a day, how many acres will 20 men mow in the same time? Here the less the number of men, the

XC. *Q.* How may the absent term in Proportion be found? 2. Why so? 1 Describe that operation which has received the name of the Rule of Three. 10. On what principle is it based? 11. What reason is given for multiplying and dividing in that rule? 11. What is the method of stating example 12, and why? 13. ☞Whenever an example is referred to in this manner, it is expected the teacher will read the example audibly to the scholar before he asks the question.☜ What is the statement of example 14, and why?

less the number of acres mowed, being a case in which it is said, *less requires less*, and is also called *Direct Proportion.* *A.* 30A

17. DIRECT PROPORTION, *then, is when one ratio increases as another increases, or decreases as another decreases, and was formerly called the* RULE OF THREE *Direct*.*

18. If 20 men mow a meadow in 10 days, how long will it take 40 men to do the same? Here the more men, the less time will be required; then double the number of men would require half as many days; or if the number of men decreases in any ratio whatever, the number of days will increase in the same ratio.

19. The ratio then of 20 men to 40 men in the last example is directly the reverse of 10 days to the answer; but if the 20 and 40 change places, that is, be inverted, the ratios will become equal, in which case we can proceed as before; thus—40 men : 20 men : : 10 days. *A.* 5 days.

20. The proportion here then may be called *Inverse*, but the method of stating the question is the same as before.

21. *That is, take of the first two terms, the greater one for the second term, when the answer requires it, otherwise, take the smaller for the first term.*

22. If 40 men mow a field in 5 days, in what time will 20 men mow the same? Here the less men, the more days will be required. it being a case in which it is said *less requires more*. The ratio then of 40 men to 20 men decreases as that of 5 days to the answer increases.

23. But if we invert the first couplet, we have 20 men : 40 men, which have the same ratio that the third term has to the answer, in which case we may proceed as before.—20 men : 40 men : : 5 days. *A.* 10 days.

24. In the foregoing example the proportion is also called *Inverse*, but the method of stating and operating corresponds exactly with the directions given above.

25. INVERSE PROPORTION, *then, is when one ratio increases as another decreases, or decreases as another increases, and was formerly called* THE RULE OF THREE INVERSE.*

Q. What is Direct Proportion? 17. What is the statement of example 23, and the reason for it? What is the proportion called? 20. What is the method of procedure? 21. What is the statement of example 22? What reasons are assigned? 22, 23. What is Inverse Proportion? 25. What distinction was formerly made in Proportion? * 25. What was the Rule of Three Direct? * 25. What, that of Inverse? * 25. Which place did the term like the answer occupy? * 25.

---

* Formerly the following distinctions obtained in respect to Direct and Inverse Proportion, viz.

THE RULE OF THREE DIRECT has three terms given to find a fourth, which shall have the same proportion (or ratio) to the third term that the second has to the first

THE RULE OF THREE INVERSE has three terms given to find a fourth, which shall have the same proportion to the second as the first has to the third.

The Rule of Three Direct is when more requires more, or less requires less. It may be known thus: more requires more when the third term is more than the first, and requires the fourth term, or answer, to be more than the second; and less requires less, when the third term is less than the first, and requires the fourth term, or answer, to be less than the second

COMPOUND PROPORTION.

XCI. 1. If a man travels 60 miles in 5 days, traveling 3 hours each day, how far will he travel in 10 days, if he travels 9 hours each day?

2. By Analysis.—If he travels 60 miles in 5 days, he travels in 1 day $\frac{1}{5}$ of 60, which is 12 miles; and if he travels 3 hours each day, he travels in 1 hour $\frac{1}{3}$ of 12, which is 4 miles. Then, if he travels 4 miles in 1 hour, he will travel in 9 hours, or 1 day, 4 times 9, which is 36 miles; and in 10 days, 10 times 36, which is 360 miles, the answer.

3. In the foregoing example the answer evidently depends on two circumstances, viz.—the number of days the man travels, and the number of hours he travels each day. These circumstances we will now consider separately, on the principles of Simple Proportion.

4. We will first enquire how far he will go in 10 days, provided he travel an equal number of hours each day; this question, then, may be expressed as follows:

5. If a man travel 60 miles in 5 days, how far will he travel at that rate in 10 days?—which will form the following proportion—

5 days : 10 days : : 60 miles : *A*. 120 miles.

6. In the next place we will consider the other circumstance, viz. the difference in the number of hours; the question will then be—

7. If a man, by traveling 3 hours a day, travels 120 miles in a certain number of days, how far will he go in the same number of days if he travel 9 hours each day?—which gives the following proportion—

3 hours : 9 hours : : 120 miles : *A*. 360 miles.

These two statements brought together stand thus—

5 days : 10 days : : 60 miles : *A*. 120 miles.
3 hours : 9 hours : : 120 miles : *A*. 360 miles.

XCI. Q. What is the solution of the first question by analysis? 2. ☞The scholar should be allowed, in cases like the last, to copy the example referred to, on his slate, and have it before him while he is performing the operation, or answering the questions respecting it.☜ On what does the answer of the first example depend? 3. What is the first object of enquiry? 4. What will be the form of the question? 5. Statement? 5. What is the next enquiry? 7.

Rule 1. *State the question, that is, place the numbers so that the first and third terms may be of the same name, and the second term of the same nome with the answer, or thing sought.*

2. *Bring the first and third terms to the same denomination, and reduce the second term to the lowest denomination mentioned in it.*

3. *Divide the product of the second and third terms by the first term; the quotient will be the answer to the question, in the same denomination with the second term, which may be brought into any other denomination required.*

The Rule of Three Inverse is when more requires less, or less requires more, and may be known thus: more requires less when the third term is more than the first, and requires the fourth term, or answer, to be less than the second; and less requires more when the third term is less than the first, and requires the fourth term to be more than the second.

Rule. *State and reduce the terms as in the Rule of Three Direct; then multiply the first and second terms together, and divide their product by the third term; the quotient will be the answer, in the same denomination with the middle term*

8. In performing these examples, we in the first place multiplied 60 by 10, and divided the product by 5, making the 120 in the second statement; then we multiplied the 120 by 9, and divided the product by 3.

9. But since the result will be the same, we may as well multiply the 60 at once, by the product of the two multipliers, 9 and 10, and divide this result, (5,400,) by the product of the two divisors, 3 and 5, n which case the two statements may be incorporated into one, and performed as follows—

```
 5 days  : 1 0 days }   miles          Then 6 0 miles
 3 hours     9 hours } : : 6 0                  9 0
 -------     -------                       ---------
 1 5         9 0                     1 5 ) 5 4 0 0
                                        A. 3 6 0 miles.
```

10. Or, since the ratio of 5 to 10 is $\frac{10}{5}$, or 2, he will travel, in 10 days, (other things being equal,) 2 times as far as in 5 days, that is, 2 times 60, or 120 miles. And since the ratio of 3 hours to 9 hours is $\frac{9}{3}$, or 3, he will travel 3 times as far, when he travels 9 hours each day, as when he travels only 3 hours each day; that is, 3 times 120, or 360 miles.

11. The last process consists in multiplying the third term first by one ratio, and that product by the other; but the effect is the same, if we multiply the 60 at once by the product of the two ratios, 2 and 3, or 6, thus—$6 \times 60 = 360$ miles, answer.

12. Hence the method of stating is the same in principle as that of Simple Proportion.

13. If a man travel 7,800 miles in 260 days, traveling 4 hours each day, how many miles would he travel in 390 days, traveling 8 hours each day?

```
Days   2 6 0 :   3 9 0 }      miles
Hours      4 :       8 } : : 7 8 0 0
      -------   -------
     1 0 4 0   3 1 2 0
```

Then $7{,}800 \times 3{,}120 \div 1{,}040 = 23{,}400$ miles, answer.

We write 7,800 for the 3d term, because it is like the answer. Then we take 2 terms of the same kind, say 260 days and 390 days, and because he would go farther in 390 days than in 260 days, we write the greater for the 2nd term and the smaller for the 1st. Next, taking the other two terms, they being of the same kind, and because he would go farther when he travels 8 hours a day than when he travels only four hours a day, we write the greater for a 2nd term, and the smaller for a first term. Then the third term, multiplied by the product of the second terms, and the result divided by the product of the first terms, gives the answer.

14. This process may be shortened, as in Simple Proportion, by

---

Q. How are the two statements performed by one operation? 8, 9. How is the same done by Ratio? 10, 11. On what principle does the statement proceed? 12. How is example 13 done by statement?—by analysis?—by abbreviating the statement? 14.

substituting the quotients arising from dividing any two terms by their greatest common divisor.

Thus 260 and 390 divided each by 130=2 and 3.

And 4 and 8 divided each by 4=1 and 2.

$$\text{Then}\left.\begin{array}{c}2:3\\ \underline{1:2}\\ 2:6\end{array}\right\}::\overset{\text{miles}}{7{,}800}.\quad \text{Next } 7{,}800\times 6\div 2=23{,}400 \text{ miles, Ans.}$$

15. Again, since the two factors 2 and 6 have a common divisor, 2, we may substitute in their stead their respective quotients, which are 1 and 3, thus—1 : 3 : : 7,800; then 7,800 × 3 = 23,400. *A.* 23,400m.

16. THE SAME BY ANALYSIS.—He would travel in 1 day $\frac{1}{260}$ of 7,800 miles, which is 30 miles; and in 1 hour $\frac{1}{4}$ of 30 miles, which is $7\frac{1}{2}$ miles; then in 390 days, 390 times $7\frac{1}{2}$, or 2,925 miles; and by traveling 8 hours each day, 8 times 2,925 miles, which is 23,400 miles, the answer.

RECAPITULATION.

17. COMPOUND PROPORTION is when the relation of the required quantity to the given quantity depends on several circumstances combined.

18. COMPOUND RATIO is that which results from multiplying two or more simple ratios together.

19. COMPOUND PROPORTION is sometimes called the DOUBLE RULE OF THREE, because it embodies in a single process all those terms which by SIMPLE PROPORTION or the SINGLE RULE OF THREE would require two or more separate statements.

RULE.

20. *Select that number which is of the same kind with the answer for the third term, and take of the remaining numbers any two of the same kind, and arrange them as in single proportion; then take two more of a kind, and arrange them in like manner; and so on, till all are used: then multiply the third term by the continued product of the second terms, and divide the result by the continued product of the first terms.*

21. When the terms of any couplet, or their products, have a common divisor, divide by it, and substitute their quotients for the terms themselves; after which, multiply and divide by the third term as above directed.

22. BY RATIO. Multiply the third term by the product of the ratios of the other terms, recollecting to cancel equal terms in all practicable cases.

23. If 5 men can build 90 rods of wall in 6 days, how many rods can 20 men build in 18 days?

$$\left.\begin{array}{l}\text{Men } 5:20\\ \text{Days } 6:18\end{array}\right\}::\overset{\text{rods}}{90}\qquad 5\times 6=30:18\times 20=360 \text{ then } 90\times 360\div 30=1080.\quad A.\ 1{,}080 \text{ rods}$$

Q. What is Compound Proportion? 17 Compound Ratio? 18. Double Rule of Three? 19 General Rule? 20.

24. Or if we divide 5 and 20 each by 5, and 6 and 18 each by 6, we shall have the following statement:—

1 : 4 } rods.
1 : 3 } : : 90. Then $3 \times 4 \times 90 = 1,080$. *A.* 1,080 rods.

25. THE SAME BY RATIO.—The ratio of 5 to 20 is $\frac{20}{5}$ or 4, and that of 6 to 18 is 3; then $4 \times 3 \times 90 = 1,080$. *A.* 1,080 rods.

26. THE SAME BY ANALYSIS.—1 man will build $\frac{1}{5}$ of 90 rods, or 18 rods, in 6 days, and in 1 day $\frac{1}{6}$ of 18 rods, or 3 rods; then 20 men will build 20 times 3 rods, or 60 rods, in 1 day, and in 18 days, 18 times 60 rods, or 1,080 rods, answer.

27. If 10 men can build a wall 360rds. long in 9 days, how many rods of wall could 75 men build in 24 days? *A.* 7,200 rds.=22m. 4fur.

28. If a man travel 100 miles in 5 days, traveling 4 hours each day, how far could he go in 12 days, provided he travels 10 hours each day? *A.* 600 miles.

29. If 40 men could cradle, in 10 days, 800 acres of rye, how many acres could 60 men cradle in 15 days? *A.* 1,800 acres.

30. If 75 men can build a wall 7,200 rods long in 24 days, how many rods of wall could 10 men build in 9 days? 75 men will build more wall than 10 men, therefore write 10 for the second term, and 75 for the first term. *A.* 360 rods.

31. If a man travel 100 miles in 5 days, traveling 4 hours each day, in how many days could he travel 600 miles, provided he travel 10 hours each day?

32. By analysis, it appears that when he travels 10 hours each day, he goes 50 miles a day; then $600 \div 50 = 12$. In stating, take notice, that the more hours he travels in a day, the less days will be required; therefore make the second term the smaller one. *A.* 12d.

33. If 10 men can build a wall 360 rods long in 9 days, how many men would be required to build a wall 7,200 rods long in 24 days? Make 9 days the second term, because the more men the less days. *A.* 75 men.

34. If a family of 8 persons spend $480 in 24 months, how much would 16 persons spend in 8 months? *A.* $320.

35. If a family of 16 persons spend, in 8 months, $320, how many persons would spend, in 24 months, $480? *A.* 8 persons.

36. If 4 men receive $24 for 6 days' work, how much would 8 men receive for 12 days' work? *A.* $96.

37. If 4 men receive $24 for 6 days work, how many men may be hired 12 days for $96? *A.* 8 men.

38. If $2,000 will support a garrison of 150 men 3 months, how long will $6,000 support 4 times as many men? Ratios 3 and $\frac{1}{4}$. *A.* $2\frac{1}{4}$ months.

39. If $100 gain $6 interest in 1 year, in what time will $900 gain $36 interest? *A.* 8 months.

Q. How is example 23 done by statement?—by ratio?—by analysis?—by abbreviating the statement? 24. (The teacher can select other examples, to be performed in a similar manner.)

40. An usurer put out \$150 at interest, and when it had been on interest 8 months, he received for principal and interest \$160. What rate per cent. did he receive; that is, how many dollars on \$100 for 12 months? *A.* 10 per cent.

41. When the amount of \$18,000 for 2y. 6m. 15d. is \$20,745, what is the rate per cent.? *A.* 6 per cent.

42. Suppose you pay \$5.712 for transporting 12cwt. 3qr. 400 miles, what must you pay for transporting 13T. 7cwt. 3qr. over a distance of only 75 miles? 13T. 7cwt. 3qr.=1,071qr. : 12cwt. 3qr.= 51qr. Ratios 21 and $\frac{3}{16}$. *A.* 22.491.

43. When you pay \$22.491 for transporting 13T. 7cwt. 3qr.,75 miles, what must you pay for transporting 12cwt. 3qr.,400 miles? *A.* \$5.712.

44. When the transportation of two boxes, each weighing 2cwt. 3qr. 5lb. 200 miles costs \$5.60, what will be the cost of transporting 2T. 4cwt. 3qr. 5lb. 150 miles? Ratios 8 and $\frac{3}{4}$. *A.* \$33.60.

45. If 45 yards of cloth, 5 quarters wide, will make 10 suits of clothes, how many pieces, each containing 25 yards, but only 3 quarters wide, will be required to make 50 suits? *A.* 15 pieces.

46. If 25 men can dig a trench 36 feet long, 12 feet broad, in 9 days, in how many days would 15 men dig a trench of the same depth, but 48 feet long, and only 8 feet broad?

15 men : 25 men				The less men, the more days.
36 length : 48 length	days :: 9	For	The more length, the more days.
12 width : 8 width				The less width, the less days.

47. The multiplying of the third term by the product of the middle terms, and dividing the result by the product of the first terms, gives $13\frac{1}{3}$ days.

THE SAME BY RATIO.—The ratio of 15 to 25 is $\frac{5}{3}$; of 36 to 48, $\frac{4}{3}$; of 12 to 8, $\frac{2}{3}$. The operation, then, may be expressed as follows—

$\frac{5}{3}\times\frac{4}{3}\times\frac{2}{3}\times\frac{9}{1}=\frac{360}{27}=13\frac{1}{3}$. *A.* $13\frac{1}{3}$ days.

48. If 8 men build a wall 80 rods long and 5 feet thick in 6 days, in how many days would 3 men build a wall of the same breadth, but 120 rods long and 2 feet thick? The ratios $\frac{8}{3}$, $\frac{3}{2}$, and $\frac{2}{5}$, (by canceling,)$=\frac{8}{5}$, then $\frac{8}{5}\times 6=\frac{48}{5}=9\frac{3}{5}$. *A.* $9\frac{3}{5}$ days.

49. If 15 men dig a trench 48 feet long and 8 feet broad in $14\frac{2}{5}$ days, in how many days would 25 men dig a trench of the same depth, but 36 feet long and 12 feet wide? *A.* $9\frac{18}{25}$ days.

50. If 25 men, by working 10 hours a day, can dig a trench 36 feet long, 12 feet broad, and 5 feet deep, in 9 days, how many hours a day must 15 men work, in order to dig a trench 48 feet long, 8 feet broad, and 3 feet deep, in 12 days?

15 men : 25 men,		The less men, the more days.
36 length : 48 length,		The more length, the more hours
12 breadth : 8 breadth,	hours. :: 10	The less breadth, the less hours
5 depth : 3 depth,		The less depth, the less hours.
12 days : 9 days,		The more days, the less hours.

51. The ratios in the last statement are $\frac{5}{3}\times\frac{4}{3}\times\frac{2}{3}\times\frac{3}{5}\times\frac{3}{4}=\frac{2}{3}$; then $\frac{2}{3}\times 10=6\frac{2}{3}$ hours. Or the product of the first terms is 388,800, and that of the second terms 259,200, which multiplied by 10, and the result divided by the first product, gives $6\frac{2}{3}$. *A.* $6\frac{2}{3}$ hours.

52. If 15 men, by working $6\frac{2}{3}$ hours a day, can dig a trench 48 feet long, 8 feet broad, and 3 feet deep, in 12 days, how many hours a day must 25 men work, in order to dig a trench 36 feet long, 12 feet broad, and 3 feet deep in 9 days? *A.* 10 hours.

53. Suppose that 50 men, by working 3 hours each day, can dig, in 45 days, 24 cellars, which are each 36 feet long, 21 feet wide, and 20 feet deep, how many would be required to dig, in 27 days, 18 cellars, which are each 48 feet long, 28 feet wide, and 15 feet deep provided they work only 5 hours each day?

24 cellars : 18 cellars,			The less cellars, the less men.
36 length : 48 length,			The more length, the more men.
21 width : 28 width,	men.		The more width, the more men
20 depth : 15 depth,	: : 50.		The less depth, the less men.
27 days : 45 days,			The less days, the more men.
5 hours : 3 hours,			The more hours, the less men.

54. In the last example, either product of the terms is 48,988,800, and the ratios just cancel each other; the third term, then, is the answer as it stands. *A.* 50 men.

55. If 80 men, by working 5 hours in a day, can dig, in 27 days, 20 cellars, which are 45 feet long, 28 feet wide, 10 feet deep, how many men would dig, in 45 days, 36 cellars, which are 30 feet long 21 feet wide, and 15 feet deep, by working only 3 hours each day? *A.* 108 men.

CONJOINED PROPORTION.

XCII. 1. Conjoined Proportion, or Chain Rule, as it is sometimes called, is the combination of several ratios, or proportions through which the ratio between the first and last term is discovered

2. This rule relates principally to the exchanges between different countries, in respect to specie, weights, and measures, but is applicable to common business transactions.

3. To find how much of the quantity mentioned first is equal to a certain portion of the quantity mentioned last.

RULE.

4. *Call the first terms in each part of the general question antecedents, and the following ones consequents; then place the antecedents in a column on the left, and the consequents in another column on the right, with the odd term under the former column.*

Q How is example 46 stated, and why?—example 50, and why?—example 53, and why? How is the last performed by cancelling? 53.

XCII. Q. What is Conjoined Proportion? 1. To what does it principally relate? 2. What is the first rule? 4, 5 What does it find? 3.

5. *Multiply the column of consequents together for a divisor, and that of the antecedents for a dividend; the quotient will be the quantity sought. Or, first reject opposite and equal terms, or opposite and equal factors, by dividing by the greatest common divisor; then proceed with the remaining terms as first directed.*

6. If 10bl. of flour may be bought for 30bu. of wheat, and 20bu. of wheat for 10 yards of broadcloth, and 60 yards of broadcloth for 120gal. of brandy, and 30 gallons of brandy for 60 barrels of cider, how many barrels of flour will purchase 12 barrels of cider?

10bl. flour = 30bu. wheat.	
20bu. wheat = 10yd. cloth.	$30\times10\times120\times60=2{,}160{,}000$; 10
60yd. cloth = 120gal. brandy.	$\times20\times60\times30\times12=4{,}320{,}000\div$
30gal. brandy = 60bl. cider.	$2{,}160{,}000=2$. *A.* 2bl. flour.
12bl. cider.	

But, by cancelling equal terms, we have left only—

20bu. wheat : 120gal. brandy.	Then $20\times12=240\div120=2$.
12bl. cider.	*A.* 2bl. flour

Again, if we reject common factors, by dividing 20 and 120 by 20 we have only 1, 12, and 6. Then $1\times12=12\div6=$ 2bl. flour.

Again, dividing the opposite terms 12 and 6 by 6, we have only 1 and 2 and 1. *A.* 2bl. flour.

7. For the proof, suppose the flour to be worth $6 a barrel; then find, on that supposition, the cost of a single one of each quantity; and lastly, find whether 2bl. of flour are worth just 12 barrels of cider. *A.* Wheat at $2; cloth, $4; brandy, 2; cider, $1. Then 2bl. of flour are worth 12bl. of cider, the above answer.

8. If 280 braces at Venice are worth 300 braces at Leghorn, and 7 braces at Leghorn are worth 4 yards at Boston, (U. S.) how many braces at Venice are worth 100 yards at Boston? *A.* 163$\frac{1}{3}$yd.

9. If 12lb. at Boston are equal to 10lb. at Amsterdam, and 10lb. at Amsterdam are equal to 12lb. at Paris, how many pounds at Boston are equal to 500 at Paris? Cancel equal terms, and the work is done. *A.* 500.

10. Suppose 200 bushels of wheat in Boston (Mass.) are worth 300bu. in New York, and 20 in New York are worth 40 in Ohio, and 60 in Ohio are worth 75 in Michigan, and 25 in Michigan are worth 30 in Illinois, how many bushels in Boston are worth 1,000 bushels in Illinois? *A.* 222$\frac{2}{9}$.

11. Suppose 20 girls in a factory can do as much work as 15 boys, and 60 boys as much as 25 men, how many girls would accomplish as much as 250 men? *A.* 800 girls.

12. If 15s. in N. England be the same value as 20s. in N. York, and 24s. in N. York the same as 22s. 6d. in N. Jersey, and 30s. in N. Jersey the same as 20s. in Canada, h[illegible] many pounds in N. England are the same value as £240. 7s. 6d. [illegible] anada? *A.* £288. 9s.

13. If 15 quarts of [illegible]ilk will make [illegible]unds of butter, and 6 pounds of butter require as m[illegible]h milk as 5[illegible]ds of cheese, and 40 pounds

of cheese be made from 2 cows in 3 days, how many quarts of milk, at that rate, would 25 cows give in 6 months? *A.* 45,000 qts.

14. If in the last example the milk sell for 5 cents a quart, the butter for 25 cents a pound, and the cheese for 10 cents a pound, which would be the most profitable, for the given time, 6 months, the selling of the milk, the making of butter, or the making of cheese?

A. The making of cheese is more profitable than the selling of milk by $\frac{1}{4}$, or $750; and more profitable than the making of butter by $\frac{1}{2}$, or $1500.

15. To find how much of the thing mentioned last, is equal to a certain quantity of the thing mentioned first.

16. RULE. *Arrange the terms as before, except the odd term, which place under the column of consequents, then cancel, multiply, and divide the columns as directed in the last rule.*

17. In the example (No. 6,) find how many barrels of cider will purchase 20 barrels of flour? *A.* 120bl. cider.

18. In the example (No. 8,) find how many yards at Boston, are worth 250 braces at Venice. *A.* $153\frac{3}{49}$.

19. In the example (No. 9,) how many pounds, at Paris, are equal to 600 pounds at Boston? *A.* 600.

20. In the example (No. 10,) how many bushels, in Illinois, are worth 500 bushels in Boston? *A.* 2,250.

21. In the example (No. 11,) find how many men would accomplish as much as 480 girls. *A.* 150 men.

22. In the example (No. 12,) how many pounds in Canada are equal to £250.12s. 6d. in New England? *A.* £208.17s. 1d.

23. In the example (No. 13,) find how many cows would give $5\frac{5}{9}$ kilderkins of milk in one day. *A.* 40 cows.

FELLOWSHIP.

XCIII. 1. FELLOWSHIP, which is another name for the Rule of Three, is employed by persons in partnership in ascertaining their respective gain or loss in trade, when these are in proportion to the stock, or stock and time together.

2. SINGLE OR SIMPLE FELLOWSHIP is when the stock of each partner is continued in trade for equal periods of time; each one's gain or loss, therefore, is evidently in proportion to his stock in trade.

3. This rule may be applied to cases of bankruptcy and taxation, in apportioning the part of each person interested.

4. RULE. *As the whole stock : is to each man's stock :: so is the whole gain or loss : to each man's gain or loss.*

Q. What is the second rule? 16. What does it find? 15.

XCIII. Q. What is Fellowship? 1. Single or Simple Fellowship? 2. What other cases does it embrace? 3. What is the Rule? 4. What abbreviation may be used? 5.

5. NOTE. The operation may oftentimes be much abridged by Analysis, or by multiplying the third term by the ratio of the other two.

6. Three men, A, B, and C, traded in company; the first put in $200, the second $400, and the third $600. They gained $300. What was each man's share of the gain?

A's stock $200 } $1200 : $200::$300 : $50 Ans. A's gain.
B's stock $400 } $1200 : $400::$300 : $100 Ans. B's gain.
C's stock $600 } $1200 : $600::$300 : $150 Ans. C's gain.

The same by ratio. There are $\frac{200}{1200}$, $\frac{400}{1200}$, $\frac{600}{1200}$, $=\frac{1}{6}$, $\frac{1}{3}$, $\frac{1}{2}$, then $\frac{1}{6}$ of $300 is $50; $\frac{1}{3}$ of $300 is $100; $\frac{1}{2}$ of $300 is $150.

A. $50; $100; $150.

The same by analysis. If $1200 gain $300, then $1 will gain $\frac{1}{1200}$ of $300, which is $$\frac{1}{4}$, and $200 will gain 200 times as much, which is the same as $\frac{1}{4}$ of $200=$50; $\frac{1}{4}$ of $400=$100; $\frac{1}{4}$ of $600=$150.

A. $50; $100; $150.

7. Three merchants, A, B, and C, gained by trading in company $200; A's stock was $150; B's stock $250, and C's $400; what was the gain on $1, and what each man's gain?

A. $$\frac{1}{4}$; then A's $37.50; B's $62.50; C's $100.

8. A, B, and C freight a ship with 270 tons; A shipped on board 96 tons, B 72, and C 102. In a storm the captain was obliged to throw 90 tons overboard. What was the loss on one ton, and what the loss of each man? *A.* $\frac{1}{3}$T.; A's 32T.; B's 24T.; C's 34T.

9. A and B traded in company, with a joint capital of $600. A put in $350.50, and B $249.50. They gained $120. What was that on $1, and what portion belonged to each?

A. $$\frac{1}{5}$; A's $70.10; B's $49.90.

10. A ship valued at $25,200 was lost at sea. A owned $\frac{1}{3}$ of it; B $\frac{1}{2}$, and C the rest. What was the loss of each man, provided an insurance of $18,000 had been effected on her?

A. $$\frac{2}{7}$ on $1; A's $2,400; B's $3,600; C's $1,200.

11. A detachment, consisting of 5 companies, was sent into a garrison, in which the duty required 228 men a day; the first company consisted of 162 men; the second, 153; the third, 144; the fourth, 117; and the fifth, 108. How many men must each company furnish in proportion to the whole number of men? (The proportion for 1 man is $\frac{1}{3}$; then, $\frac{1}{3}$ of 162=54, first company; the second, 51; the third ,48; the fourth, 39; and the fifth, 36 men.)

A. 54; 51; 48; 39; 36.

12. Two men, A and B, traded in company, with a joint capital o $1,000: they gained $400, of which A took $300 and B the remainder. What was each person's stock?

A. $1 gain on $$2\frac{1}{2}$ stock; A's $750; B's $250.

13. A bankrupt is indebted to A $350, to B $1,000, to C $1,200, to D $420, to E $85, to F $40, and to G $20; his whole estate is worth no more than $1,557.50. What will be each creditor's part of the property?

14. NOTE. In adjusting such claims, it is the general practice to find how much the debtor pays on $1, first. *A.* $1=$$\frac{1}{2}$; A, $175, B, $500; C, $600; D, $210; E, $42.50; F, $20; G, 10.

15. A wealthy merchant at his death, left an estate of 30,000 to be divided among his children in such a manner that their shares should be to each other as their ages, which were 7, 10, 12, 15 and 16 yr's. What was the share of each? *A.* $3,500; $5,000; $6,000; $7,500; 8,000.

16. A and B invest equal sums in trade and clear $220, of which A is to have 8 shares, because he is to transact the business, and B only 3 shares: what is each man's gain, and what allowance is made A for his time? *A.* $60; A $100 for his time.

17. ASSESSMENT OF TAXES. A TAX is a rate or sum of money which is paid for the support of government by the citizens, in proportion to their property, except that on their heads, which is called a poll tax.

18. RULE. *Having taken an inventory or valuation of all the taxable property of the town, and the number of polls, deduct from the whole tax, the poll tax (assessed equally on all) then find how much the remainder is on one dollar of the said inventory for the multiplier of each person's inventory, and to the product add his poll tax; the sum of which will be his whole tax.*

19 Suppose that a certain town which has $500,000 of taxable property, and 2,000 polls, which are taxed $.70 each, is assessed at $21,400:—

What is A's tax, whose list is $1,400 and 2 polls?* *A.* $57.40.
What is B's tax, whose list is $1,200 and 2 polls? *A.* $49.40.
What is C's tax, whose list is $1,265 and 1 poll? *A.* $51.30.
What is D's tax, whose list is $2,125 and 3 polls? *A.* $87.10.
What is E's tax, whose list is $3521 and 2 polls? *A.* $142.24.
What is F's tax, whose list is 825\frac{1}{2}$ and 3 polls? *A.* 35.12.
What is G's tax, whose list is 800\frac{40}{100}$ and 2 polls? *A.* 33.41$\frac{6}{10}$.
What is H's tax, whose list is 375\frac{1}{4}$ and 1 poll? *A.* 15.71.
What is I's tax, whose list is 265\frac{3}{10}$ and 2 polls? *A.* 12.01\frac{2}{10}$.

* 2,000 polls at 70 cts. each=$1,400 from $21,400 leaves $20,000 to be assessed on$500,000; which on 1 dollar is $$\frac{20000}{500000}$, or $\frac{1}{25}$ of $1=4 cts.; that is 4 cents on a dollar. Then A's list being $14,00 ×4 cents=$56 which added to $140, A's tax on 2 polls at 70 cents each, makes $57.40 for A's whole tax, as above.

COMPOUND FELLOWSHIP.

XCIV. 1. COMPOUND FELLOWSHIP is when the stock of each partner is employed for *unequal periods* of time: each one's gain or

Q. What is a tax? 17. What is the first requisite? 18. What deduction is first to be made? 18. How is the poll tax apportioned? 18. Describe the rest of the process in finding an individual's tax. 18.

loss therefore is in proportion to both his stock and the time it is continued in trade.

2. Two men hired a pasture for $9; A put in 2 oxen for six months, and B 3 oxen for 5 months; what ought each to pay for the pasture?

3. Two oxen for 6 months is the same as (2×6=) 12 oxen for 1 month, and 3 oxen for 5 months is the same as (3×5=) 15 oxen for 1 month: thus,—

2×6=12 } 27 : 12 : $9 : $4 A's Ans.
3×5=15 } 27 : 15 : $9 : $5 B's Ans.

RULE.

4. *Having multiplied each man's stock by the time it was in trade, then say as the sum of these products is to each man's product, so is the whole gain or loss, to each man's gain or loss.*

5. Three merchants, A, B, and C, enter into partnership; A puts in $60 for 4 mo.; B $50 for 10 mo., and C $80 for 12mo.; but by misfortune they lose $50; how much loss must each man sustain?
A. A's $7.058+; B's $14.705+; C's $28.235+.

6. Three butchers hire a pasture for $48; A puts in 80 sheep for 4mo.; B 60 sheep for 2mo, and C 72 sheep for 5mo.; what share of the rent must each man pay? *A.* A's $19.20; B's $7.20; C's $21.60.

7. Two merchants entered into partnership for 16mo.; A at first put in stock to the amount of $600, and, at the end of 9 months, put in $100 more; B put in at first $750, and, at the expiration of 6 months, took out $250; with this stock they gained $386: what was each man's part? *A.* A's, $200.797; B's, $185.202.

8. On the first of January, A began to trade with $760, and, on the first of February following, he took in B with $540; on the first of June following, they took in C with $800; at the end of the year, they found they had gained $872; what was each man's share of the gain? *A.* A's, $384.929; B's, $250.71; C's, $236.36.

XCIV. *Q.* What is Compound Fellowship? 1. In example 2, what number of oxen for 1 month is equal to the given number for the given time? What is the Rule? 4.

APPENDIX.

PART THIRD.

PRACTICE.

XCV. 1. PRACTICE is a concise method of answering questions in the Rule of Three, when the first term happens to be unity.

2. Operations in Practice are conducted principally by supposing a price, and taking aliquot or even parts of the same for the true price. [XLII. 1.]

3. What will 50 bushels of rye cost at 5s. a bushel? Suppose the price were £1 per bushel, then the 50 bushels would cost £50; but at 5s. per bushel only $\frac{1}{4}$ as much for 5s. = £$\frac{1}{4}$.

5s. = £$\frac{1}{4}$)£50
£12.10s.

4. What will be the cost of 8640 yards of cloth at the following prices.

At 10 shillings per yard? = £$\frac{1}{2}$	A. £4,320.
At 6s. 8 pence per yard? = £$\frac{1}{3}$.	A. £2,880.
At 4 shillings per yard? = £$\frac{1}{5}$.	A. £1,728.
At 3s. 4 pence per yard? = £$\frac{1}{6}$.	A. £1,440.
At 2s. 6 pence per yard? = £$\frac{1}{8}$.	A. £1,080.
At 1s. 8 pence per yard? = £$\frac{1}{12}$.	A. £720.
At 1s. 3 pence per yard? = £$\frac{1}{16}$.	A. £540.
At 1 shilling per yard? = £$\frac{1}{20}$.	A. £432.
At 10 pence per yard? = £$\frac{1}{24}$.	A. £360.
At 8 pence per yard? = £$\frac{1}{30}$.	A. £288.
At 5 pence per yard? = £$\frac{1}{48}$.	A. £180.
At 2$\frac{1}{2}$ pence per yard? = £$\frac{1}{96}$.	A. £ 90.

6. What is the cost of the following quantities at the prices annexed?

3,150 gallons of oil at 2s. 6d. per gallon?	A. £393. 15s.
4,235 yards of cloth at 3s. 4d. per yard?	A. £705. 16s. 8d.
2,434 bushels of oats at 1s. 8d. per bushel?	A. £202.16s.. 8d.
2,678 dozen of oranges at 5 pence for each?	A. £669. 10s.
4,595 quarts of strawberries at 3d. per quart?	A. £57. 8s. 9d.

7. When the price is not an aliquot part, we may take the one nearest to it first, then take an aliquot part of that part, and so on.

XCV. Q. What is Practice? How is it performed? How is example 3 performed? 3. What are the divisor and dividend when the quantity is 8640 and the price 5s.?—is 10s.?—6s. 8d.—4s.?—3s. 4d.?—2s. 6d.?—1s. 3d.?—5d.? When the price is not an aliquot part of the given quantity, what is the direction? 7.

8. What will 51 barrels of cider cost at 7s. 6d. per barrel? (2s. 6d. =£$\frac{1}{8}$ or $\frac{1}{2}$ of 5s.)

$\frac{1}{8}$, $\frac{1}{4}$)	£ 5 1 at £ 1 per bl.	Or $\frac{1}{4}$)	£ 5 1
	£ 1 2 . 1 5 s. at 5 s.	$\frac{1}{2}$)	£ 1 2 . 1 5 s
	£ 6 . 7 s. 6 d. at 2 s. 6 d.		£ 6 . 7 s. 6 d
A.	£ 1 9 . 2 s. 6 d. at 7 s. 6 d.	A.	£ 1 9 . 2 s. 6 d

9. What will the following articles cost at the prices annexed?—

724 desks at 12s 6d. each?	*A.* £452 . 10s.
140 chairs at 15s. 3d. each?	*A.* £106 . 15s.
936 razors at 8s. 6d. each?	*A.* £397 . 16s.
812 books at 3s. 9d. each?	*A.* £152 . 5s.
715 bonnets at 17s. 6d. each?	*A.* £625 . 12s. 6d

10. What will 11cwt. 3qr. 13lb. of rice cost at $9.60 per cwt.?

2 qr. =$\frac{1}{2}$	$ 9 . 6 0	=cost of 1 cwt.
	1 1	
	1 0 5 . 6 0	=cost of 11 cwt.
1 qr. =$\frac{1}{2}$)	4 . 8 0	=cost of 2 qr.
10 lb. =$\frac{2}{5}$)	2 . 4 0	=cost of 1 qr.
2lb. =$\frac{1}{5}$)	. 9 6	=cost of 10 lb.
1 lb. =$\frac{1}{2}$)	. 1 9 2	=cost of 2lb.
	. 0 9 6	=cost of 1 lb.
Ans.	$ 1 1 4 . 0 4 8	=cost of 11cwt. 3qr. 13lb.

11. What would be the cost of the following quantities?—

25 yards 2 quarters at $2.40 per yard?	*A.* $61.20
18 bushels 3 pecks at $3.60 per bushel?	*A.* $67.50.
5cwt. 3qr. 5lb. at $4.20 per cwt.?	*A.* $24.36.
3cwt. 1qr. 24lb. at $3.60 per cwt.?	*A.* $12.564.
4T. 15cwt. 3qr. at $10.50 per ton?	*A.* 50.268\frac{3}{4}$.
5 e. E. 2qr. 3na. at $2.75 per ell?	*A.* 15.262\frac{5}{10}$.

12. Suppose a merchant buys 7hhd. 7gal. 2qt. of molasses at 10.62\frac{1}{2}$ per hhd., and sells $\frac{1}{3}$ of it for 11\frac{1}{4}$ per hhd.; $\frac{2}{3}$ of it for $12 per hhd. and the balance for $15 per hhd.; how much profit does he make on the whole? *A.* $12.457.+

DUODECIMALS.

XCVI. 1. DUODECIMALS are so called from *duodecim*, the Latin for *twelve*, because they decrease by *twelves* from the left hand towards the right.

2. In Duodecimals, the foot is divided first into twelve equal parts,

---

Q. How is example 8 performed? When the quantity is 11cwt. 3qr. 13lb., what are the several divisors? [See 10.]

XCVI. Q. What are Duodecimals? 1. What is the integer and its divisions? 2

called inches or primes; each prime into 12 equal parts, called seconds; each second into 12 equal parts, called thirds, and so on.

3. That is, 1 inch or prime is $\frac{1}{12}$ of a foot.

1 second is $\frac{1}{12}$ of $\frac{1}{12}$, that is $\frac{1}{144}$ of a foot.

1 third is $\frac{1}{12}$ of $\frac{1}{12}$ of $\frac{1}{12}=\frac{1}{1728}$ of a foot.

1 fourth is $\frac{1}{12}$ of $\frac{1}{12}$ of $\frac{1}{12}$ of $\frac{1}{12}=\frac{1}{20736}$ of a foot.

4. These fractions are distinguished usually by marks called accents; thus $8'=8$ inches or primes; $8''=\frac{8}{144}$ or 8 seconds; $8'''=\frac{8}{1728}$ or 8 thirds, &c., each additional mark denoting an inferior denomination.

5. Since feet stand in the place of units, feet multiplied by feet must give feet; feet multiplied by 12ths must give 12ths, that is inches or primes, and so on as follows :—

6. Feet multiplied by feet give feet.
Feet multiplied by primes give primes.
Feet multiplied by seconds give seconds.
Primes multiplied by primes give seconds.
Primes multiplied by seconds give thirds.
Seconds multiplied by seconds give fourths.
Seconds multiplied by thirds give fifths.
Thirds multiplied by thirds give sixths, &c.

7. That is, the product will always be of that denomination which is indicated by the sum of the accents; thus, $7''' \times 5''''=35'''''''$ or 35 sevenths.

TABLE OF SOME OF THE HIGHER DENOMINATIONS.

$12''''''$	(sixths)	make $1'''''$	(fifth.)
$12'''''$	(fifths)	make $1''''$	(fourth.)
$12''''$	(fourths)	make $1'''$	(third.)
$12'''$	(thirds)	make $1''$	(second.)
$12''$	(seconds)	make $1'$	(prime.)
$12'$	(primes)	make 1ft.	(foot.)

8. The operations of addition, subtraction, multiplication, division and reduction of duodecimals are the same as of other compound numbers, 12 of an inferior denomination invariably making one of the next higher denomination, as in the foregoing Table.

9. How many feet are there in 1,685′? *A.* 140ft. 5′.

10. How many primes are there in 140ft. 5′? *A.* 1685′.

11. How many feet are there in 31,049″? *A.* 215ft. 7′ 5″.

12. How many thirds in 23ft. 4′7″8‴? *A.* 40,412‴.

13. How many feet in 2,985,984′′′′′′? *A.* 1 foot.

14. How many feet in 1,504,935,936′′′′′′′′? *A.* $3\frac{1}{2}$ feet.

15. How many ninths in 7 feet? *A.* 36,118,462,464′′′′′′′′′.

Q. What parts of a foot are these sub-divisions? 3. How are these denominations distinguished? 4. What do feet, primes, &c., multiplied by each other form? 6. How can the denomination of the product be determined? Repeat the Table.

16. Add together 425ft. 4′ 8″ 7‴ 5⁗ 11′′′′′ 9′′′′′′. 125ft. 3′ 4″ 9‴ 2⁗ 3′′′′′ 7′′′′′′ and 43ft. 2′ 5″ 11‴ 3⁗ 6′′′′′ 10′′′′′′.

A. 593ft. 10′ 7″ 3‴ 11⁗ 10′′′′′ 2′′′′′′.

17. If a stick of timber which contains 39ft. 2′ 3″ 9‴ be divided into two parts, one of which shall contain 23ft. 8′ 1″ 10‴, what will the other part contain? *A*. 15ft. 6′ 1″ 11‴.

18. Suppose that a person agreed to furnish at a certain price, 15 sticks of round timber, each to contain in solid measure 90ft. 3′ 5″ 6‴; also 30 other sticks, each measuring 101ft. 2′ 6″ 9‴; but on its delivery $\frac{2}{5}$ of the whole was rejected on the ground that it did not answer the description in the contract; what was the quantity received?

A. 52T. 34ft. 5′ 3″.

CROSS MULTIPLICATION OF DUODECIMALS.

19. Duodecimals are principally used by workmen and artificers in ascertaining the square or solid contents of their work.

20. The square content, we have seen, [VII. 46.] is the product of the length by the breadth; and the solid content, the square contents, multiplied by the depth or thickness. [VII. 60.]

22. The principle illustrated in (6.) which see, forms the basis of the following rule.

RULE.

22. *Having written feet under feet, primes under primes, &c., multiply by each denomination separately, beginning with the highest of the multiplier and the lowest of the multiplicand.*

23. *Place those products that are of the same denomination under each other, which will carry the first denomination in each successive product after the first, one place farther toward the right than the former; then the sum of these partial products will form the required product.*

24. In a stick of timber 20ft. 9′ long, 2ft. 5′ wide and 2ft. 3′ thick how many solid feet does it contain?

```
    2 0 ft.  9′
      2  .  5′
    ----------------
    4 1  .  6′
      8  .  7′ . 9″
    ----------------
    5 0  .  1′ . 9″
      2  .  3′
    ----------------
  1 0 0  .  3′   6″
    1 2  .  6′ . 5″ . 3‴
    ---------------------
A.1 1 2  .  9′ 1 1″ . 3‴
    =====================
```

* For the value of each product see 8; recollecting always to carry by 12; thus, 2ft.×9′=18′÷12=1ft. 6 primes; then 20ft.×2ft.=40ft.+1ft. (to carry)=41ft. Next 5′×9′=45″=3′ and 9″; 20ft×5′=100′+3′ (to carry) 103′=8ft. 7′ and add the products together. To multiply by 2ft. 3′, say 2ft.×9″=18″=1′ 6″ : 2ft.×1′=2′+1′=3′ : 2ft×50ft.=100ft. : 3′×9″=27‴=2″ 3‴ : 3′×1′=3″+2″=5 : 3′×50ft.=150′=12ft. 6′.

```
  2 0 ft.  9′
    2  .  5′
  -------------
    8  .  7′ . 9″
  4 1  .  6′
  -------------
  5 0  .  1′ . 9″
  =============
```

We begin on the left of the multiplier instead of the right, because it is more convenient, as may be seen by comparing the adjacent operation with the one above, with which it corresponds, except that we begin to multiply as usual.†

† Lacroix's method of illustration.

25. How many square ft. in a board 10ft. 8′ long and 1ft. 5′ broad? *A.* 15ft. 1′ 4″.

26. In a load of wood 8ft. 4′ long, 2ft. 6′ high, and 3ft. 3′ wide, how many solid feet? *A.* 67ft. 8′ 6″.

NOTE.—Artificers compute their work by different measures Glazing and masons' flat work are computed by the square foot; painting, paving, plastering, &c. by the square yard; flooring, roofing, tiling, &c. by the square of 100 feet; brick work by the rod of 16½ feet, whose square is 272¼; the contents of bales, cases, &c. by the ton of 40 cubic feet; and the tonnage of ships by the ton of 95ft

27. What will be the expense of plastering the walls of a room 8ft 6′ high, and each side 16ft. 3′ long, at 62½ cents per square yard? *A.* $38.368.

28. How many cubic feet in a block 4ft. 3′ wide, 4ft. 6′ long, and 3ft. thick? *A.* 57ft. 4′ 6″.

29. How much will a marble slab cost, that is 7ft. 4′ long and 1ft. 3′ wide, at $1.25 per foot? *A.* $11.458.

30. How many cubic feet of wood in a load 6ft. 7′ long, 3ft. 5′ high, and 3ft. 8′ wide? *A.* 82ft. 5′ 8″ 4‴.

31. What will the paving of a court-yard, which is 70ft. long and 56ft. 4′ wide, come to, at $.20 per square yard? *A.* $87.63, nearly.

32. How many solid feet are there in a stick of timber 70ft. long, 15′ thick, and 18′ wide? *A.* 131ft. 3′.

33. A man built a house consisting of 3 stories; in the upper story there were 10 windows, each containing 12 panes of glass, each pane 14′ long, 12′ wide: the first and second stories contained 28 windows, each 15 panes, and each pane 16′ long, 12′ wide: how many square feet of glass were there in the whole house? *A.* 700 sq. ft.

INVOLUTION.*

XCVII. 1. INVOLUTION is the process of finding powers. POWERS are the several products arising from multiplying any number by itself, and that product by the same number again, and so on.

2. Any number is called a first power of itself; but when it becomes repeatedly a factor in producing other powers, it is called their root,† because they seem, as it were, *to grow* out of it.

Q. By whom are Duodecimals used, and for what purpose? 19. How are the square and solid contents of any thing ascertained? 20. What is the Rule? 22, 23.
XCVII. *Q.* What is Involution? 1. What are meant by powers? 1. What, by first powers, second powers, &c.? 2.

---

* INVOLUTION, from the Latin *in*, for *in*, and *volvo*, *to roll*, signifies the act of enrolling, enwrapping, or involving; the state of being mixed or complicated; the raising of powers, because a given number thereby becomes, by repeated multiplications, *involved* in other numbers.

† ROOT. The part of a plant in the ground. Figuratively, the bottom or lower part; the origin, cause, ancestry; a primitive word or theme. The first power, because it forms the basis of all the succeeding powers.

3. The first product, because the same factor is used twice, is called the *second* power, or square; the next product, because the same factor is used three times, is called the *third* power, or cube; and so on, as follows—

Thus $3 = 3$, 1st power, or root.
$3\times3 = 9$, 2d power, or square.
$3\times3\times3 = 27$, 3d power, or cube.
$3\times3\times3\times3 = 81$, 4th power, or biquadrate.*
$3\times3\times3\times3\times3 = 243$, 5th power.

4. The Index or Exponent of a power denotes the number of times the root must be used as a factor to produce that power; consequently, *index* is only another name for the number of the power

5. Powers are frequently expressed by writing their indices in smaller figures on the right of their respective roots, as—

The 8th power of 2 is $2^8 = 2\times2\times2\times2\times2\times2\times2\times2 = 256$.
The 2d power of 5 is $5^2 = 5\times5 = 25$.
The 3d power of 4 is $4^3 = 4\times4\times4 = 64$.

RULE

6. *Involve the given number or root, that is, multiply it by itself, and the product by the root again, and so on till the root has been used as a factor as many times as are indicated by the given power or its exponent.*

7. What is the second power or square of 13? *A.* 169.
8. What is the third power or cube of 18? *A.* 5,832.
9. What is the fourth power or biquadrate of 11? *A.* 14,641.
10. What is the fifth power of 7?—of 9? *A.* 16,807; 59,049.
11. What is the sixth power of 5?—of 4? *A.* 15,625; 4,096.
12. What number is meant by 3^2?—by 5^3?—by 20^5?—by 7^4?—by 6^5? *A.* 9; 125; 3,200,000; 2,401; 7,776.
13. What is the numerical difference between 2^6 and 8^3?—between 9^5 and 4^8?—between 10^{10} and 20^5? *A.* 448; 6,487; 9,996,800,000.
14. What is the 2d power of $\frac{3}{4}$?—of .75?—of $\frac{3}{5}$? What is the 3d power of $\frac{2}{7}$?—of 4.22? *A.* $\frac{9}{16}$ .5625; $\frac{9}{25}$; $\frac{8}{343}$; 75.151448.
15. What is the square of $5\frac{1}{2}$? *A.* $30\frac{1}{4}$.
16. What is the square of $16\frac{1}{2}$? *A.* $272\frac{1}{4}$.
17. What is the difference between the cube of $\frac{\frac{2}{5}}{8}$ and the biquadrate of $\frac{3}{4}$ of $\frac{2}{3}$? *A.* $\frac{499}{8000}$.
18. What is the numerical value of $\frac{3}{5}^4$?—of $5\frac{1}{2}^3$? *A.* $\frac{81}{625}$; $166\frac{3}{8}$.
19. What number is equal to $3^2\times4^3$?—to $3^5\times2^4$? *A.* 576; 3,888.
20. What is the difference between 4^5 and $4^3\times4^2$? *A.* 0.

Q. What is meant by Index or Exponent? 4. Give an example. What is the rule? 6. What are the second, third and fourth powers sometimes called? 3 What is the square of 20?—cube of 3?—biquadrate of 3?—fifth power of 2?—seventh power of 2?

---

* Biquadrate, from two Latin words, *bis, twice*, and *quadra, a square*, is so called because that number which is used *twice* as a factor in producing a square is used *twice* more in producing the biquadrate or fourth power

21. In the last example the exponents of 4^3 and 4^2 added together make 5, the exponent of 4^5; therefore—

22. *The powers of the same root are multiplied by adding their exponents.*

23. $215^3\times215^4\times215^6$ are equal to what? *A.* 215^{13}.

24. Involve 2^4; that is, raise it to the power denoted by its exponent. *A.* 16. Involve 2^6. *A.* 64. Involve 2^2. *A.* 4.

25. What is the quotient of 2^6 (=64) divided by 2^2 (=4)?
A. $16=2^4$.

26. *Hence powers of the same root may be divided by subtracting the exponent of the divisor from the exponent of the dividend.*

27. Divide 315^{21} by 315^{16}, and 82^{39} by 82^{18}. *A.* 315^5; 82^{21}.

28. What is the square of the 9 digits?
A. 15,241,578,750,190,521.

29. What is the sum of the squares of all the composite numbers between 1 and 20? *A.* 1,442.

30. What is the sum of the cubes of all the prime numbers between 1 and 20. *A.* 15,803.

31. Suppose there is a pile of wood, whose dimensions, that is, its length, breadth, and depth, are each 17 feet; how many cords does the pile contain? *A.* 38c. 49ft.

32. Suppose a piece of land lies in the form of a square, and each side measures 135 rods; how many acres does it contain?
A. 113A. 145rd.

33. Suppose a pile of wood, whose dimensions are each 18 feet, be sold for \$8$\frac{3}{4}$ per cord; what will the pile bring?
A. \$398.672, nearly.

34. If the amount of \$1 at compound interest for 1 year is \$1.06, what is the amount for 4 years?—for 5 years?—for 7 years?—for 10 years? *A.* \$1.262477+; \$1.338225+; \$1.50363+; \$1.790848+.

35. What is the difference in value, at \$18$\frac{3}{4}$ per acre, between a quantity of land containing 250 square miles and one which is 250 miles square? [See VII. 44.] *A.* \$747,000,000.

36. If a solid block of granite, 27 feet long, 13$\frac{1}{2}$ feet wide, and 13$\frac{1}{2}$ feet thick, be halved, what will be the value of .19 of each part, at the rate of 18$\frac{3}{4}$ cents for 3 solid feet? *A.* \$29.2169+.

Q. What is the numerical difference between 5 whose index shall be 3, and 12 whose index shall be 2? What is the product of 15 whose index shall be 8, if multiplied by 23 whose index shall be 5? What is the rule for it? 22. What is the quotient of 4^6 divided by 4^2? What is the rule? 26. How many square rods in a plat of ground 12 rods square? What is the difference in square yards, between 11 square yards and 11 yards square? [See VII. 43, 44.]

EVOLUTION.*

XCVIII. 1. EVOLUTION is the finding of the root from having the power given, and is therefore the converse of Involution, which is the finding of the power from having the root given.

2. Thus, the second or square root of 36 is 6, because the second power or square of 6 is 36; the third or cube root of 27 is 3, because the third power or cube of 3 is 27; the fourth root of 16 is 2, because the fourth power of 2 is 16, &c.

3. A ROOT, then, of any number is that factor which, multiplied into itself a certain number of times, will produce the given number. The process of finding it is called its EXTRACTION.

4. The number or name of the root corresponds with the number or name of its power.

5. That is, if 4 be the second power or square of 2, then 2 is the second or square root of 4; and if 27 be the cube of 3, then 3 is the cube root of 27.

---

6. Find by trial the square root of 64?—of 144?—of 3,600?—of 25?—of 42.25?—of $\frac{9}{16}$?—of $\frac{1}{4}$ of $\frac{1}{9}$?—of $6\frac{1}{4}$?—of $3\frac{1}{16}$?

A. 8; 12; 60; .5; 6.5; $\frac{3}{4}$; $\frac{1}{6}$; ($6\frac{1}{4}=\frac{25}{4}$) $\frac{5}{2}$, or $2\frac{1}{2}$; $1\frac{3}{4}$.

7. Find by trial the cube root of 1?—of 8?—of 27?—of 64?—of .125?—of $\frac{8}{27}$?—of $\frac{8}{1000}$?—of $3\frac{3}{8}$?

A. 1; 2; 3; 4; .5; $\frac{2}{3}$; $\frac{2}{10}$ ($=\frac{1}{5}$); $1\frac{1}{2}$.

8. Find by trial the biquadrate or fourth root of 16?—of 10,000?—of $\frac{16}{81}$?—of $\frac{1}{16}$?—of .0016?—of $\frac{1}{10}$ of $\frac{1}{1000}$?

A. 2; 10; $\frac{2}{3}$; $\frac{1}{2}$; .2; $\frac{1}{10}$.

9. In Involution, the required power of any number may be exactly

---

XCVIII. *Q.* What is Evolution? 1. Give an example. What is the root of any number? 3. Whence their names? 4. Give an example. Give the answers to the examples (on being read aloud by the teacher) in No. 6—in No. 7—in No. 8. Are all numbers susceptible of exact powers and roots? 9, 10. What classification is made in reference to such numbers? 11, 12. Give an example. 13.

---

* EVOLUTION. [*e, from,* and *volvo, to roll.*] The act of unfolding; the diverse figures, motions, &c. of a body of soldiers. Evolution is so called, because the root, by the process, becomes evolved or disentangled from other numbers

* TABLE OF POWERS AND ROOTS.

1st Power.	2d Power.	3d Power.	4th Power.	5th Power.	6th Power.	7th Power.	8th Power.	9th Power.
1	1	1	1	1	1	1	1	1
2	4	8	16	32	64	128	256	512
3	9	27	81	243	729	2187	6561	19683
4	16	64	256	1024	4096	16384	65536	262144
5	25	125	625	3125	15625	78125	390625	1953125
6	36	216	1296	7776	46656	279936	1679616	10077696
7	49	343	2401	16807	117649	823543	5764801	40353607
8	64	512	4096	32768	262144	2097152	16777216	134217728
9	81	729	6561	59049	531441	4782969	43046721	387420489

ascertained, because it is done by multiplication, which produces an exact product.

10 On the contrary, in Evolution there are many numbers whose roots cannot be accurately expressed, as the square root of 2, there being no factor that, multiplied into itself, will produce it.

11. Numbers whose roots can be exactly ascertained are called PERFECT POWERS, and their roots RATIONAL NUMBERS.

12. But other numbers are called IMPERFECT POWERS, and their oots IRRATIONAL NUMBERS, or SURDS.

13. Thus 16 is a perfect square, because its root is a rational number; but 16 is an imperfect cube, because there is no factor the third power of which is that number. Its root, then, is a *surd*.

14. By the means of decimals, however, we can come nearer and nearer to the desired root; that is, approximate towards it to any assignable degree of exactness; as the square root of 2, which is nearly 1.41421356+.

15. Roots are often indicated after the manner of powers in Involution, the numerators of which show the powers of the given numbers, and the denominators the required roots; thus—

$4^{\frac{1}{2}}$ means the square root of 4^1 or 4; then $4^{\frac{1}{2}}=2$.

$27^{\frac{1}{3}}$ means the cube root of 27, which is 3; then $27^{\frac{1}{3}}=3$.

$16^{\frac{1}{4}}$ means the fourth root of 16, which is 2; then $16^{\frac{1}{4}}=2$.

$3^{\frac{4}{2}}$ means the square root of the fourth power of 3; then $3^{\frac{4}{2}}=9$.

$2^{\frac{6}{3}}$ means the cube root of the sixth power of 2; then $2^{\frac{6}{3}}=4$.

16. The square root is also indicated by the radical sign $\sqrt{}$, and other roots by placing before the same sign their respective indices.

17. Thus, $\sqrt{9}$, $\sqrt[3]{8}$, $\sqrt[4]{16}$, denote the square, cube, and fourth roots, respectively.

18. Since $\sqrt{25}=5$, therefore $\sqrt{25}\times\sqrt{25}=25$.

19. Since $\sqrt[3]{8}=2$, therefore $\sqrt[3]{8}\times\sqrt[3]{8}\times\sqrt[3]{8}=8$.

20. Since $\sqrt[4]{16}=2$, therefore $\sqrt[4]{16}\times\sqrt[4]{16}\times\sqrt[4]{16}\times\sqrt[4]{16}=16$.

21. But $\sqrt[4]{16}\times\sqrt[4]{16}\times\sqrt[4]{16}=8$, since $\sqrt[4]{16}=2$, and $2\times2\times2=8$.

22. When numbers have a line, called a vinculum, drawn over them, or are enclosed in a parenthesis, they are to be taken together.

23. Thus, $\sqrt[3]{\overline{30-3}}$ or $\sqrt[3]{(30-3)}$ means that 3 is first to be taken from 30, leaving 27, of which the cube root is to be extracted.

---

24. Find by trial the difference between the square of 81 and the square root of 81. *A.* 6,552.

Q. How may surd roots be expressed with a tolerable degree of exactness? 14. How are roots indicated? 15. What is meant by $4^{\frac{1}{2}}$? [Read 4 with the index $\frac{1}{2}$.] What by $27^{\frac{1}{3}}$?—by $16^{\frac{1}{4}}$?—by $3^{\frac{4}{2}}$?—by $2^{\frac{6}{3}}$? What other indications of roots are there? 16. Give an example. See 17. When are numbers to be taken jointly? 22

25. Find the difference between 125^3 and $125^{\frac{1}{3}}$;—between 16^4 and $\sqrt[4]{16}$. *A.* 1,953,120; 65,534.

26. What is the sum of $3^{\frac{4}{2}}$ and $\sqrt{16}$?—of $9^{\frac{1}{2}}$ and $\sqrt[4]{16}\times\sqrt[4]{16}\times\sqrt[4]{16}\times\sqrt[4]{16}$? *A.* 13; 19.

27. What is the difference between $64^{\frac{1}{3}}$ and $\sqrt[6]{64}$?—between $\sqrt{\frac{1}{121}}$ and $\frac{25}{81}^{\frac{1}{2}}$? *A.* 2; $\frac{46}{99}$.

28. What is the amount of 32^5, $32^{\frac{1}{5}}$, and $\sqrt[6]{64}$?—also $\frac{2}{3}$ of $\frac{3}{4}$ of 54^3 and $\sqrt[3]{\frac{8}{27}}\times\sqrt{25}$? *A.* 33,554,436; 787,35 $\frac{1}{3}$.

29. Add together $36^{\frac{1}{2}}$, $\sqrt[5]{32}$, 5^4, $\sqrt[3]{\frac{1}{8}}$; $\sqrt[4]{\frac{1}{2}}$ of $\frac{1}{8}$ and $\sqrt[3]{10\frac{3}{4}-2.75}$. *A.* 636.

30. Suppose an orchard has 2,500 trees, and that it is in the form of a square; how many trees are there in each row? *A.* 50 trees.

31. A man desirous of appropriating $2\frac{1}{2}$ acres of a certain lot of land for a vegetable garden in the form of a square; what would be the distance round the garden? *A.* 80 rods.

32. A countryman in returning from market, said he received for his butter $4.41, and that he got as many cents a pound as there were pounds; how many pounds had he, and what was the price per pound? *A.* 21lb. at 21 cents.

33. If in digging a cellar of equal length, breadth and depth, there was thrown out 1,331 solid feet, how deep must the cellar have been? *A.* 11 feet.

34. Suppose that there are two square floors, one containing 121 square feet, and the other 400; now what is the sum of all the sides of both squares? *A.* 124 feet.

35. If a pile of wood in the form of a cube, sold at $7.50 per cord, comes to $3,750, what must be either the length, breadth or depth of the pile? *A.* 40 feet.

36. The foregoing process, when the numbers are large, is so tedious, that rules have been invented, by which the extraction of the required root is rendered comparatively easy.

EXTRACTION OF THE SQUARE ROOT.

XCIX. 1. The SQUARE of any given number, is the product of that number multiplied by itself. (XCVII. 3.)

2. The SQUARE ROOT of any given number, is such a number, as will, on being multiplied by itself, produce the given number. (XCVIII. 2.)

3. A square figure has, as we have seen, four equal sides and four equal angles. (VII. 36). The length of a square, multiplied by its breadth, produces its square content or superficies, sometimes called its area. (VII. 46).

4. The length and breadth of a square being equal, the square of

---

XCIX. *Q.* What is a square? 1. What is the square root? 1. Describe a square figure. 3. What is meant by the area of a square? 3. How is it found? 4. Of what use is the area in finding the side of a square? 4.

either of its sides is equal to its area; of course the square root of its area is equal to the length of either of its sides.

5. When a garden, which is laid out in the form of a square, contains 1,296 square rods, what is the length of each side? that is, what is the square root of 1296?

OPERATIONS.

```
        1st.                        2d.
     Square Rods.               Square Rods.
     3 0)1 2 9 6(3 0         3)1 2 9 6(3 6
          9 0 0                  9
6 0+6=6 6)3 9 6(6          6 6)3 9 6
          3 9 6                  3 9 6
          =====                  =====
```

In this example, we know that the root or the length of one side of the garden must be greater than 30, for $30^2=900$, and less than 40, for $40^2=1600$, which is greater than 1296; therefore we take 30, the less, and, for convenience' sake, write it at the left of 1296 as a kind of divisor likewise at the right of 1296, in the form of a quotient in division; (See Operation 1st.;) then subtracting the square of 30, =900 sq rods, from 1296 square rods, leaves 396 square rods.

FIGURE.

```
        30 rods.                  6 rods.
   ┌──────────────────────────┬──────────┐
 6 │              3 0         │     6    │ 6
 r │      B         6         │   C 6    │ r
 o │            ────          │   ──     │ o
 d │            1 8 0         │   3 6    │ d
 s │──────────────────────────┼──────────│ s
   │              A           │     D    │
   │  Rods.                   │          │
30 │  3 0,  length of A.      │   3 0    │ 30
 r │   3 0, breadth of A.     │     6    │ r
 o │  ────                    │   ──     │ o
 d │  9 0 0, sq. rods in A.   │  1 8 0   │ d
 s │                          │          │ s
   └──────────────────────────┴──────────┘
        30 rods.                  6 rods.
```

The pupil will bear in mind that the FIG on the left hand is the form of the garden and contains the same number of square rods, viz. 1296. This figure is divided into parts, called A, B, C, and D. It will be perceived that the 900 sq. rds. which we deducted, are found by multiplying the length of A, being 30 rds. by the breadth, being also 30 rods, that is, $30^2=900$.

To obtain the square rods in B, C, and D, the remaining parts of the figure, we may multiply the length of each by the breadth of each, thus; $30\times6=180$; $6\times6=36$; and $30\times6=180$; then $180+36+180=396$ square rods; or, add the length of B, that is, 30, to the length of D, which is also 30, making 60; or which is the same thing, we may double 30, making 60; to this add the length of C, 6 rods, and the sum is 66. Now, to obtain the square rods in the whole length of B, C, and D, we multiply their length, 6 rods, by the breadth of each side, thus, $66\times6=396$ square rods, the same as before.

We do the same in the operation; that is, we first double 30 in the quotient, and add the 6 rods to the sum, making 66 for a divisor; next, multiply 66, the divisor, by 6 rods, the width, making 396; then taking 396 from 396 leaves 0.

The pupil will perceive, the only difference between the 1st and 2d operation (which see) is, that in the 2d we neglect writing the ciphers at the right of the numbers, and use only the significant figures. Thus, for 30+6, we write 3 (tens,) and 6 (units,) which, joined together, make 36; for 900, we write 9 (hundreds). This is obvious from the fact, that the 9 retains its place under the 2 (hundreds). Instead of 60+6, we write 66. Omitting the ciphers in this manner cannot possibly make any difference, and, we see, it does not, for the result is the same in both.

6. By either of the foregoing operations, then, we find that the length of each side of the garden is 36 rods; or, that the square root of 1296 is 36.

7. PROOF. All the parts of the above figure make as follows,—

A contains 9 0 0 square rods.
B contains 1 8 0 square rods.
C contains 3 6 square rods.
D contains 1 8 0 square rods.
The given sum 1 2 9 6 square rods.

Or by Involution, thus, 36 rods ×36 rods=1296 square rods.

8. If then the square of the root, found from the operation, be equal to the given sum, the work is right.

9. Since the square of 99, the greatest factor of two figures, is 9801, which has the same number of figures as both its factors, or only double the number of figures in the root 99, therefore,—

10. The square of any root cannot have more figures than double the number of figures in the root.

11. Since the square of 10, the least factor of two places, is 100, which has only one figure less than both its factors, or only one less than double the number of figures in the root, therefore,—

12. The square of any root can never have but one figure less than double the figures of the root.

13. Hence, if we divide any given number into periods of two figures each, the number of periods will equal the number of figures of which the root will consist.

RULE.

14. *Point off the given number into periods of two figures each, by puttting a dot over the units, another over the hundreds, and so on; and if there are decimals, point them in the same manner, from the units towards the right hand.*

15. *Find the greatest square in the last period on the left, write its root on the right, as a quotient, subtract the square from the said period, and to the remainder bring down the next period for a dividend.*

16. *Double the root (quotient) for a partial divisor, and on its right, place, for the total divisor, such a figure as will express the greatest number of times that the true divisor is contained in the dividend, which figure will be the second in the root, or quotient.*

17. *Multiply the divisor by the last quotient figure; subtract the product from the dividend; and to the remainder bring down the next period for a new dividend, with which proceed as before, by doubling all the figures in the quotient, or root, &c.*

Q. How is the operation proved? 8. What is the greatest number of figures which any root can have? 10. What is the least number? 12. What reason is given for each? 9, 11. What is the inference? 13. What is the rule for pointing off the given number? 14. How is the first dividend obtained? 15. What is the direction for finding the second figure in the root? 16. What for finding the next dividend? 17. Repeat the entire rule. 14, 15 16. 17.

18. OPERATION.

```
        8)7 5 6 9(8 7
          6 4
    1 6 7)1 1 6 9
          1 1 6 9
Proof. 8 7×8 7=7 5 6 9
```

Find the sq. root of 7569. *A.* 87.
Find the sq. root of 9025. *A.* 95.
Find the sq. root of 4225. *A.* 65.
Find the sq. root of 1369. *A.* 37
Find the sq. root of 2304. *A.* 48.
Find the sq. root of 6561. *A.* 81.

19. Recollect to double all the quotient figures for a divisor.

```
    2)6 5 5 3 6(2 5 6
      4
  4 5)2 5 5
      2 2 5
5 0 6)3 0 3 6
      3 0 3 6
```

Find the sq. root of 65536. *A.* 256.
Find the sq. root of 470596. *A.* 686.
Find the sq. root of 123201. *A.* 351.
Find the sq. root of 801025. *A.* 895.
Find the sq. root of 412164. *A.* 642.
Find the sq. root of 966289. *A.* 983.
Find the sq. root of 765625. *A.* 875.

20. Extract the square root of 2125764. *A.* 1458.
21. Extract the square root of 6718464. *A.* 2592.
22. Extract the square root of 4294967296. *A.* 65536.
23. When the divisor is too large, increase the dividend by bringing down the next period of the given sum, then place a cipher in the root, and find a new divisor as before.

```
    2)4 2 0 2 5(2 0 5
      4
4 0 5)2 0 2 5
      2 0 2 5
```

Find the sq. root of 42025. *A.* 205.
Find the sq. root of 651249. *A.* 807.
Find the sq. root of 49126081. *A.* 7009.
Find the sq. root of 25806400. *A.* 5080.

24. What is the square root of 6480.25? [See R. 14.] *A.* 80.5.

```
      8)6 4 8 0.2 5(8 0.5
        6 4
1 6 0 5)8 0 2 5
        8 0 2 5
```

Find the sq. root of 913.8529. *A.* 30.23.
Find the sq. root of 9.3025. *A.* 3.05.
Find the sq. root of .00015625. *A.* .0125.
Find the sq. root of 196.5604. *A.* 14.02.

25. Extract the square root of .0000000001018081. *A.* .00001009.
26. When the last divisor leaves a remainder, the operation may be continued by annexing successively periods of decimal ciphers.

```
       3)1 0(3.1 6+
         9
   6 1)1.0 0
         6 1
 6 2 6)3 9.0 0
       3 7.5 6
Remainder, 1.4 4
```

Find the sq. root of 10. *A.* 3.162+
Find the sq. root of 175. *A.* 13.228+
Find the sq. root of 90. *A.* 9.4868+
Find the sq. root of 5. *A.* 2.23606+
Find the sq. rt. of 2. *A.* 1.41421356237+

27. When the last period of a decimal consists of only one figure, annex a cipher to complete the period.
28. What is the square root of 11.7? *A.* 3.4205+
29. What is the square root of 8.003? *A.* 2.828+
30. What is the square root of .018? *A.* .1341+

Q. What is the direction when the divisor is not contained in the dividend? 23. What is to be done with the final remainder? 26. What with an imperfect decimal period? 27.

31. When fractions have terms that are perfect powers, [XCVIII. 11.] extract the roots of the most simple terms.

32. What is the square root of $\frac{64}{144}$? *A.* $\frac{8}{12}$ or $\frac{2}{3}$.

33. What is the square root of $\frac{450}{2048}$? *A.* $\frac{15}{32}$.

34. What is the square root of $\frac{121}{1234321}$? *A.* $\frac{1}{101}$.

35. When the terms are either of them imperfect powers, [XCVIII. 12.] reduce them first to a decimal.

36. What is the square root of $\frac{70}{84}$? *A.* .9128+

37. What is the square root of $\frac{11}{13}$? *A.* .9198+

38. What is the square root of $\frac{9}{13}$? *A.* .83205.

39. What is the square root of $\frac{3}{4}$? *A.* .866+

40. What is the square root of $\frac{1}{2}$? $\frac{2}{3}$? $\frac{4}{5}$? $\frac{5}{6}$? $\frac{1}{4}$? $\frac{4}{7}$? $\frac{5}{9}$? $\frac{1}{3}$?
A. .707+ ;.816+ ;.894+ ;.912+ ; $\frac{1}{2}$;.755+ ;.745+ ;.577+.

41. A mixed number may first be reduced to an improper fraction, and its roots be expressed again by a mixed number, unless its terms are imperfect powers, in which case the operation must be conducted decimally.

42. Extract the square root of $420\frac{1}{4}$. *A.* $20\frac{1}{2}$.

43. Extract the square root of $912\frac{1}{25}$. *A.* $30\frac{1}{5}$.

44. Extract the square root of $272\frac{1}{4}$. *A.* $16\frac{1}{2}$.

45. What is the square root of $17\frac{23}{100}$? *A.* 4.1509+

46. What is the square root of $87\frac{471}{1112}$? *A.* 9.35+

47. What is the root indicated by $\sqrt{234\frac{9}{13}}$? *A.* 15.3196+

48. What is the root indicated by $81\frac{16}{1031}$? *A.* 9.000862+

49. If the next divisor or double of the root be written under the final remainder, the fraction will express very nearly the radical remainder, which should be first reduced, either to its lowest terms, or to a decimal, and annexed to the root.*

50. How much is $\sqrt{10946}$? $=104$ and 130 rem.: $104\times2=208$ he next divisor. *A.* $104\frac{130}{208}=104\frac{5}{8}$ or 104.625.

51. How much is $\sqrt{43256789101}$? *A.* $207982\frac{276777}{415964}$.

52. How much is $\sqrt{543010940567}$? *A.* $736892\frac{1120903}{1473784}$.

53. How much is $\sqrt{6732100100954}$? *A.* $2594629\frac{453313}{5189258}$.

54. How much is $\sqrt{1000900010007}$? *A.* $1000449\frac{1808406}{2000898}$.

55. From $\sqrt{729}$ take $144^{\frac{1}{2}}+2^{\frac{4}{2}}$. *A.* 11.

56. From $\sqrt{729}$ take $256^{\frac{1}{2}}+2^{\frac{4}{2}}$. *A.* 7.

57. From $20^{\frac{6}{2}}$ take $\sqrt{46311.04}$. *A.* 7784.8.

58. From $4^{\frac{5}{2}}$ take $\sqrt{915\frac{1}{16}}+1.1025^{\frac{1}{2}}$. *A.* .7.

59. How much are $\sqrt{529}+\sqrt{1764}+144^{\frac{1}{2}}+1459264^{\frac{1}{2}}+\sqrt{6724}$
A. 1367.

Q. What is the rule for extracting the roots of fractions? 31, 35. What for mixed numbers? 41.

---

* **Although the remainder is a little too great in the square root, and a little too small in the cube root, they are nevertheless sufficiently exact for most purposes, and much more convenient than the operation by annexing ciphers.**

60. From $\sqrt{152399025}$ take $(\sqrt{4120900}+\sqrt{\frac{5}{80}}+.00060025^{\frac{1}{2}}.)$
A. 10314.7255

61. What is the square root of 15241578750190521?
A. The 9 digits.

62. Find the sum of the roots or numbers involved in all the per fect squares between 1 and 100. *A.* 44.

63. Find the sum of the *squares* whose roots are surds, between 1 and 20? *A.* 160.

64. Suppose that a commandant of an army has 180625 effective men, and would form them into a solid square, how many would there be in each rank and file? *A.* $\sqrt{180625}=425$.

65. Suppose a town proposes to levy a poll tax of $216.09 so that each man shall pay as many cents as there are men to be taxed; what is each man's tax on his head? *A.* $1.47.

66. Suppose there are two portions of land each in the form of a square, and that one is $30\frac{1}{4}$ miles square, and the other contains $30\frac{1}{4}$ square miles; what is the sum of the distances round both squares?
A. 143 miles.

67. If the surface of the earth, which is computed to contain 196,-000,000 square miles were in the form of a square, what would be the distance round it? *A.* 56,000 miles.

68. If a tract of land $6\frac{1}{4}$ miles long, and 4 miles wide, which cost $\$1\frac{1}{4}$ per acre be exchanged for the same quantity in the form of a square, and subsequently be divided into one hundred equal and square farms, $\frac{3}{8}$ of which should bring at auction $\$11\frac{3}{4}$ per acre; $\frac{2}{5}$ of them $12 per acre, and the rest $\$10\frac{1}{5}$ per acre; what would be the profit in the transaction, and what the sum of the distances round all the squares? *A.* 200 miles; $164020 profit.

PROPORTIONS INVOLVING ROOTS AND POWERS.

69. *The product of the square roots of any two numbers, is equal to the square root of their product.*

70. Prove $\sqrt{81}\times\sqrt{225}=\sqrt{81\times225}$.

71. To find a mean proportional between any two numbers:— *Extract the square root of their product.*

72. For in the proportion 2 : 10 : : 10 : 50; of which the 10 is a mean proportional between 2 and 50; we have on geometrical principles, $2\times50=10^2$.

73. What is the mean proportional between 3 and 12? 4 and 36? 24 and 96? 16 and 64? *A.* 6; 12; 48; 32.

74. What is the mean proportional between 25 and 289? 25 and 156.25? *A.* 85; $62\frac{1}{2}$.

75. What is the mean proportional between 7 and $1\frac{3}{4}$? $10\frac{3}{10}$ and $41\frac{1}{5}$? *A.* $3\frac{1}{2}$: 20.6.

Q. To what is the product of the square root of any two numbers equal? 69. How is a mean proportional between any two numbers found? 71. What is the mean proportional between 4 and 9? between 2 and 18?

76. *The mean proportional between any two numbers, has the same ratio to those numbers, that the square roots of those numbers have to each other.*

77. Find the mean proportional between 25 and 36, and the ratio between it and those numbers, and see if it is the same as the ratio between $\sqrt{25} : \sqrt{36}$.

78. * To find any two numbers from having their sum and product given:—*From the square of half their sum, subtract their product: extract the square root of the remainder, and add it to half their sum, for the larger number; or subtract it therefrom for the smaller number.*

79. A certain field contains an area of 30 acres 2 roods and 20 rods: required its length and breadth, the sum of these being 148 rods. *A*. 98rd. : 50rd.

80. A gentleman having purchased a certain quantity of flour, for $1935, found that if he added the number of dollars it cost per barrel to the number of barrels, the sum would be 224. How many barrels must he have bought? *A*. 215 barrels.

81. To find any two numbers from having their sum and the sum of their squares given:—*Find the difference between the square of their sum, and the sum of their squares: half this difference subtract from the square of half their sum, and add the square root of the remainder to their half sum for the greater number, or subtract it therefrom for the smaller number.*

82. Suppose that two square fields contain together 9A. 2R. 5rd and that the sum of either their length or breadth is 55 rods; pray what is the length of each lot? *A*. 25rd. : 30rd.

EXTRACTION OF THE CUBE ROOT.

C. 1. The CUBE of any given number is the product of that number multiplied by its square. [XCVII. 3.]

2. The CUBE ROOT of any given number, is such a number as will, on being multiplied by its square, produce the given number. [XCVIII. 2.]

3. A body in the form of a cube is a solid of six equal sides, each containing an exact square. [See the block accompanying this work.]

4. A CUBE then has three dimensions, viz., length, breadth, and thickness or depth; the product of which multiplied into each other is called its solid content. [VII. 60.]

5. The length, breadth, and thickness of a cube being equal, the cube of either of its sides must be equal to its solid contents; of course the cube root of its solid contents must be equal to the length of either of its sides.

Q. To what is the ratio of any mean proportional equal? 76. How are any two numbers found from having their sum and product given? 78.—from having their sum and the sum of their squares given? 81.

C. *Q.* 1. What is the cube of any number? 1. What is the cube root? 2. What is a cubical body? 3. What its dimensions? 4. How are its solid contents found? 5. How either of its dimensions? 5.

* This and the following proportion are deduced from Algebraic processes

6 * The blocks which accompany this work for the purpose of illustrating the operation of the following example are eight in all, and when put together, they should form a perfect cube of 24,389 sd. feet.

7. These blocks are marked by the letters A, B, C, and D, whose proportional dimensions are supposed to be as follows :

A is a cube, 20 feet long, 20 feet wide, and 20 feet thick.

Three B's, each 20 feet long, 20 feet wide, and 9 feet thick.

Three C's each 20 feet long, 9 feet wide, and 9 feet thick.

D is a cube 9 feet long, 9 feet wide, and 9 feet thick.

8. If a cubical block which is formed by the 8 small ones above, contains 24,389 solid ft. ; what must be the length of each of its sides?

$24389(20$

$20^3=8000$

$2^2\times300=1200)16389(9$

Div. $1200\times$quo. $9=10800$

$2\times30\times\ 9^2=\ 4860$

$9^3=\ 729$

16389

The same without the ciphers.

$24389(29$ Ans.

$2^3=8$

$2^2\times300=1200)16389$ dividend.

$1200\times9=\ 10800$

$2\times30\times\ 9^2=\ 4860$

$9^3=\ 729$

16389 subtrahend.

In this example, we know that one side cannot be 30ft., for $30^3=27000$ solid feet, are more than 24389, the given sum—therefore, we will take 20 for the length of one side of the cube.

Then $20\times20\times20=8000$ solid feet, which we must, of course, deduct from 24389 leaving 16389

These 8000 solid feet the pupil will perceive, are the solid contents of the cubical block marked A. This corresponds with the operation; for we write 20 feet, the length of the cube A, at the right of 24389, in the form of a quotient; and its square 8000, under 24389; from which subtracting 8000, leaves 16389 as before.

As we have 16389 cubic feet remaining, we find the sides of the cube A are not so long as they ought to be; consequently we must enlarge A; but in doing this we must enlarge three sides of A, in order that we may preserve the cubical form of the block. We will now place the three blocks each of which is marked B, on these three sides of A. Each of these blocks, in order to fit, must be as long and as wide as A; and, by examining them, you will see that this is the case; that is, they are 20 feet long and 20 feet wide; then $20\times20=400$, the square contents in one B; and $3\times400=1200$, square contents in three Bs; then it is plain, that 16389 solid contents, divided by 1200, the sq. contents will give the thickness of each block. But an easier method is to square the 2, (tens,) in the root 20, making 4, and multiply the product 4, by 300, making 1200, a divisor, the same as before.

We do the same in the operation (which see); we multiply the square of the

*This rule is best illustrated by means of blocks which may be supposed to contain a certain proportional number of feet, inches, &c., corresponding with the operation of the rule. They may be made in a few minutes, from a small strip of pine board, with a common penknife, at the longest, in less time than the teacher can make the pupil comprehend the reason, from merely seeing the picture on paper. This method of demonstrating the rule will be an amusing and instructive exercise, both to teacher and pupil, and may be comprehended by any pupil, however young, who is so fortunate as to have progressed as far as this rule. It will give him distinct ideas respecting the different dimensions of square and cubic measures, and indelibly fix on his mind the reason of the rule, and consequently the rule itself. But, for the convenience of teach- [illegible] illustrative of the operation, [illegible] foregoing example, accompany this work

quotient figure, 2, by 300, thus $2 \times 2 = 4 \times 300 = 1200$; then the divisor, 1200 (the square contents) is contained in 16389 (solid contents) 9 times, that is, 9ft. is the thickness of each block marked B. This quotient figure, 9, we place at the right of 16389, and then 1200 square feet × 9 feet, the thickness,=10800 s. ft.

If we now examine the block, thus increased by the addition of the 3 Bs, we shall see that there are yet three corners not filled up; these are represented by the three blocks, each marked C, and each of which, you will perceive, is as long as either of the Bs, that is, 20 ft., being the length of A, which is 20 in the quotient. Their thickness and breadth are the same as the thickness of the Bs, which we found by dividing, to be 9 feet, the last quotient figure. Now, to get the solid contents of each of these Cs, we multiply their thickness (9 feet) by their breadth, (9 feet,)=81 square feet; that is, the square of the last quotient figure, 9=81; these square contents must be multiplied by the length of each, (20 feet,) or, as there are 3, by $3 \times 20 = 60$; or, which is easier in practice, we may multiply the 2, (tens) in the root, 20, by 30, making 60, and this product by $9^2 = 81$, the square contents = 4860 solid feet.

We do the same in the operation, by multiplying the 2 in 20 by $30 = 60 \times 9 \times 9 = 4860$ solid feet, as before; this 4860 we write under the 10800, for we must add the several products together by and by, to know if our cube will contain all the required feet.

By turning over the block with all the additions of the blocks marked B and C, which are now made to A, we shall spy a little square space, which prevents the figure from becoming a complete cube. The little block for this corner is marked D, which the pupil will find, by fitting it in, to exactly fill up this space. This block D, is exactly square, and its length, breadth and thickness are alike, and, of course, equal to the thickness and width of the Cs, that is, 9 feet, the last quotient figure; hence 9ft. × 9ft. × 9ft.=729 solid feet in the block D; or, in other words, the cube of 9, (the quotient figure,) which is the same as $9^3 = 729$, as in the operation. We now write the 729 under the 4860, that this may be reckoned in with the other additions.

We next proceed to add the solid contents of the Bs, Cs, and D, together, thus, $10800 \times 4860 \times 729 = 16389$, precisely the number of solid feet which we had remaining after we deducted 8000 feet, the solid contents of the cube A.

If, in the operation, we subtract the amount, 16389, from the remainder, or dividend, 16389, we shall see that our additions have taken all that remained, after the first cube was deducted, there being no remainder.

The last little block, when fitted in, as you saw, rendered the cube complete, each side of which we have now found to be $20 + 9 = 29$ feet long, which is the cube root of 24389 (solid feet); but let us see if our cube contains the required number of solid feet.

9. *Proof.*—8000 s. ft. in A+10800 s. ft. in 3 Bs+4860 s. ft. in 3 Cs × 729 s. ft. in D=24389 s. ft. in the given sum which because they are equal to 29^3 form a perfect cube, then, 29 is the length of the required side; therefore,—

10. If by Involution the cube of the root found from the operation be equal to the given sum, the operation is correctly performed.

11. By reasoning similar to that employed in XCIX. 9, it may be shown that the product of any three numbers into each other never has more figures than all its factors, nor fewer than that same number less two.

12. We infer also from the same reasoning, that if we point off any sum into periods of three figures each, the number of periods will equal the number of figures in its root. Hence the direction in the rule

RULE.

13. *Divide the given number into periods of three figures each, by placing a point over the unit figure, and over every third one from the place of units to the left in whole numbers, and to the right in decimals.*

Q. Of how many figures will every root consist? 12. What is the reason for it? 11. What is the rule for pointing off the given number? 13.

14. *Find the greatest cube in the first left hand period, and place its root in the quotient. Subtract the cube thus found from this period, and to the remainder bring down the next period, and the result will be the dividend.*

15. *Multiply the square of the root or quotient by* 300 *for a divisor. Divide the dividend by the divisor for the next figure in the root.*

16. *Multiply the divisor by the quotient figure; multiply the former quotient figure or figures by* 30 *times the square of the last quotient figure; finally, cube the last quotient figure; then add these three results together for a subtrahend.*

17. *Subtract the subtrahend from the dividend, and to the remainder bring down the next period for a new dividend, with which proceed as before, and so on till all the periods are brought down.**

18. NOTE. When the subtrahend happens to be larger than the dividend, the quotient figure must be made one less, and we must find a new subtrahend. The reason why the quotient figure will be sometimes too large, is, because this quotient figure merely shows the width of the three first additions to the original cube; consequently, when the subsequent additions are made, the width (quotient figure) may make the solid contents of all the additions more than the cubic feet in the dividend, which remain after the solid contents of the original cube are deducted.

19. When we have a remainder after all the periods are brought down, we may continue the operation by annexing periods of ciphers, as in the square root. When it happens that the divisor is not contained in the dividend, a cipher must be written in the quotient (root,) and a new dividend formed by bringing down the next period in the given sum.

$$
\begin{array}{rl}
 & \dot{9}66\dot{3}59\dot{7}(213, \text{ Ans.} \\
2^3 = & 8 \\
2^2\times300=1200) & 1663 \text{ dividend.} \\
1200\times1= & 1200 \\
2\times30\times1^2= & 60 \\
1^3= & 1 \\
 & \overline{1261} \text{ subtrahend.} \\
21^2\times300=132300) & 402597 \text{ dividend.} \\
132300\times3= & 396900 \\
21\times30\times3^2= & 5670 \\
3^3= & 27 \\
 & \overline{402597} \text{ subtrahend.}
\end{array}
$$

20. What is the cube root of 9663597? *A.* 213.

21. What is the cube root of 91125? *A.* 45.

22. What is the cube root of 970299? *A.* 99.

23. What is the cube root of 778688? *A.* 92.

24. What is the cube root of 2000376? *A.* 126.

25. What is the cube root of 3796416? *A.* 156.

26. What is the cube root of 94818816? *A* 456.

27. What is the cube root of 175616000? 560.

Q. What for finding the first dividend? 14. What for finding the subtrahend? 16. Describe the rest of the process. 17. What is the whole rule? 13, 14, 15, 16, 17. What is to be done when the subtrahend is too large? 18 What when the divisor is too large? 19. What is to be done with the final remainder? 19.

* The root of the first period take,
And of that root a quotient make:
Which root must now a cube become,
To be a period taken from;
To the remainder then you must
Bring down another period just;
Which being done, you then must see,
This number straight divided be,
By just three hundred times the square
Of what the quotient figures are;
The last squared, multiplied by the rest,
The product thirty times exprest;
The cube of the last figure, too,
You must put in, if right, you do;
Add these, subtract them; so descend,
From point to point unto the end.

28. What is the cube root of 1,879,080,904? *A.* 1,234.

29. Where the divisor is larger than the dividend. [See 19.]

```
                     748613312(908
                     729
90²×300=2430000)19613312
    2430000×8 =19440000
     90×30×8²=   172800
           8³=      512
                 19613312
```

30. What is the cube root of 748,613,312? *A.* 908.

31. What is the cube root of 8,365,427? *A.* 203.

32. What is the cube root of 517,781,627? *A.* 803.

33. What is the cube root of 731,189,187,729? *A.* 9,009.

34. Find the cube root of 8,096,384,512,000,000,000? *A.* 2008000.

35. What is the cube root of .000,015,625? *A.* .025.

36. What is the cube root of 12.167? *A.* 2.3.

37. What is the cube root of 26.2? [See xcix. 27. *A.* 2.97 +

38. What is the cube root of 15.32? *A.* 2.483+

39. What is the cube root of $\frac{27}{64}$? [See xcix. 31.] *A.* $\frac{3}{4}$.

40. What is the cube root of $\frac{216}{343}$? *A.* $\frac{6}{7}$.

41. What is the cube root of $\frac{4913}{9261}$? *A.* $\frac{17}{21}$.

42. What is the cube root of $\frac{1}{400}$? [See xcix. 35.41.] *A.* .13+

43. What is the cube root of $\frac{64}{343}$? $\frac{729}{140608}$? $49\frac{8}{27}$? $7,558\frac{197}{512}$? *A.* $\frac{4}{7}$; $\frac{9}{52}$: $3\frac{2}{3}$; $19\frac{5}{8}$.

44. What is the cube root of $15\frac{5}{8}$? *A.* $2\frac{1}{2}$

45. What is the cube root of $1,242\frac{19}{64}$? *A.* $10\frac{3}{4}$.

46. What is the cube root of $1,984\frac{21}{83}$? *A.* 12.566+.

47. What is the cube root of $200\frac{45}{39}$? *A.* 5.859+.

48. What is the cube root of $183,457\frac{3}{7}$? *A.* 56.82+.

49. How much is $\sqrt[3]{8,000}+1,728^{\frac{1}{3}}$? *A.* 32.

50. What is $132,651^{\frac{1}{3}}-3^{\frac{6}{3}}$? *A.* 42.

51. Add into one sum the cube roots of 274,625; 2,197; 6,859 *A.* 97.

52. How much are $\sqrt[3]{4,096}+1,000^{\frac{1}{3}}+\sqrt[3]{166\frac{3}{8}}+512^{\frac{1}{3}}+\sqrt[3]{9\frac{305}{512}}$? *A.* $41\frac{5}{8}$.

53. To find two mean proportionals between any two given numbers, as the 30 and 150 in 6 : 30 : : 150 : 750: in which 5 is the common ratio. Now $6\times5^3=750$; of course $750\div6$ and this quotient by $125=5^3=5$ the ratio again, then 6×5 the ratio$=30$ the smaller mean, and $750\div5=150$ the greater mean; therefore—

54. *Divide the greater of the two numbers by the smaller, and extract the cube root of the quotient for the common ratio, with which*

Q. What is the method of procedure with an imperfect decimal period? 37. With a fraction? 39, 42. With a mixed number? 42. What is the rule for finding the mean proportionals between two numbers? 54. What is the illustration? 55

multiply the smaller of the given numbers for the smaller mean proportional, and divide the greater of the given numbers by the same ratio for the greater mean proportional.

55. What are the two mean proportionals between 32 and 16,384? *A.* 256 : 2,048.

56. What are the two mean proportionals between $\frac{3}{10}$ and $37\frac{1}{2}$?—between $\frac{3}{5}$ of $\frac{1}{27}$ and 6.3 times $\frac{1}{7}$ of $\frac{2}{3}$? *A.* 1.5 and $7\frac{1}{2}$; $\frac{1}{15}$ and $\frac{1}{3}$.

57. What are the two mean proportionals between .000625 and 625? *A.* $\frac{1}{16}$ and $6\frac{1}{4}$.

58. Find the sum of the cube roots of all the numbers under 1,001, which are perfect powers of those roots. *A.* 55.

59. Find the sum of all the *powers* under 20, the cube roots of which are surds. *A.* 181.

60. If the amount of a certain sum at compound interest for 3 years be $1.191016; what is the amount for the first year, the rate per cent., and the principal? *A.* $1.06; rate 6; $1 principal.

61. Suppose that in making an excavation, there were thrown out 838,561,807 solid feet of earth; what would be the length of one side of a cube of equal contents? *A.* 943 feet.

62. If a pile of wood which is 2,565ft. long, 40ft. wide, and 40ft. high, be thrown into the form of a cube and sold for $\frac{3}{4}$ of as many dollars as the cube would be feet long, what sum would the cube bring? *A.* $96.

63. If 5,375 tons of round stones of equal size, 1,849 of which just weigh one ton, be thrown into a cubical pile, and all be sold for what the number of stones that will reach across one side of the cube would bring at the rate of $5 for 7 stones; what would be the purchase price? *A.* 153.57\frac{1}{7}$.

EXTRACTION OF THE ROOTS OF ALL POWERS.

CI. 1. When the index of the given power is a composite number—*Resolve it into as many indices or factors as is possible; then extract the roots of the given power successively as their indices require.*

2. *That is, extract the root denoted by one index, then the root of that root as denoted by another index, and so on till the number of extractions shall equal the number of indices.*

3. Thus the 4th root (2×2) = the square root of the square root; the sixth root (3×2) = the cube root of the square root, or the square root of the cube root.

4. The 8th root $(2\times2\times2)$ = the square root of the square root of the square root; the ninth root (3×3) = the cube root of the cube root; the 10th root, $(5\times2,)$ the fifth root of the square root, &c.

5. What is the biquadrate or fourth root of 20,736? $\sqrt{20,736}=144$, and $\sqrt{144}=12$. *A.* 12.

CI. *Q.* How are the roots of most powers extracted? 1. What is meant by that process? 2. How, for example, are the fourth and sixth roots extracted? 3 How, the 8th and 9th roots? 4

6. What is the biquadrate root of 2,998,219,536? *A.* 234.
7. What is the sixth root of 1,178,420,166,015,625? *A.* 325.
8. What is the eighth root of 722,204,136,308,736? *A.* 72.
9. What is the ninth root of 387,420,489? *A.* 9.
10. What is the twelfth root of 282,429,536,481? *A.* 9.

GENERAL RULE FOR EXTRACTING ALL ROOTS.

11. *Point off, from the unit's place, the periods, as the required root directs; that is, for the fourth root point off periods of four figures each; for the fifth root, periods of five figures, &c.*

12. *Find by trial the greatest root in the left hand period, and subtract its power from the said period.*

13. *To the remainder bring down the next figure in the next period, for a dividend.*

14. *Involve the root to the power next inferior to that which is given, and multiply the result by the index of the given power for a divisor.*

15. *Divide the dividend by the divisor, and consider the quotient the next figure of the root.*

16. *Involve the whole root to the given power, and subtract it from as many left hand periods as the root has places of figures.*

17. *To the remainder bring down the next period for a new dividend, to which find a new divisor as before, and so on till the periods are all brought down.*

18. What is the sursolid or 5th root of 701,583,371,424?

7 0 1 5 8 3 3 7 1 4 2 4 (2 3 4
$2^5 =$ 3 2 subtrahend.
$2^4 \times 5 =$ divisor 8 0) 3 8 1 dividend.
$23^5 =$ 6 4 3 6 3 4 3 subtrahend.
$23^4 \times 5 =$ 1 3 9 9 2 0 5) 5 7 9 4 9 0 7 dividend.
$234^5 =$ 7 0 1 5 8 3 3 7 1 4 2 4 subtrahend.

19. Observe that only one figure is brought down to form the dividend, and that the subtrahend is in each instance taken directly from the periods in the top line.

20. What is the fifth root of 1,934,917,632? *A.* 72.

21. What is the seventh root of 10,030,613,004,288? *A.* 72.

22. What is the tenth root of 3,486,784,401? The better method is to extract the 5th root of the square root. *A.* 9.

23. If the amount of $100 for 8 years at compound interest be $159.38480745308416, what is the amount for the first year, and what is the rate per cent.? *A.* $106; 6 per cent.

Q. What is the fourth or biquadrate root of 256?—of 10,000? In the rule which is applicable to all powers, what is the direction for pointing off? 11. What is the rule for obtaining the dividend? 12, 13. What, for finding the divisor? 14. What, for finding the second figure in the root? 15. Describe the rest of the process? 16, 17

ALLIGATION.

CII. 1. ALLIGATION is the method of mixing several simples of different qualities, so that the compound or composition may be of a mean or middle quality.

2. When the quantities and prices of the several things or simples re given, to find the mean price or mixture compounded of them, the rocess is called

ALLIGATION MEDIAL.

3. A farmer mixed together two bushels of rye, worth 50 cents a bushel, 4 bushels of corn, worth 60 cents a bushel, and 4 bushels of oats, worth 30 cents a bushel; what is a bushel of this mixture worth?

```
2 bushels at $ . 5 0 cost $ 1 . 0 0
4 - - - - - - - $ . 6 0 - - - $ 2 . 4 0
4 - - - - - - - $ . 3 0 - - - $ 1 . 2 0
1 0 - - - - - - - - - - - - - ) $ 4 . 6 0 ( 4 6
```

4. In this example, it is plain, that if the cost of the whole be divided by the whole number of bushels, the quotient will be the price of one bushel of the mixture. *A.* 46 cents.

RULE.

5. *Divide the whole cost by the whole number of bushels, &c.; the quotient will be the mean price or cost of the mixture.*

6. A grocer mixed 10cwt. of sugar at $10 per cwt., 4 cwt. at $4 per cwt., and 8cwt. at $7\frac{1}{2}$ per cwt.; what is 1cwt. of this mixture worth?—what is 5cwt. worth?

A. 1cwt. is worth $8, and 5cwt. is worth $40.

7. A composition was made of 5lb. of tea at $1\frac{1}{4}$ per lb., 9lb. at $1.80 per lb., and 17lb. at $1\frac{1}{2}$ per lb.; what is a pound of it worth?

A. $1.546\frac{7}{10}$+.

8. If 20 bushels of wheat, at $1.35 per bushel, be mixed with 15 bushels of rye, at 85 cents per bushel, what will a bushel of this mixture be worth? *A.* $1.135\frac{7}{10}$+.

9. If 4lb. of gold, of 23 carats fine, be melted with 2lb. 17 carats fine, what will be the fineness of this mixture? *A.* 21 carats.

ALLIGATION ALTERNATE.

CIII. 1. ALLIGATION ALTERNATE is the process of finding the proportional quantity of each simple, from having the mean price or rate, and the mean prices or rates of the several simples given; consequently, it is the reverse of ALLIGATION MEDIAL, and may be proved by it.

2. A farmer has oats worth 25 cents a bushel, which he wishes to mix with corn worth 50 cents per bushel, so that the mixture may be worth 30 cents per bushel, what proportion or quantities of each must he take?

CII. *Q.* What is Alligation? 1. Alligation Medial? 2. Rule? 5
CIII *Q.* What is Alligation Alternate? 1.

3. In this example, it is plain, that if the price of the corn had been 35 cents, that is, had it exceeded the price of the mixture (30 cents) just as much as it falls short, he must have taken equal quantities of each sort; but, since the difference between the price of the corn and the mixture price is 4 times as much as the difference between the price of the oats and the mixture price, 4 times as much oats as corn must be taken, that is, 4 to 1, or 4 bushels of oats to 1 of corn. But since we determine this proportion by the differences, these differences will represent the same proportion.

4. These are 20 and 5, that is, 20 bushels of oats to 5 of corn, which are the quantities or proportions required. In determining those differences, it will be found convenient to write them down in the following manner:

```
       cts.  bushels.
30 { $.25 ─┐  −20 }
   { $.50 ─┘  − 5 } Ans.
```

5. It will be recollected, that the difference between 50 and 30 is 20, that is, 20 bushels of oats, which must stand at the right of the 25, the price of the oats, or, in other words, opposite the price, that is connected or linked with the 50; again, the difference between 25 and 30=5, that is, 5 bushels of corn opposite the 50, the price of the corn.

6. The answer, then, is 20 bushels of oats to 5 bushels of corn, or in that proportion.

7. By this mode of operation, it will be perceived that there is precisely as much gained by one quantity as there is lost by another, and therefore the gain or loss on the whole is equal.

8. The same will be true of any two ingredients mixed together in the same way. In like manner, the proportional quantities of any number of simples may be determined; for, if a less be linked with a greater than the mean price, there will be an equal balance of loss and gain between every two; consequently an equal balance on the whole.

9. It is obvious that this principle of operation will allow a great variety of answers; for, having found one answer, we may find as many more as we please by only multiplying or dividing each of the quantities found by 2, or 3, or 4, &c.; for if two quantities of two simples make a balance of loss and gain, as it respects the mean price, so will also the double or treble, the $\frac{1}{2}$ or $\frac{1}{3}$ part, or any other ratio of these quantities, and so on to any extent whatever.

10. *Proof.*—We will now ascertain the price of the mixture by the last rule, thus:

```
20 bushels of oats at 25 cents per bushel=$5.00
 5  - - -  corn at 50  - - - - -        =$2.50
25 - - - - - - - - - - - - - - - -       )7.50(30 cts. A.
```

RULE.

11. *Having reduced the several prices to the same denomination,*

Q. Why does not the operation affect the total value of the commodity? 7, 8 Why is not the result confined to one answer? 9. Rule? 11, 12, 13 *connect by a line each price that is less than the mean rate with one or more that is greater, and each price greater than the mean rate with one or more that is less.*

12. *Place the difference between the mean rate and that of each of the simples opposite the price with which they are connected.*

13. *Then, if only one difference stands against any price, it expresses the quantity of that price; but if there be several, their sum will express the quantity.*

14. A merchant has several sorts of tea, some at 10s., some at 11s., some at 13s. and some at 24s. per lb.; what proportions of each must be taken to make a composition worth 12s. per lb.?

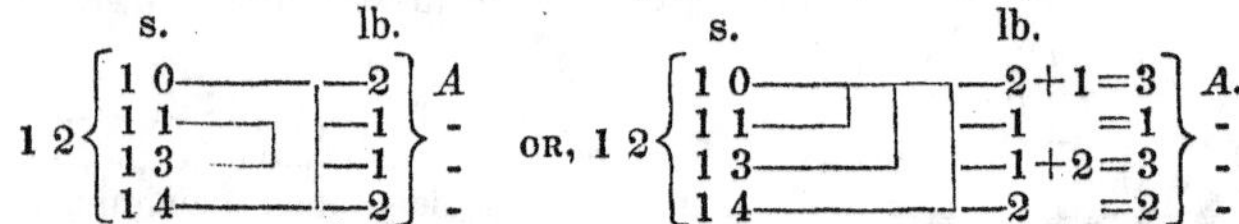

15. How much wine, at 5s. per gallon and 3s. per gallon, must be mixed together, that the compound may be worth 4s. per gallon?

A. 1 gallon of each.

16. How much corn, at 42 cents, 60 cents, 67 cents, and 78 cents, per bushel, must be mixed together, that the compound may be worth 64 cents per bushel? *A.* 14bu. at 42c.; 3 at 60; 4 at 67; 22 at 78.

17. A grocer would mix different quantities of sugar, viz.—one at 20, one at 23, and one at 26 cents per lb.; what quantity of each sort must be taken to make a mixture worth 22 cents per lb.?

A. 5lb. at 20 cents; 2 at 23; 2 at 26.

18. A jeweller wishes to procure gold of 20 carats fine from gold of 16, 19, 21, and 24 carats fine; what quantity of each must he take?

A. 4, 1, 1, 4.

19. We have seen that we can take 3 times, 4 times, $\frac{1}{3}$, $\frac{1}{4}$, or any proportion of each quantity, to form a mixture.

20. Hence, when the quantity of one simple is given, to find the proportional quantities of any compound whatever, after having found the proportional quantities by the last rule, we have the following

RULE.

21. *As the proportional quantity of that piece whose quantity is given is to each proportional quantity, so is the given quantity to the quantities or proportions of the compound required.*

22. A grocer wishes to mix one gallon of brandy, worth 15s. per gallon, with rum worth 8s., so that the mixture may be worth 10s per gallon; how much rum must be taken?

23. By the last rule, the differences are 5 to 2; that is, the proportions are 2 of brandy to 5 of rum; hence, he must take $2\frac{1}{2}$ gallons of rum for every gallon of brandy. *A.* $2\frac{1}{2}$ gallons.

24. A person wishes to mix 10 bushels of wheat, at 70 cents per bushel, with rye at 48 cents, corn at 36 cents, and barley at 30 cents per bushel, so that a bushel of this mixture may be worth 38 cents;

what quantity of each must be taken? We find by the last rule, that the proportions are 8, 2, 10, and 32.

Then, as 8 : 2 : : 1 0 : 2 $\frac{1}{2}$ bushels of rye.
8 : 1 0 : : 1 0 : 1 2 $\frac{1}{2}$ bushels of corn. } Answer.
8 : 3 2 : : 1 0 : 4 0 bushels of barley.

25. How much water must be mixed with 100 gallons of rum, worth 90cts. per gallon, to reduce it to 75cts. per gallon. *A.* 20gal.

26. A grocer mixes teas at $1.20, $1, and 60 cents, with 20lb. at 40c. per lb.; how much of each sort must he take to make the composition worth 80c. per lb. *A.* 20 at $1.20, 10 at $1, 10 at 60c.

27. A grocer has currants at 4 cents, 6 cents, 9 cents, and 11 cents per lb.; and he wishes to make a mixture of 240lb., worth 8 cents per lb.; how many currants of each kind must he take? In this example we can find the proportional quantities by linking, as before; then it is plain that their sum will be in the same proportion to any part of their sum, as the whole compound is to any part of the compound, which exactly accords with the principle of Fellowship.

RULE.

28. *As the sum of the proportional quantities found by linking, as before : is to each proportional quantity : : so is the whole quantity or compound required : to the required quantity of each.*

We will now apply this rule in performing the last question.

8 { 4——3, 6——1, 9——2, 1 1——4 } Then, { 1 0 : 3 : : 2 4 0 : 7 2 lb. at 4 cts.
1 0 : 1 : : 2 4 0 : 2 4 lb. at 6 cts.
1 0 : 2 : : 2 4 0 : 4 8 lb. at 9 cts.
1 0 : 4 : : 2 4 0 : 9 6 lb. at 1 1 cts. } *A.*

29. A grocer, having sugars at 8c., 12c., and 16c. per lb., wishes to make a composition of 120lb., worth 13c. per lb.; what quantity of each must be taken? *A.* 30lb. at 8, 30lb. at 12, 60lb. at 16.

30. How much water, at 0 per gal., must be mixed with wine, at 80c. per gal., so as to fill a vessel of 90gal., which may be offered at 50c per gal.? *A.* 56$\frac{2}{8}$ gallons of wine, and 33$\frac{6}{8}$ gallons of water.

31. How much gold, of 15, 17, 18, and 22 carats fine, must be mixed together, to form a composition of 40 ounces of 20 carats fine?
A. 5oz. of 15, of 17, of 18, and 25oz. of 22.

ARITHMETICAL PROGRESSION.

CIV. 1. Arithmetical Progression, or Series, is any rank of numbers more than two, that increase by a constant addition, or decrease by a constant subtraction, of the same number.

2. The Common Difference is the number added or subtracted as above.

3. An Ascending Series is one formed by a continual addition of the common difference, as 2, 4, 6, 8, 10, &c.

CIV. *Q.* What is Arithmetical Progression? 1. What the Common Difference? 2. An Ascending Series? 3. A Descending Series? 4. Give an example of each 3, 4. What are the terms? 5

4. A DESCENDING ARITHMETICAL SERIES, is one formed by a continual subtraction of the common difference, as 10, 8, 6, 4, 2, &c.

5. THE TERMS are those numbers that form the series, the first and last of which are called the EXTREMES, and the other the MEANS.

6. In Arithmetical Progression there are reckoned five terms, any three of which being given, the remaining two may be found, viz.—

7. 1. *The first term;* 2. *The last term;* 3. *The number of terms;* 4. *The common difference;* 5. *The sum of all the terms.*

8. The first term, the last term, and the number of terms, being given, to find the Common Difference;—

9. A man had 6 sons, whose several ages differed alike: the youngest was 3 years old, and the oldest 28; what was the common difference of their ages?

10. The difference between the youngest son and the eldest, evidently shows the increase of the 3 years by all the subsequent additions, till we come to 28 years; and, as the number of these additions are, of course, 1 less than the number of sons (5), it follows, that, if we divide the whole difference (28−3=), 25, by the number of additions (5), we shall have the difference between the ages of each, that is, the common difference. Thus, 28−3=25; then, 25÷5=5 years, the common difference. *A.* 5 years.

11. Hence, to find the common difference, —*Divide the difference of the extremes by the number of terms, less* 1, *and the quotient will be the common difference.*

12. If the extremes be 3 and 23, and the number of terms 11, what is the common difference? *A.* 2.

13. A man is to travel from Boston to a certain place in 6 days, and to go only 5 miles the first day, increasing the distance traveled each day by an equal excess, so that the last day's journey may be 45 miles; what is the daily increase, that is, the common difference? *A.* 8 miles.

14. If the amount of $1 for 20 years, at simple interest, be $2.20, what is the rate per cent.? In this example, we see the amount of the first year is $1.06 and the last year $2.20, consequently, the extremes are 106 and 220, and the number of terms 20. *A.* $.06=6 per cent.

15. A man bought 60 yards of cloth, giving 5 cents for the first yard, 7 for the second, 9 for the third, and so on to the last; what did the last cost? Since, in the last example, we have the common difference given, it will be easy to find the price of the last yard: for, as there are as many additions as there are yards, less 1, that is, 59 additions of 2 cents to be made to the first yard, it follows, that the last yard will cost 2×59=118 cents more than the first, and the whole cost of the last, reckoning the cost of the first yard, will be 118 +5=$1.23. *A.* $1.23.

16. Hence, when the common difference, the first term, and the

Q. What is the rule for finding the common difference? 11. For finding the last term? 16

number of terms, are given, to find the last term.—*Multiply the common difference by the number of terms, less* 1, *and add the first term to the product.*

17. If the first term be 3, the common difference 2, and the number of terms 11, what is the last term? *A.* 23.

18. A man went from Boston to a certain place in 6 days, traveling the first day 5 miles, the second 8 miles, and each successive day 3 miles farther than the former; how far did he go the last day? *A.* 20 miles.

19. What will $1, at 6 per cent., amount to, in 20 years, at simple interest? The common difference is the 6 per cent.; for the amount of $1, for 1 year, is $1.06, and 1.06+$.06=$1.12, the second year, and so on. *A.* $2.20.

20. A man bought 10 yards of cloth, in arithmetical progression; for the first yard he gave 6 cents, and for the last yard he gave 24 cents; what was the amount of the whole? In this example, it is plain that half the cost of the first and last yards will be the average price of the whole; thus, 6 cts.+24 cts.=30÷2=15 cts., average price; then, 10 yds.×15=$1.50, whole cost. *A.* $1.50.

21. Hence, when the extremes, and the number of terms, are given, to find the sum of all the terms.—*Multiply half the sum of the extremes by the number of terms, and the product will be the answer*

22. If the extremes be 3 and 273, and the number of terms 40, what is the sum of all the terms? *A.* 5520.

23. How many times does a clock strike in 12 hours? *A.* 78.

24. A butcher bought 100 oxen, and gave for the first ox $1, for the second $2, for the third $3, and so on to the last; how much did they come to at that rate? *A.* $5050.

25. What is the sum of the first 1000 numbers, beginning with their natural order, 1, 2, 3, &c.? *A.* 500500.

26. If a board, 18 feet long, be 2 feet wide at one end, and come to a point at the other, what are the square contents of the board? *A.* 18 feet.

27. If a piece of land, 60 rods in length, be 20 rods wide at one end, and at the other terminate in an angle or point. what number of square rods does it contain? *A.* 600.

28. A number of flat stones were laid, 2 yards distant, for the space of 1 mile, from each other, and the first, 2 yards from a certain basket. How far will that man travel who gathers them up singly, and returns with them one by one to the basket? *A.* 881 miles.

29. A person traveling into the country, went 3 miles the first day, and increased every day's travel 5 miles, till at last he went 58 miles in one day; how many days did he travel?

30. We found, in the example 1, the difference of the extreme divided by the number of terms, less 1, gave the common difference: consequently, if, in this example, we divided (58−3=) 55, the difference of the extremes, by the common difference, 5, the quotient 11,

---

Q. The sum of all the terms? 21.

will be the number of terms, less 1; then, $1+11=12$, the number of terms. *A.* 12.

31. Hence, when the extremes and common difference are given, to find the number of terms:—*Divide the difference of the extremes by the common difference, and the quotient, increased by* 1, *will be the answer.*

32. If the extremes be 3 and 45, and the common difference 6 what is the number of terms? *A.* 8.

33. A man being asked how many children he had, replied, that the youngest was 4 years old, and the eldest 32, the increase of the family having been 1 in every 4 years; how many had he? *A.* 8

GEOMETRICAL PROGRESSION.

CV. 1. Geometrical Progression, is any rank or series of numbers, which increases by a constant multiplier, or decreases by a constant divisor.

2. Thus, 3, 9, 27, 81, &c., is an increasing geometrical series; and 81, 27, 9, 3, &c., is a decreasing geometrical series.

3. There are five terms in Geometrical Progression, and, like Arithmetical Progression, any three of them being given, the other two may be found, viz:—

4. 1. *The first term.* 2. *The last term.* 3. *The number of terms* 4. *The sum of all the terms.* 5. *The ratio.*

5. A man purchased a flock of sheep, consisting of 9; and by agreement, was to pay what the last sheep came to, at the rate of $4 for the first sheep, $12 for the second, $36 for the third, and so on, trebling the price to the last; what did the flock cost him?

6. We may perform this example by multiplication; thus, $4\times3\times3\times3\times3\times3\times3\times3\times3=\$26{,}244$. *A.* But this process, you must be sensible, would be, in many cases, a very tedious one; let us see if we cannot abridge it and make it easier.

7. In the above process, we discover that 4 is multiplied by 3 eight times, one time less than the number of terms; consequently, the 8th power of the ratio 3, expressed thus, 3^8, multiplied by the first term, 4, will produce the last term. But, instead of raising 3 to the 8th power in this manner, we need only raise it to the 4th power, then multiply this 4th power into itself; for, in this way, we do, in fact, use the 3 eight times, raising the 3 to the same power as before; thus, $3^4=81$; then $81\times81=6561$; this, multiplied by 4, the first term, gives $26,244, the same result as before. *A.* $26,244.

8. Hence, when the first term, ratio, and number of terms, are given, to find the last term.

RULE.

9. *Write down some of the leading powers of the ratio, with the*

Q. Number of terms? 31.

CV. Q. What is Geometrical Progression? 1. What are the terms? 4 Give ex[illegible] ascending and a descending series. 2.

numbers 1, 2, 3, *&c., over them, being their several indices. Add together the most convenient indices to make an index less by* 1 *than the number of terms sought.*

10. *Multiply together the powers, or numbers standing under those indices; and their product, multiplied by the first term, will be the term sought.*

11. If the first term of a geometrical series be 4, and the ratio 3, what is the 11th term?

1, 2, 3, 4, 5, indices. } *Note.*—The pupil will notice that
3, 9, 27, 81, 243, powers. } the series does not commence with the first term, but with the ratio. The indices 5+3+2=10, and the powers under each, 243×27×9=59,049; which, multiplied by the first term, 4, makes 236,196, the 11th term required.

A. 236,196.

12. The first term of a series, having 10 terms, is 4, and the ratio 3; what is the last term? *A.* 78,732.

13. A sum of money is to be divided among 10 persons; the first to have $10, the second $30, and so on, in threefold proportion: what will the last have? *A.* $196,830.

14. A boy purchased 18 oranges, on condition that he should pay only the price of the last, reckoning 1 cent for the first, 4 cents for the second, 16 cents for the third, and in that proportion for the whole; how much did he pay for them? *A.* $171,798,691.84.

15. What is the last term of a series having 18 terms, the first of which is 3, and the ratio 3? *A.* $387,420,489.

16. A butcher meets a drover, who has 24 oxen. The butcher inquires the price of them, and is answered, $60 per head; he immediately offers the drover $50 per head, and would take all. The drover says he will not take that; but, if he will give him what the last ox would come to, at 2 cents for the first, 4 cents for the second, and so on, doubling the price to the last, he may have the whole. What will the oxen amount to at that rate? *A.* $167,772.16.

17. A man was to travel to a certain place in 4 days, and travel at whatever rate he pleased; the first day he went 2 miles, the second 6 miles, and so on to the last, in a threefold ratio; how far did he travel the last day, and how far in all?

18. In this example, we may find the last term as before, or find it by adding each day's travel together, commencing with the first, and proceeding to the last, thus: 2+6+18+54=80 miles, the whole distance traveled, and the last day's journey is 54 miles. But this mode of operation, in a long series, you must be sensible, would be very troublesome. Let us examine the nature of the series, and try to invent some shorter method of arriving at the same result.

19. By examining the series 2, 6, 18, 54, we perceive that the last term (54,) less 2 (the first term,)=52, is 2 times as large as the sum of the remaining terms; for 2+6+18=26; that is, 54−2=52÷2=26; nence, if we produce another term, that is, multiply 54, the last term,

Q. Give the rule for finding the last term. 9, 10

by the ratio 3, making 162, we shall find the same true of this also; for 162−2 (the first term) =160÷2=80, which we at first found to be the sum of the four remaining terms; thus, 2+6+18+54=80. In both of these operations it is curious to observe, that our divisor, (2,) each time, is 1 less than the ratio (3).

20. Hence, when the extremes and ratio are given, to find the sum of the series, we have the following

RULE.

21. *Multiply the last term by the ratio, from the product subtract the first term, and divide the remainder by the ratio, less* 1; *the quotient will be the sum of the series required.*

22. If the extremes be 5 and 6,400, and the ratio 6, what is the whole amount of the series?

$$\frac{\overline{6400 \times 6 - 5}}{6 - 1} = 7679, \text{ Ans.}$$

23. A sum of money is to be divided among 10 persons in such manner, that the first may have $10, the second $30, and so on, in three-fold proportion; what will the last have, and what will the whole have?

24. The pupil will recollect how he found the *last* term of the series by a foregoing rule: and, in all the cases in which he is required to find the *sum* of the series, when the last term is not given, he must first find it by that rule, and then work for the sum of the series by the present rule. *A* The last, $196,830; and the whole, $295,240

25. A hosier sold 14 pair of stockings, the first at 4 cents, the second at 12 cents, and so on in geometrical progression; what did the last pair bring him, and what did the whole bring him?

A. Last, $63,772.92; whole, $95,659.36.

26. A man bought a horse, and, by agreement, was to give a cent for the first nail, three for the second, &c.; there were four shoes, and in each shoe eight nails; what did the horse come to at that rate?

A. $9,265,100,944,259.20.

27. At the marriage of a lady, one of the guests made her a present of a half-eagle, saying that he would double it on the first day of each succeeding month throughout the year, which, he said, would amount to something like $100; how much did his estimate differ from the true amount? *A*. $20,375.

28. If our pious ancestors, who landed at Plymouth, A. D. 1620, being 101 in number, had increased so as to double their number in every 20 years, how great would have been their population at the end of the year 1840? *A*. 206,747.

ANNUITIES AT SIMPLE INTEREST.

CVI. 1. An ANNUITY is a sum of money, payable every year, for a certain number of years, or forever.

CVI. *Q*. What is an annuity? 1.

2. When the annuity is not paid at the time it becomes due, it is said to be in *arrears*.

3. The sum of all the annuities, such as rents, pensions, &c. remaining unpaid, with the interest on each, for the time it has been due, is called the *amount* of the annuity.

4. Hence, to find the amount of an annuity—*Calculate the interest on each annuity for the time it has remained unpaid, and find its amount; then the sum of all these several amounts will be the amount required.*

5. If the annual rent of a house, which is $200, remain unpaid (that is, in arrears) 8 years, what is the amount?

6. In this example, the rent of the last (8th) year being paid when due, of course there is no interest to be calculated on that year's rent.

The amount of $200 for 7 years=$284
The amount of $200 for 6 years=$272
The amount of $200 for 5 years=$260
The amount of $200 for 4 years=$248
The amount of $200 for 3 years=$236
The amount of $200 for 2 years=$224
The amount of $200 for 1 year =$212
The eighth year, paid when due,=$200
$1,936 *A.*

7. If a man, having an annual pension of $60, receive no part of it till the expiration of 8 years, what is the amount then due? *A.* $580.80.

8. What would an annual salary of $600 amount to, which remains unpaid (or in arrears) for 2 years?—1,236. For 3 years?—1,908. For 4 years?—2,616. For 7 years?—4,956. For 8 years?—5,808. For 10 years?—7,620. *A.* Total, $24,144.

9. What is the present worth of an annuity of $600, to continue 4 years? The present worth [LXXXIII.] is such a sum as, if put at interest, would amount to the given annuity; hence,

$600÷$1.06=$566.037, present worth, 1st year.
$600÷$1.12=$535.714, present worth, 2d year.
$600÷$1.18=$508.474, present worth, 3d year.
$600÷$1.24=$483.870, present worth, 4th year.
$2,094.095, present worth required.

10. Hence, to find the present worth of an annuity.—*Find the present worth of each year by itself, discounting from the time it becomes due, and the sum of all these present worths will be the answer.*

11. What sum of ready money is equivalent to an annuity of $200, to continue 3 years, at 4 per cent.? *A.* $556.063.

12. What is the present worth of an annual salary of $800, to continue 2 years?—1,469.001. 3 years?—2,146.967? 5 years?—3,407.512. *A.* Total, $7,023.48.

Q. What is meant by *arrears* and *amount?* 2, 3. Rule for finding the amount? 4. For finding the present worth? 10.

ANNUITIES AT COMPOUND INTEREST

CVII. 1. *The amount of an annuity at simple and compound interest is the same, excepting the difference in interest.*

2. Hence, to find the amount of an annuity at compound interest.—*Proceed as in* CVI., *reckoning compound instead of simple interest.*

3. What will a salary of $200 amount to, which has remained unaid for 3 years?

The amount of $200 for 2 years=$224.72
The amount of $200 for 1 year =$212.00
The 3d year - - - - - - - =$200.00
A. $636.72

4. If the annual rent of a house, which is $150, remain in arrears for 3 years, what will be the amount due for that time? *A.* 477.54.

5. Calculating the amount of the annuities in this manner, for a long period of years, would be tedious. This trouble will be prevented, by finding the amount of $1, or £1, annuity, at compound interest, for a number of years, as in the following

TABLE I.

Showing the amount of $1, or £1, annuity, at 6 per cent., compound interest, for any number of years, from 1 to 50.

Y.	6 per cent.	Y.	6 per cent.	Y.	6 per cent.	Y.	6 per cent.	Y.	6 per cent.
1	1.0600	11	14.9716	21	39.9927	31	84.8016	41	165.0467
2	2.0600	12	16.8699	22	43.3922	32	90.8897	42	175.9495
3	3.1836	13	18.8821	23	46.9958	33	97.3431	43	187.5064
4	4.3746	14	21.0150	24	50.8155	34	104.1837	44	199.7568
5	5.6371	15	23.2759	25	54.8645	35	111.4347	45	212.7423
6	6.9753	16	25.6725	26	59.1563	36	119.1208	46	226.5068
7	8.3938	17	28.2123	27	63.7057	37	127.2681	47	231.0972
8	9.8974	18	30.9056	28	68.5281	38	135.9042	48	245.9630
9	11.4913	19	33.7599	29	73.6397	39	145.0584	49	261.7208
10	13.1807	20	36.7855	30	79.0581	40	154.7619	50	278.4241

It is evident, that the amount of $2 annuity is 2 times as much as one of $1; and one of $3, 3 times as much.

6. Hence, to find the amount of an annuity, at 6 per cent.—*Find, by the table, the amount of* $1, *at the given rate and time, and multiply it by the given annuity, and the product will be the amount required.*

7. What is the amount of an annuity of $120, which has remained unpaid 15 years? The amount of $1, by the table, we find to be $23.2759; therefore, $23.2759 × 120=$2,793.108. *A.*

8. What will be the amount of an annual salary of $400, which has been in arrears 2 years?—824. 3 years?—1,273.44. 4 years?—

CVII. *Q.* What is meant by the amount of an annuity at compound interest? 1. How is it found? 2.

1,749.84. 6 years?—2,790.12. 12 years?—6,747.96. 20 years?—14,714.2. *A.* Total, $28,099.56.

9. If you lay up $100 a year, from the time you are 21 years of age till you are 70, what will be the amount at compound interest? *A.* $26,172.08.

10. What is the present worth of an annual pension of $120, which is to continue 3 years?

11. In this example, the present worth is evidently that sum which, at compound interest, would amount to as much as the amount of the given annuity for the three years. Finding the amount of $120 by the table, as before, we have $382.032; then, if we divide $382.032 by the amount of $1, compound interest, for 3 years, the quotient will be the present worth. This is evident from the fact, that the quotient, multiplied by the amount of $1, will give the amount of $120, or, in other words, $382.032. The amount of $1 for 3 years at compound interest is $1.19101; then, $382.032÷$1.19101= $320.763, *A.*

12. Hence, to find the present worth of an annuity.—*Find its amount in arrears for the whole time; this amount, divided by the amount of $1 for said time, will be the present worth required.*

13. NOTE.—The amount of $1 may be found, ready calculated, in the table of compound interest, [LXXXII.]

14. What is the present worth of an annual rent of $200, to continue 5 years? *A.* $842.472.

15. The operations in this rule may be much shortened by calculating the present worth of $1 for a number of years, as in the following

TABLE II.

Showing the present worth of $1, or £1, annuity, at 6 per cent., compound interest, for any number of years, from 1 to 32.

Y.	6 per cent.	Y.	6 per cent.	Y.	6 per cent.	Y.	6 per cent.
1	0.94339	9	6.80169	17	10.47726	25	12.78335
2	1.83339	10	7.36008	18	10.82760	26	13.00316
3	2.67301	11	7.88687	19	11.15811	27	13.21053
4	3.46510	12	8.38384	20	11.46992	28	13.40616
5	4.21236	13	8.85268	21	11.76407	29	13.59072
6	4.91732	14	9.29498	22	12.04158	30	13.76483
7	5.58238	15	9.71225	23	12.30338	31	13.92908
8	6.20979	16	10.10589	24	12.55035	32	14.08398

16. To find the present worth of any annuity by this table, we have only to multiply the present worth of $1, found in the table, by the given annuity, and the product will be the present worth required.

17. What sum of ready money will purchase an annuity of $300, to continue 10 years? The present worth of $1 annuity, by the Table, for 10 years, is $7.36008; then 7.36008×300=$2,208.024, *A.*

Q. How is the present worth found? 12

18. What is the present worth of a yearly pension of $60, to continue 2 years?—110.0034. 3 years?—160.3806. 4 years?—207.906 8 years?—372.5874. 20 years?—688.1952. 30 years?—825.8898 Total, $2,364.9624.

19. What salary, to continue 10 years, will $2,208.024 purchase? This example is the 17th example reversed; consequently, $2.208.024 ÷7.36008=300, the annuity required. *A.* $300.

20. Hence, to find that annuity which any given sum will purchase.—*Divide the given sum by the present worth of* $1 *annuity for the given time, found by Table* II.; *the quotient will be the annuity required.*

21. What salary, to continue 20 years, will $688.195 purchase? *A.* $60+.

22. What annuity, to continue 10 years, is equivalent to $3,680.04? *A.* $500.

23. To divide any sum of money into annual payments, which, when due, shall form an equal amount at compound interest.—*First find an equivalent annuity as above,* (20,) *then its present worth for each required period of time.* (15.)

24. A certain manufacturing establishment in Massachusetts was actually sold for $27,000, and the sum divided into four notes, payable annually, so that the principal and interest of each, when due, should form an equal amount, at compound interest, and the several principals, when added together, should make $27,000; now, what were the principals of said notes?*

A.
The first note is $7,350.915; amount for 1 year, $7,791.97032.
The second note is $6,934.825; amount for 2 years, $7,791.97032.
The third note is $6,542.288; amount for 3 years, $7,791.97032.
The fourth note is $6,171.970; amount for 4 years, $7,791.97032.
Proof—$27,000, lacking 2 mills

PERMUTATION.

CVIII. 1. **Permutation** is the method of finding how many different ways any number of things may be changed.

2. How many changes may be made of the first three letters of the alphabet? In this example, had there been but two letters, they could only be changed twice; that is, a, b, and b, a; that is, 1×2=2: but, as there are three letters, they may be changed 1×2×3=6 times, as follows—1, a, b, c; 2, a, c, b; 3, b, a, c; 4, b, c, a; 5, c, b, a; 6, c, a, b.

Q. What is the rule for finding what sum a given annuity will purchase? 20. CVIII. *Q.* What is Permutation? 1. How many changes can be made with the first three letters of the alphabet? 2. What are they? 2. Rule? 3.

*The annuity which $27,000 will purchase, found as before, is 7,791.97032+. To obtain an exact result, we must reckon the decimals, which were rejected in forming the tables. This makes the last divisor 3.4651056.

3. Hence, to find the number of different changes or permutations which may be made with any given number of different things.—*Multiply together all the terms of the natural series, from* 1 *up to the given number, and the last product will be the number of changes required.*

4. How many different ways may the first five letters of the alphabet be arranged? *A.* 120.

5. How many changes may be rung on 15 bells, and in what time may they be rung, allowing 3 seconds to every round?

A. 1,307,674,368,000 changes; 3,923,023,104,000 seconds.

6. What time will it require for 10 boarders to seat themselves differently every day at dinner, allowing 365 days to the year?

A. 9,941$\frac{335}{365}$ years.

7. Of how many variations will the 26 letters of the alphabet admit? *A.* 403,291,461,126,605,635,584,000,000.

POSITION.

Position is a rule which teaches, by the use of supposed numbers, to find true ones. It is divided into two parts, called Single and Double.

SINGLE POSITION.

CIX. 1. Single Position teaches to resolve those questions whose results are proportional to their suppositions.

2. A schoolmaster, being asked how many scholars he had, replied, "If I had as many more as I now have, one half as many more, one third and one fourth as many more, I should have 296." How many had he?

Suppose he had	24
As many more=	24
$\frac{1}{2}$ as many=	12
$\frac{1}{3}$ as many=	8
$\frac{1}{4}$ as many=	6
	74

We have now found that we did not suppose the right number. If we had, the amount would have been 296. But 24 has been increased in the same manner to amount to 74, that some unknown number, the true number of scholars, must be, to amount to 296. Consequently, it is obvious, that 74 has the same ratio to 296 that 24 has to the true number. The question may, therefore, be solved by the following statement: As 74 : 296 : : 24 : 96, *A.*

3. This answer we prove to be right by increasing it by itself, one half of itself, one third of itself, and one fourth of itself, as, 96+96+48+32+24=296.

RULE.

4. *Suppose any number you choose, and proceed with it in the same manner you would with the answer, to see if it were right; then say, as this result : the result in the question : : the supposed number : number sought.*

CIX. Q. What is Position? Single Position? 1. Rule? 4.

5. James lent William a sum of money on interest, and in 10 years it amounted to $1,600; what was the sum lent? *A.* $1,000.

6. Three merchants gained, by trading, $1,920, of which A took a certain sum, B took 3 times as much as A, and C four times as much as B; what share of the gain had each?

A. A, $120; B, $360; C, $1,440.

7. A person having about him a certain number of crowns, said if a third, a fourth, and a sixth, of them were added together, the sum would be 45; how many crowns had he? *A.* 60.

8. What is the age of a person, who says, that if $\frac{6}{10}$ of the years he has lived be multiplied by 7, and $\frac{2}{3}$ of them be added to the product, the sum would be 292? *A.* 60 years.

9. What number is that, which, being multiplied by 7, and the product divided by 6, the quotient will be 14? *A.* 12

DOUBLE POSITION.

CX. 1. Double Position teaches to solve questions by means of two supposed numbers.

2. In Single Position, the number sought is always multiplied or divided by some proposed number, or increased or diminished by itself, or some known part of itself, a certain number of times. Consequently, the result will be proportional to its supposition, and but one supposition will be necessary; but, in Double Position, we employ two, for the results are not proportional to the suppositions.

3. A gentleman gave his three sons $10,000, in the following manner; to the second $1000 more than to the first, and to the third as many as to the first and second? What was each son's part?

Let us suppose the share of the first 1,000
Then the second=2,000
Third=3,000
Total, 6,000
This subtracted from 10,000, leaves 4,000

The shares of all the sons will, if our supposition be correct, am't to 10,000; but, as they amount to $6,000 only we call the error 4000.

Suppose again, that the share of the first was 1,500
Then the second=2,500
Third=4,000
8,000
2,000

We perceive the error in this case to be 2000.

4. The first error, then, is $4,000, and the second $2,000. Now, the difference between these errors would seem to have the same relation to the difference of the suppositions, as either of the errors would have to the difference between the supposition which produced it, and the true number. We can easily make this statement, and ascertain whether it will produce such a result:

5. As the difference of errors, 2,000 : 500 difference of suppositions :: either of the errors (say the first,) 4,000 : 1,000, the difference

between its supposition and the true number Adding this difference to 1,000, the supposition, the amount is 2,000 for the share of the first son: then $3,000 that of the second, $5,000 that of the third, Ans. For 2,000+3,000+5,000=10,000, the whole estate.

6. Had the supposition proved too great, instead of too small, it is manifest that we must have subtracted this difference. The differences between the results and the result in the question are called *errors;* these are said to be *alike*, when both are either too great or too small: *unlike*, when one is too great, and the other too small.

RULE.

7. *Suppose any two numbers, and proceed with each according to the manner described in the question, and see how much the result of each differs from that in the question.*

8. *Then say, as the difference* of the errors : the difference of the suppositions :: either error : difference between its supposition and the number sought.*

9. Three persons disputing about their ages, says B, "I am 10 years older than A;" says C, "I am as old as you both:" now, what were their several ages, the sum of them all being 100?

A. A's, 20; B's, 30; C's, 50.

10. Two persons, A and B, have the same income: A saves $\frac{1}{4}$ of his yearly; but B, by spending $150 per annum more than A, at the end of 8 years, finds himself $400 in debt; what is their income, and what does each spend per annum?

A. A's income $400; A spends $300; B $450.

11. There is a fish whose head is 8 feet long, his tail is as long as his head and half his body, and his body is as long as his head and tail; what is the whole length of the fish. *A.* 64 feet.

12. A laborer contracted to work 80 days for 75 cents per day, and to forfeit 50 cents for every day he should be idle during that time He received $25: now how many days did he work, and how many days was he idle? *A.* 52 days; idle 28.

MENSURATION.

CXI. 1. MENSURATION is the measuring of Surfaces and Solids.

OF ANGLES.

2. AN ANGLE is the inclination or opening of two lines that meet each other, as in the Figures on next page. The point of intersection is called the *Angular point;* and in common language, the *Angle*

3. An ANGLE is greater or less, not according to the length of the

CXI. *Q.* What is Mensuration? 1. An Angle? 2. The point of intersection? 2. How is the size of an angle determined? 3.

* The difference of the errors, when alike, will be one *subtracted* from the other when unlike, one *added* to the other.

lines, but according as they are more or less inclined or opened; thus, the angle at C, below, is the greatest of the three.

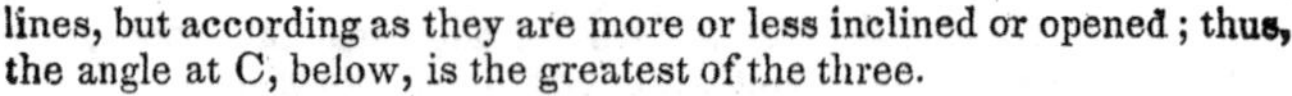
FIG. 1 FIG. 2. FIG. 3.

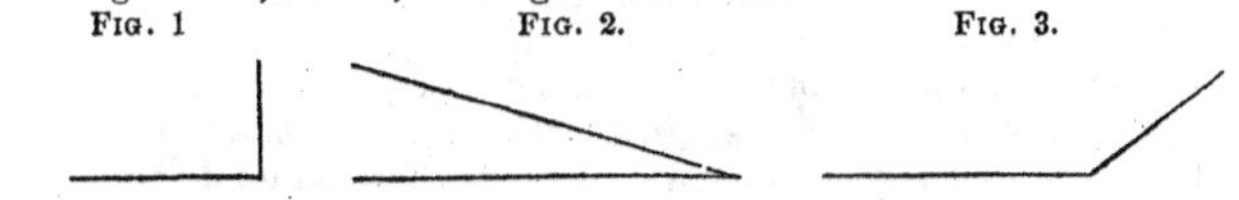
Right Angle, A Acute Angle. B Obtuse Angle, C

4. A RIGHT ANGLE is one formed by a line drawn perpendicular to another: as A, in Fig. 1.

5. OBLIQUE ANGLES are those formed by oblique lines, and are either Acute or Obtuse; as B and C.

6. An OBTUSE ANGLE is greater, and an ACUTE ANGLE is less than a right angle.

OF TRIANGLES.

7. A TRIANGLE is a plane* figure that has three sides and three angles; as in the following Figures.

FIG. 4. FIG. 5. FIG. 6. FIG. 7.

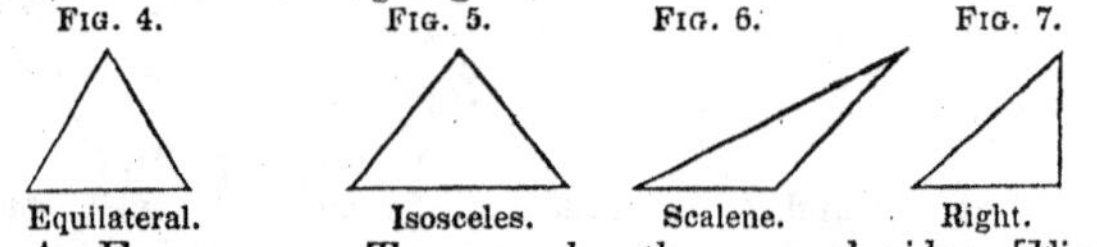
Equilateral. Isosceles. Scalene. Right.

8. AN EQUILATERAL TRIANGLE has three equal sides. [Fig. 4.]

9. AN ISOSCELES TRIANGLE has two equal sides. [Fig. 5.]

10. A SCALENE TRIANGLE has three unequal sides. [Fig. 6.]

11. A RIGHT-ANGLED TRIANGLE has one right angle. [Fig. 7.]

12. AN OBTUSE-ANGLED TRIANGLE has an obtuse angle. [Fig. 6.]

13. AN ACUTE-ANGLED TRIANGLE has three acute angles. [Fig. 4.]

14. In a right-angled triangle the longest side is the HYPOTHENUSE and the other two sides the LEGS, or the BASE and PERPENDICULAR. In other triangles the longest side is usually considered the Base.

FIG. 8.

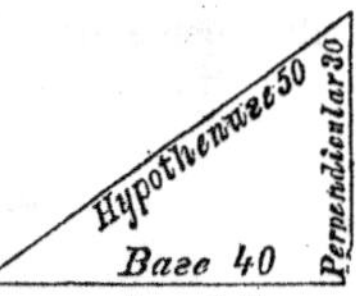

15. In every right-angled triangle,—*The square of the hypothenuse is equal to the sum of the squares of the other two sides; as,* $50^2 = 40^2 + 30^2$. [Fig. 8.]

16. Hence, to find the different sides, we may proceed as follows

To find the hypothenuse.—*Add the squares of the two legs together, and extract the square root of that sum.* To find either leg *From the square of the hypothenuse subtract the square of the given leg, and the square root of the remainder will be the other leg.*

Q. How is a right angle formed? 4. What are oblique angles? 5. Obtuse? 6. Acute? 6. What is a triangle? 7. An equilateral triangle? 8. An Isosceles? 9. Scalene? 10. Right angled triangle? 11. Obtuse angled triangle? 12. Acute angled triangle? 13. What are the names of the sides in a right angled triangle? 14. How are each found? 16. On what principle is each operation based? 15.

---

* PLANE, [L. *Planus.*] An even or level surface, like *plain* in common language. An instrument used in smoothing boards.

☞ The learner will discover the application of the rule better by drawing a triangle on his slate, like Fig. 8, and noting the sides which are intended to correspond with those which are given in the question. Indeed, without some such illustration, he will scarcely be able to apply the rule at all, except in cases where the particular sides are designated.

17. Required the hypothenuse of a right-angled triangle whose legs are 24 and 32 feet. *A.* 40 feet.

18. Required the base of a right-angled triangle the other sides of which are respectively 15 and 25 feet. *A.* 20 rods.

19. Suppose a lot of land lies in the form of a right-angled triangle, and that the longest side is 100 rods and the shortest 60; what is the distance around it? *A.* 240 rods.

20. A river 80 yards wide passes by a fort, the walls of which are 60 yards high; now what is the distance from the top of the wall to the opposite bank of the river. *A.* 100 yards.

21. A ladder 40 feet long may be so placed upon one side of a street that it will reach a window 32 feet from the ground; and without moving it at the bottom, it will reach a window on the other side 24 feet high; what is the width of the street? *A.* 56 feet.

22. There is a certain elm, 2 feet in diameter, growing in the centre of a circular island, the distance from the top of the tree in a straight line to the water is 120 feet, and the distance from the foot 89ft.; what is the height of the tree? (89 + 1 base.) *A.* 79.372ft.+.

23. When two ships, which sailed from the same port, have gone, one due north 40 leagues, and the other due east 30 leagues, how far are they apart then? *A.* 50 leagues.

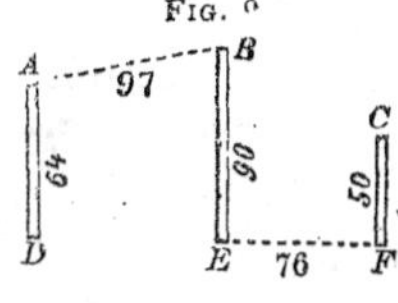

Fig. 9

24. There are three towers, A, B, and C, standing in a direct line, the heights of which are 64, 90, and 50 feet respectively. The distance between the top of the tower A and that of B, is 97 feet, and the distance between the bottom of the tower B and that of C is 76 feet. From these data please inform me what are the several distances from the top of A to the bottom of B, from the top of B to the bottom of A, from the bottom of A to the bottom of B, from the bottom of B to the top of C, from the bottom of C to the top of B, and from the top of B to the top of C. *A.* D E, 93.45+; A E, 113.26+; B D, 129.74+; C E, 90.97+; B F, 117.79+; B C, 85.883+

25. A castle wall there was, whose height was found
To be 100 feet from th' top to th' ground;
Against the wall a ladder stood upright,
Of the same length the castle was in height:
A waggish youngster did the ladder slide
(The bottom of it) 10 feet from the side;
Now I would know how far the top did fall,
By pulling out the ladder from the wall? *A* 6 in. nearly.

26. A gentleman has a garden in the form of an equilateral triangle, the sides of which are each 50 feet; at each corner of the garden stands a tower; the height of the tower A is 30 feet, that of B 34 feet, and that of C 28 feet. At what distance from the bottom of each of these towers must a ladder of the same length with each side be placed, that it may just reach the top of each tower, allowing the ground of the garden to be horizontal?

A. 40ft.; 36.66ft. +; 41.42ft.

OF SURFACES.

27. A QUADRILATERAL has four sides and four angles. PARALLELOGRAM is a general name for all *quadrilateral* figures, that have at least their opposite sides and angles equal; as below.

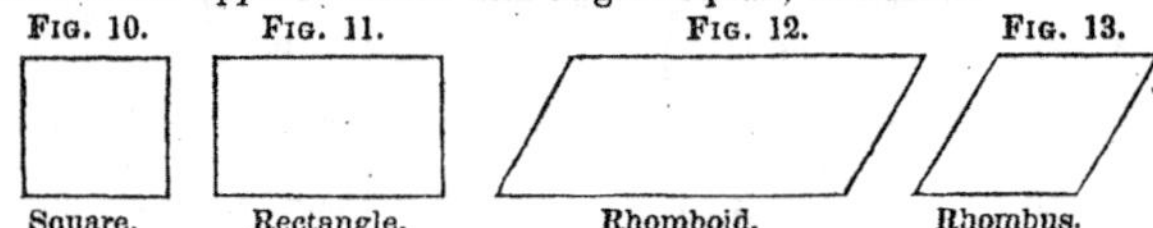

FIG. 10. Square. FIG. 11. Rectangle. FIG. 12. Rhomboid. FIG. 13. Rhombus.

28. A SQUARE is a quadrilateral that has its sides equal and its angles right angles, as Fig. 10.

29. A RECTANGLE is a quadrilateral that has its angles right angles, and only its opposite sides equal, as Fig. 11.

30. RHOMBOID is a quadrilateral which has its opposite sides and angles equal, two of its angles being acute, and two obtuse, as Fig. 12.

31. A RHOMBUS is a quadrilateral that has its sides equal, but its angles like those of a rhomboid, as Fig. 13.

32. A POLYGON is a rectilineal figure of more than four sides, which when they are all equal, form regular polygons. *Squares* are also regular figures; so are *Triangles*, when they are equilateral.

33. The PERIMETER of any plane *rectilineal** figure, is the entire distance round it; *and is found by adding together all the sides that bound it.*

34. A CIRCLE is a plane figure bounded by a curved line, called the Circumference or Periphery; which is every where equally distant from a certain point within it, called the CENTRE. An ARC is any part of the circumference.

Fig. 14.

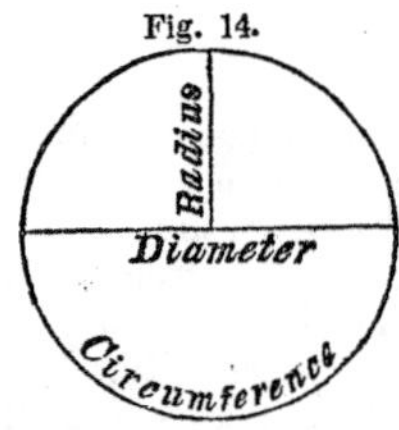

35. The DIAMETER of a circle, is a straight line drawn through the centre and terminating in the circumference on each side. A

Q. What is a Quadrilateral? 27. Parallelogram? 27. Square? 28. Rectangle? 29. Rhomboid? 30. Rhombus? 31. Polygon? 32. What other regular figures are mentioned? 32. What is the perimeter of a figure? 33. How is it found? 33. What is a circle? 34. Circumference? 34. An Arc? 34. Diameter? 35. Chord? 35. Segment? 35.

* RECTILINEAR or RECTILINEAL, [L. *rectus*, *right* or *straight*, and *linea*, a line.] consisting of right or straight lines; straight.

CHORD is a straight line shorter than the diameter, and joins the extremities of an arc. The arc and chord together form a Segment.

36. Every chord of a circle divides it into two unequal parts, and every diameter into two equal parts called SEMICIRCLES, that is, half circles. A RADIUS is half the diameter, or a right line drawn from the centre to the circumference; two or more such lines are called Radii.

FIG. 15. FIG. 16. FIG. 17.

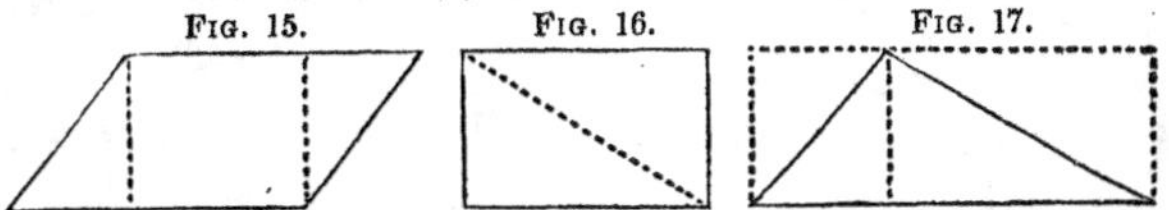

37. A PERPENDICULAR of a quadrilateral or triangle is a straigh line drawn to a point in the base from the angle opposite to that point, as the dotted lines in Fig. 15.

38. From an inspection of Fig. 15, it appears that the area or surface of a Rhombus or Rhomboid, is equal to the area of a square or parallelogram of the same length, but whose breadth is its perpendicular height.

39. A DIAGONAL is a line that passes across a quadrilateral from one angle to its opposite one.

40. The diagonal of every parallelogram divides it into two equal parts, as in Fig. 16.

41. Hence the area of every right-angled triangle is just half as much as the area of that square or rectangle whose length and breadth are equal to the base and perpendicular of the triangle.

42. Every oblique-angled triangle may, by drawing a perpendicular to its base, from its opposite angle, be formed into two right angles.

43. Hence the area of every oblique-angled triangle is just half as much as the area of that square or rectangle, whose length and breadth are equal to the longest side and perpendicular of the triangle.

RULES FOR FINDING THE AREAS OF SUPERFICES.

44. To find the area of a square or rectangle.—*Multiply the length by the breadth.*

45. To find the area of a rhomboid or rhombus.—*Multiply it length by its perpendicular height.* (See 38.)

46. To find the area of a right-angled triangle.—*Multiply the base by half the perpendicular, or the base by the whole perpendicular, and take half of the product.* (See 43.)

47. To find the area of an oblique-angled triangle.—*Multiply the base by half the perpendicular, drawn from the opposite angle.* (See 41, 42.)

48. Or, *halve the sum of the three sides, subtract the three sides*

Q. With what is the area of a rhomboid or rhombus compared? 38. What is a diagonal? 39. What comparison is made between the area of a right-angled triangle and that of a square? 41. What two equal divisions may be made of an oblique-angled triangle? 42. What is the inference? 43. What is the rule for finding the area of a square? 44.—of a rhomboid? 45.—of a right-angled triangle? 46.—of an oblique-angled triangle? 47, 48.

severally from this half sum, multiply the four results together and find the square root of the product.

COMMON RULES RESPECTING CIRCLES.

* 49. The diameter is to the circumference nearly as 7 : 22, or more accurately as 113 : 355, or decimally, as 1 : 3.14159 nearly, therefore,

50. To find the circumference,—*Either multiply the diameter by* 22, *and divide by* 7 ; *or multiply by* 355 *and divide by* 113 ; *or simply multiply by* 3.14159.

51. To find the diameter,—*Reverse the foregoing processes.*

52. To find the area of a circle :—*Multiply half the circumference by half the diameter, or the whole circumference by half the radius.*†

53. Suppose one field is 60 rods square, and another contains 60 square rods : what is the difference in their areas expressed in acres? *A.* 22A. 20rd.

54. If a site for a house is in the form of a square with 150 feet front, what is the area and what its perimeter? *A.* $82\frac{78}{121}$sq. rd.: $36\frac{4}{11}$rd. round.

55. Suppose you contract to have four floors made at $.75 per square yard, one to be 50 feet square and the other three each 20 feet square. What will be the difference between the cost of the first and that of the others? *A.* $$108\frac{1}{3}$.

56. What will be the length of each side of a square formed from an area of 10 acres? *A.* 40 rods.

57. A gentleman has two valuable building lots, one containing 40 square rods, and the other 60; for which his neighbor offers him a square field containing four times as much as his lots. How many rods in length must each side of the square be? *A.* 20 rods.

58. What are the contents of 27 boards, each 13 feet long, and 18 inches wide? *A.* $526\frac{1}{2}$.

59. Suppose that ten boards are each 15 feet long, and together contain 155 sq. feet, what may be the average width of each board? *A.* $1\frac{1}{30}$ ft.

Q. What for finding the circumference of a circle? 50.—the diameter of a circle? 51.—area of a circle? 52.

---

* MORE ACCURATE RULES.

1. To find the circumference of a circle :—*Multiply the diameter by* 3.14159 : *or the area by* 12 56636217, *and extract the square root of the product.*

2. To find the side of a square equal to a given circle :—*Multiply the diameter by* 886227, *or the circumference by* .282094.

3. To find the side of an equilateral triangle inscribed in a circle :—*Multiply the diameter by* .866024, *or the circumference by*.2756646.

4. To find the side of a square inscribed in a circle :—*Multiply the diameter by* .707016, *or the circumference by* .225079.

5. To find the area of a circle :—*Multiply the square of the diameter by* .785398, *or the square of the circumference by* .079577525.

6. To find the diameter of a circle :—*Multiply the circumference by* .31831 ; *or the area by* 1.273241, *and extract the square root of the product.*

† The exact ratio of the diameter to the circumference of a circle has never yet been ascertained, though some have exhibited an approximation, which is supposed not to vary one millionth part of a hair's breadth, in the sun's distance from the earth.

60. When a board is six inches wide, how long must it be to contain 1 sq. ft.?—3 sq. feet?—7 sq. ft.?—12 sq. ft.? *A.* Total 46ft.

61. If a road 150 miles long and 4 rods wide, would cost, when completed, $2 per square rod, what would the land cost by the acre, allowing the cost of making the road to be $2 per rod, (linear measure)? *A.* $240 per acre.

62. Suppose a square has an area of 7500 square yards; what is the breadth of a walk round it that shall take up just two thirds of the square? *A.* 18.3013yd. nearly.

63. If a rhomboid is 50 feet long and 40 feet wide, what is its area? *A.* 200ft.

64. If one rhombus be 60 feet long, with a breadth of 15 feet, and another 45 feet long, with a breadth of 20 feet, what is the difference in their areas? *A.* Nothing.

65. If a rhomboid be 80 feet long and 60 feet wide, what is the sum of the areas of the two ends which when cut off will leave the remainder in the shape of a square? *A.* 1,200 feet.

66. Herodotus estimated the largest and most remarkable of the Egyptian pyramids to be 800 feet square at its base. Now, how long a road 4 rods wide would occupy as much land as the base of the pyramid? *A.* 1m. 6fur. $27\frac{757}{1089}$rd.

67. Suppose the hypothenuse and perpendicular of a right angled triangle be 50 and 30 feet, what is the area? *A.* 600 feet.

68. If the sides of an oblique angled triangle be 40, 50, and 80 feet, what is the area? *A.* 818+sq. ft.

69. If the sides of a triangle be 16.6; 18.32, and 28.6, what is its area? *A.* 143 nearly.

70. Suppose a field has one right angle, and its hypothenuse and base are 100 and 80 rods; how many acres does it contain? *A.* 15A.

71. Suppose a piece of land in the form of a right angled triangle, whose angles are respectively 120 and 160 rods: what is the area? *A.* 60A.

72. What is the circumference of a circle whose diameter is 15? ($15\times355\div113$.) *A.* 47.12+.

73. What is the diameter of a circle whose circumference is 350? *A.* 111.4 +.

74. What is the area of a circle whose diameter is 24, and circumference 75? *A.* 450.

75. What is the area of a circle whose diameter is 24? For the method of solving several questions, see reference from 49, above. *A.* 452.3904+.

76. What is the area of a circle whose circumference is 75? *A.* 447.61875+.

77. If the diameter of a circle be 24, what is the length of one side of a square equal to the circle? *A.* 21.269+.

78. When the circumference of a circle is 75, what is the side of a square equal to the circle? *A.* 21.157+

79. What is the diameter of a circle whose area is 115? *A.* 12.1+.

80. When the side of a square is 10.5 what is the diameter of a circle which is equal to the square? *A.* 11.847+.

81. When the side of a square is 10.5 what is the circumference of a circle equal to the square? *A.* 37.224+.

82. When the diameter of a circle is 12, what is the area of a semi-circle formed from that circle? *A.* 56.548+.

83. Suppose a tract of land is 5 miles long and 3 miles wide, what is the distance round a square of an equal area? *A.* 15m. 3fur. $37\frac{2}{5}$rd.

84. If a field be 48 rods long and 10 rods wide, what will be the diameter of a circle of equal area? Having found the area of a circle, find the sum of the areas of a square inscribed, and one circumscribed;* also the side of a triangle inscribed. *A.* Diameter, 24.721+ rods; areas, 916.635 + sq. rd.; side, 21.408+rd.

85. How long will it take a man, going at the rate of 10 miles in 2 hours, to travel round an area of 256,000 acres, laid out so that the circumference shall be the shortest distance possible that will contain the given area? *A.* 14h. $10\frac{97}{125}$m.

86. A circle has an area of 308 square rods, and is to be divided into four equal concentric[1] circles; what will be the width of each circular part? *A.* 9.9+rd.; 2.05+rd.; 1.57+rd.; 1.33+rd.

87. The radii of two concentric circles are 10 and 12 yards. What is the area included between them? *A.* 138.22996+yd.

88. There is a meadow of 10 acres in the form of a square, and a horse tied equidistant from each angle or corner. What must be the length of the rope that will permit the horse to graze over every part of the meadow? *A.* 28.284+rd.

89. In the midst of a meadow well stored with grass,
I've taken just two acres to tether my ass;
Then how long must the cord be, that, feeding all round,
He mayn't graze less or more than two acres of ground?
A. 10.0925+rd.

90. What is the perimeter of Fig. 18, and what the area of its lots, *a* and *b*? *A.* Perimeter, 101.4159rd.; *a*=300 sq. rd.; *b*=150 sq. rd.

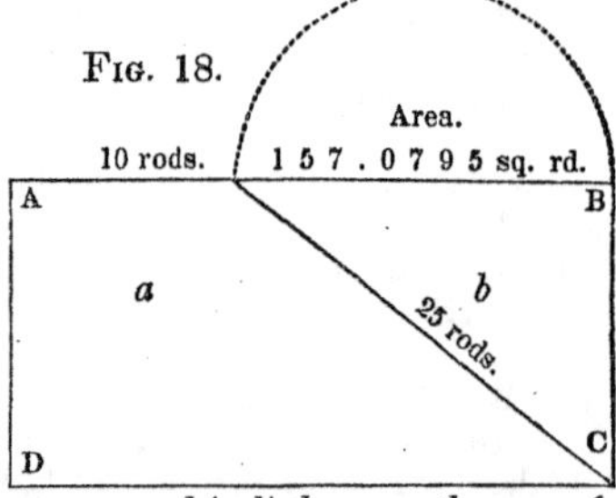

Fig. 18.

91. Since an acre is equal to a rectangle, which is 40 poles=10 chains=1,000 links in length, and 4 poles=1 chain=100 links in breadth, it will contain 1,000 × 100 = 100,000 square links, therefore—

92. *If the linear dimensions be expressed in links, and the superfi-*

* The diameter of the circle is of course the length of one side of a circumscribed square.

1 CONCENTRIC, [L. *concentrico*.] Having a common centre.

cial contents be found, these results, when divided by 100,000, *or with five figures cut off towards the right, will give the number of acres and parts of an acre, expressed in decimals.*

93. The length of a rectangular field being 25 chains 8 links, and its breadth 14 chains 75 links, what number of acres does it contain? 25 chains 8 links=2,508 links, and 14 chains 75 links=1,475 links; then 2,508 × 1,475=36.99300 acres=36 acres 3 roods 38.88 poles.

A. 36A. 3r. $38\frac{22}{25}$rd.

94. Find the area of a square field whose side is $10\frac{1}{2}$ chains.

A. 11A. 4rd.

95. The base of a triangular field is 16 chains 3 poles, and its perpendicular 6 chains 2 poles; what number of acres does it contain?

A. 5A. 1R. 31rd.

96. What is the length of the side of a square field comprising 2 acres and 4 poles? *A.* $4\frac{1}{2}$ chains.

97. Two acres of land are to be cut from a rectangular field whose breadth is 2 chains 50 links, by a line parallel with either end; what is the length of the plot? *A.* 8 chains.

FIG. 19.

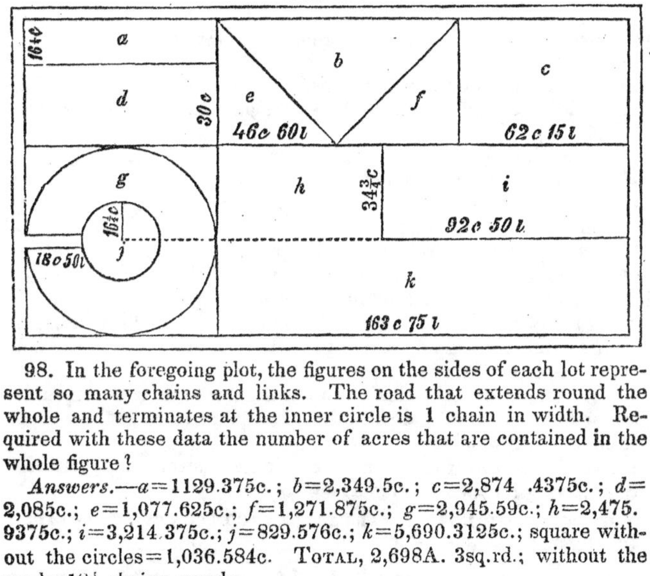

98. In the foregoing plot, the figures on the sides of each lot represent so many chains and links. The road that extends round the whole and terminates at the inner circle is 1 chain in width. Required with these data the number of acres that are contained in the whole figure?

Answers.—*a*=1129.375c.; *b*=2,349.5c.; *c*=2,874 .4375c.; *d*=2,085c.; *e*=1,077.625c.; *f*=1,271.875c.; *g*=2,945.59c.; *h*=2,475.9375c.; *i*=3,214.375c.; *j*=829.576c.; *k*=5,690.3125c.; square without the circles=1,036.584c. TOTAL, 2,698A. 3sq.rd.; without the road=$18\frac{1}{2}$ chains, nearly.

OF SOLIDS.

CXII. 1. A SOLID is any thing that has three dimensions, length, breadth, and thickness.

2. A PRISM is a solid with two equal and parallel bases or ends, and sides that are parallelograms. The sides are called the *lateral surfaces.*

3. *Prisms* receive particular names, according to the figure of their bases, as triangular, circular, square, pentagonal, and so on.

4. A CUBE is a solid or prism bounded by six equal and square sides.

5. A PARALLELOPIPED is a prism of four sides and two ends, whose length is more than its breadth, as a hewn stick of timber.

6. A CYLINDER is a round prism, whose bases or ends are of course circular, like a round column, or a common round rule.

7. A PYRAMID is a solid whose sides taper gradually from the base to one common point, called the vertex of the pyramid.

8. *Pyramids* take their names according to the figure of their bases or ends, as circular, triangular, square, and so on.

9. A CONE is a round pyramid; of course its base is circular, as a sugar loaf, if it comes to a point at the top.

10. A SPHERE or GLOBE is a round, solid body, that has a centre equally distant from every part of the surface, as an orange.

11. The DIAMETER and PERIPHERY of a sphere are the same as those of a circle of equal circumference. A HEMISPHERE is half a globe.

12. A FRUSTRUM or trunk of a pyramid is a portion of the solid next the base, cut off so that its bases are parallel. The other part is called a segment.

13. Thus the top of a sugar loaf of pyramidal form, *cut off square,* is a segment, and what remains is a frustrum.

14. The AXIS of a solid is a straight line passing from one end to the other, through the centre.

15. The *Axis* of a sphere is the same as the diameter of a circle.

16. The ALTITUDE or HEIGHT of a pyramid is the perpendicular distance from the apex or top to the centre of the base.

17. The *Slant* height of a regular pyramid is the distance from the vertex to the middle of one of the sides of the base, or, if it be a cone, to the circumference of the base.

18. A WEDGE is a solid that has a rectangular base, two triangular sides, and two quadrilateral sides that meet in an edge, as the wedge used in splitting wood.

19. A PRISMOID differs from a prism or a frustrum of a pyramid only in having its ends dissimilar.

RULES FOR FINDING THE AREAS AND CONTENTS OF SOLIDS.

20. To find the content of a cube.—*Cube either side.*

CXII. Q. What is a solid? 1. Prism? 2. Their names? 3. Cube? 4. Parallelopiped? 5. Cylinder? 6. Pyramid? 7. Their names? 8. Cone? 8. Sphere? 10. Diameter? 11. Frustrum? 12. Axis of a solid? 14. Altitude of a solid? 16. Wedge? 18. Prismoid? 19. What is the rule for finding the solidity of a cube? 20.

21. To find the solidity of a prism, or cylinder, as of round timber.—*Multiply the area of its base by its length.**

22. To find the area of a prism or cylinder.—*Add together the areas of the different sides and ends.*

23. To find the solidity of a parallelopiped, as of square timber.—*Multiply the length by the breadth, and that product by the depth.*

24. To find the solidity of a pyramid.—*Multiply the area of the base by $\frac{1}{3}$ of its height.*

25. To find the area of a pyramid.—*Multiply half the slant height by the perimeter of the base for the lateral surface, to which add the area of the base.*

26. To find the solidity of a sphere.—*Multiply the cube of the diameter by .5236. Or multiply the square of the diameter by $\frac{1}{6}$ of the circumference. Or multiply the surface by $\frac{1}{6}$ of the diameter.*

27. To find the area or surface of a sphere.—*Multiply the diameter by the circumference.*

28. To find the solidity of a frustrum of a pyramid.—*Add together the areas of the two ends, and the mean proportional between these areas; then multiply the sum by $\frac{1}{3}$ of the perpendicular height.*

29. To find the area of a frustrum of a pyramid.—*Add together the areas of the sides and ends.*

30. To find the solidity of a wedge.—*Multiply half its length into the length and breadth of its base.*

31. To find how large a cube may be inscribed in a given sphere, or be cut from it.—*Divide the square of the diameter of the sphere by 3, and extract the square root of the qaotient.*

32. The side of a cube is 24 feet. Required its content.
A. 13,824 feet.

33. When the side of a cube is 25.5 inches, what is the solidity?
A. 16,581.375 inches.

34. A prism is $20\frac{2}{3}$ feet high, with a base $2\frac{1}{4}$ feet square. Required its content. *A.* 104.625.

35. A stick of timber is 20 feet long, 1 foot 8 inches broad, and 10 inches thick. Required its solidity. *A.* $27\frac{7}{9}$ feet.

Q. Of a cylinder? 21. Of square timber? 23. Of a pyramid? 24. Of a sphere? 26. Of a frustrum of a pyramid? 28. Of a wedge? 30.

---

* The dimensions of round timber are found by girting the tree and taking $\frac{1}{4}$ of the girt for the side of a square.

When the tree tapers regularly, the girt may be taken at the middle, or at both ends, in which case, half their sum will be the mean girt. When the timber is very irregular, the girt may be taken at several places, equally distant, and their sum divided by the number of girts; or divide the tree into several lengths, according to its irregularity, find the content of each length separately, and their sum will be the whole content of the tree.

In measuring oak timber with the bark on, a deduction of $\frac{1}{10}$ or $\frac{1}{12}$ of the circumference is often made to the buyer; in respect to elm, beech, ash, &c. the deduction is less, because the bark is not so thick.

GENERAL RULE.—*Multiply the square of the quarter girt, or the square of $\frac{1}{4}$ of the mean circumference, by the length of the timber.*

This method, though very generally used, gives the content about $\frac{1}{4}$ less than that found by considering the tree as a cylinder, or the content will be nearly the same as if the tree were hewn square.

36. A granite column is 50 feet high, and each end 8 feet in circumference. Required its solidity. *A.* 254.65 nearly.

37. A stick of timber is 19 feet long, with trigonal ends, and sides each 2 feet wide. Required its solidity. *A.* 32.9sq. ft. nearly.

38. The largest Egyptian pyramid has, according to Herodotus, an altitude of 800 feet, and a square base whose perimeter is 3200 feet. Required its content. *A.* 170,666,666$\frac{2}{3}$ ft.

39. The same author says the construction of the above pyramid occupied 100,000 men nearly 20 years. What then would it have cost at the rate of \$1$\frac{1}{2}$ per day for each man, (he boarding himself,) allowing 26 working days to the month? *A.* \$936,000,000.

40. What is the solidity of a globe 12in. in diameter? *A.* 904$\frac{7}{11}$+

41. The earth is about 25,000 miles in circumference. Required its solidity.

A. 198,943,750sq. m. : 263,857,570,390 cubic miles, nearly.

42. The diameter of the moon is about 2,180 miles. What is its solidity? *A.* 5,424,617,475+ sq. miles.

43. A frustrum of a pyramid is 12 feet high, the base 9 feet square, and the other end 6 feet square. Required its solidity. (See XCIX. 69.) *A.* 684 s. ft.

44. If a round stick of timber be 30 feet long, and its extreme peripheries 4ft. 6in., and 3ft. 9 in., what is its solidity? The ratio of 4ft. 6in. to 3ft. 9in. is $\frac{5}{6}$, or which is more convenient $\frac{11}{12}$. (XCIX. 76.)

A. 40.0732+ s. ft.

45. One end of a rectangular frustrum is 60 feet by 40, the other end 40 feet by 30, and 120 feet in length. Required its solidity.

A. 212,000 s. ft. nearly.

46. A stick of hewn timber is 45 feet long, and its ends are 24in. by 20, and 30in. by 24. Required its solidity. *A.* 186$\frac{1}{4}$ s. ft. nearly.

47. If the base of a wedge be 30 by 8, and the length 60, what is its solidity? *A.* 7,200.

48. Allowing the earth's diameter to be 8,000 miles, what is the side of the largest cube that can be inscribed in it?

A. 4,618.8 miles nearly.

49. To find the solidity of any irregular body whose dimensions cannot be ascertained.—*Immerse the solid in a regular vessel of water, and carefully note the difference between the height of the water before the immersion, and afterwards; for the requisite dimensions, with which proceed according to previous rules.*

50. A solid immersed in a vessel 18 inches square, raised the water 9in. Required the content of the given solid. *A.* 1.6875 s. ft.

51. A boy boasting of his knowledge of arithmetic, was asked by his father "If he had got so far that he could measure a brush heap?" Oh, certainly, says he, only chop it up fine and throw it into the cider vat, and it is done. But, rejoined the father, suppose the vat is 6ft. square, and the cider is raised by the brush 2 inches, let us see, after all, if you can calculate the contents of the brush. *A.* 6 s. ft

SIMILAR FIGURES.

52. When two figures vary in size, but are alike in shape or form; they are called Similar Figures.

53. Similar Figures have the angles of the one equal to the corresponding angles of the other, each to each. The sides opposite to equal angles are called *homologous* sides.

RULES OF PROPORTION.

54. *The homologous sides of all similar figures are proportional.*

55. *All similar figures, whether they be triangles, quadrangles, or polygons, are in proportion to each other as the squares of their homologous sides.*

56. *The circumferences of circles are in proportion to each other as the radii or diameters of the circles. The same is true of the arcs and chords of similar segments.*

57. *Circles, or their areas are to each other as the squares of their radii, diameters or circumferences.*

58. *To find the area of a regular polygon, or any regular figure: Multiply the square of one of its sides by the area of a similar figure of which the side is a unit, as in the following:—*

TABLE OF REGULAR POLYGONS.

Names.	Sides.	Perpendiculars.	Areas.
Trigon,	3	0.2886752	0.4330127
Tetragon,	4	0.5000000	1.0000000
Pentagon,	5	0.6881910	1.7204774
Hexagon,	6	0.8660254	2.5980762
Heptagon,	7	1.0382601	3.6339124
Octagon,	8	1.2071069	4.8284271
Nonagon,	9	1.3737385	6.1818242
Decagon,	10	1.5388418	7.6942088
Undecagon,	11	1.7028439	9.3656399
Dodecagon,	12	1.8660252	11.1961524

59. *Cubes, globes, and all similar solids are to each other as the cubes of their similar dimensions.*

60. In figure 20 the perpendicular is 40 rods, and the base 30 rods; what is the base of figure 21, the perpendicular being 28 rods; what is the hypothenuse of each figure, and what the sum of the areas of both? [See 54, 55.] *A*. B C =50rd.; a c=21rd. : b c=25rd. Areas 5A. 2R. 14sq. rd.

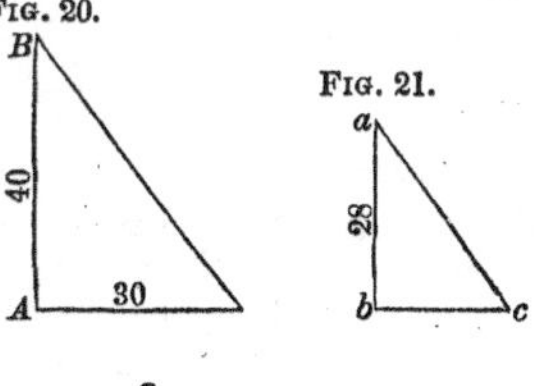

Q. What are similar figures? 52. Wherein are they equal? 53. Which are the homologous sides? 53. Which sides are proportional? 54. What proportion have similar and rectilineal figures, or their areas, to each other? 55. What proportion have the circumference of different circles? 56.

61. Wanting to know the height of the cathedral at York, I measured the length of its shadow, and found it to be 200 feet. At the same time a staff 5 feet long cast a shadow of 4 feet: required the height of that elegant and magnificent structure:—[See 55.] *A.* 250 feet.

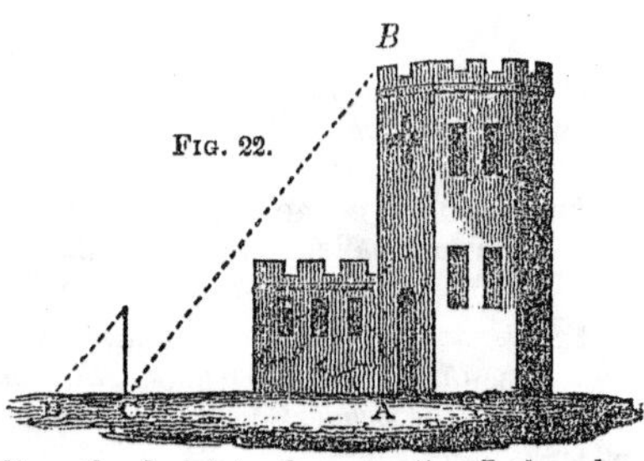

FIG. 22.

62. Being desirous of finding the height of a steeple, I placed a looking glass at the distance of 100 feet from its base on the horizontal plane, and walking backwards 5 feet, I saw the top of the steeple appear in the centre of the glass; required the steeple's height, my eye being 5 feet 6 inches from the ground? [See 54.] *A.* 110ft.

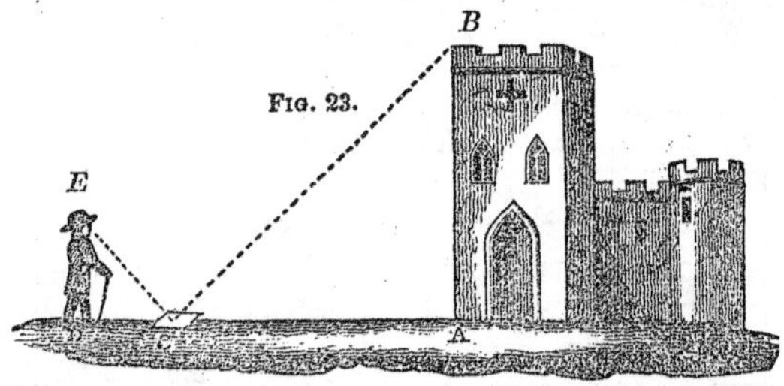

FIG. 23.

63. When the sides of a figure are each 25 rods, what would be its area in square rods, if it were—

A Pentagon?	*A.* 1075.298+.	A Nonagon?	*A.* 3863.640+.
A Hexagon?	*A.* 1623.797+.	A Decagon?	*A.* 4808.880+.
A Heptagon?	*A.* 2271.195+.	An Undecagon?	*A.* 5853.524+.
An Octagon?	*A.* 3017.766+.	A Dodecagon?	*A.* 6997.595+.

64. There is a circle whose diameter is 6 inches, required the diameter of one two times as large?—of one three times as large?—of one ten times as large? Ratios 2, 3, and 10; therefore, $6^2\times2=72$; and $\sqrt{72}=8.485$in. the diameter of one two times as large.

A. Total, 37.85in.+.

65. There is a circle with a diameter of 12 inches,—what is the diameter of one only half as large?—$\frac{1}{4}$ as large?—$\frac{1}{5}$ as large?—$\frac{1}{6}$ as large?—$\frac{2}{3}$ as large?—$\frac{3}{4}$ as large? *A.* Total, 44.938in.+

66. If 113.097 be the area of a given circle, what will be the area of one 4 times as large, and the area of one whose diameter is 4 times as large? (Retain 3 decimal figures.) *A.* 452.388: 1809.552.

67. If a ball 3 inches in diameter weigh 4lb., what will a ball of the same metal weigh, whose diameter is 6in? $3^3 : 6^3$: 4lb. *A.* 32lb.

Q. What, the area of circles? 57. What, cubes and all similar figures? 59 How is the area of a regular polygon found? 58.

68. There are two little globes, one of them is 1 inch in diameter and the other two inches; how many of the smaller globes will make one of the larger? *A.* 8 globes.

69. If the diameter of the planet Jupiter is 12 times as great as the diameter of the earth, how many globes of the earth would it take to make one as large as Jupiter? *A.* 1728 globes.

70. If the sun be 1,000,000 times as large as the earth, and the earth 8,000 miles in diameter, what is the diameter of the sun? *A.* 800,000 miles.

GAUGING.

CXIII. 1. GAUGING is the process of ascertaining the capacity of any regular vessel, in bushels, gallons, &c.

2. The ale gallon contains 282 cubic inches.

3. The wine gallon contains 231 cubic inches.

4. The bushel contains 2,150.4 cubic inches.

5. A cubic foot of pure water weighs 1,000 ounces=$62\frac{1}{2}$ pounds avoirdupois.

6. To find what weight of water may be put into a given vessel.— *Multiply the cubic feet by* 1000 *for the ounces, or by* $62\frac{1}{2}$ *for the pounds, avoirdupois.*

7. What weight of water can be put into a cistern $7\frac{1}{2}$ feet square? *A.* 26,367lb. 3oz.

8. What weight of water will fill a circular fish pond that is 15 rods in circumference, and has an uniform depth of 4 feet? *A.* 609T. 6cwt. 2qr. 4lb. 10oz. 12dr.

9. To find the number of gallons or bushels that a given vessel may contain.—*Calculate the content in inches, which divide by* 282 *for the ale gallons; by* 231 *for the wine gallons, and by* 2,150.4 *for the bushels.*

10. How many barrels of ale will a vat 8 feet square hold? *A.* 87bl. $5\frac{51}{141}$gal.

11. What will the oats cost at $62\frac{1}{2}$ cents a bushel, that will fill 25 bins, 12 of which are cylindrical, being 18 feet in circumference, and 7 feet deep; and the rest 7 feet square? *A.* \$3327.18+

12. A cellar 20 feet long, 15 feet wide, and 8 feet deep, became, during a heavy rain, filled with water. What would be the expense, when labor is $\$1\frac{1}{4}$ per day of 8 hours, of removing the water, allowing that one man can empty three buckets, in 2 minutes, each bucket to hold $2\frac{2}{3}$ gallons, (wine measure)? *A.* \$11.688+.

13. To find the number of gallons in a cask, or to gauge it:—

CXIII. Q. What is Gauging? 1. What is the number of cubic inches in an ale or wine gallon? 2, 3. In a bushel? 4. What is the rule for finding what weight of water may be put into a given vessel? 6. What weight of water may be put into a vessel 1 foot square? Into one 2 feet square? Into one 3 feet deep and 2 feet square at each end? What is the rule for ascertaining the capacity of vessels? 9.

RULE.

14. *Take the dimensions in inches, viz., the diameter of the bung and head, and the length of the cask, and find the difference between the bung and the head diameter.*

15. *If the staves of the cask be much curved between the bung and the head, multiply the difference (found above) by .7; if not quite so much curved, by .65; if they bulge yet less, by .6; and if they are almost straight, by .55; add the product to the head diameter; the sum will be the mean diameter, by which the cask is reduced to a cylinder.*

16. *Square the mean diameter thus found, then multiply it by the length; divide the product by* 359 *for ale or beer gallons, and by* 294 *for wine.*

17. There is a certain cask, whose bung diameter is 35 inches, head diameter 27 inches; and length 45 inches. Required its capacity in ale and wine gallons.

18. Thus, 35−27=8×.7=5.6+27=32.6×32.6=1062.76×45 =47824.20÷359 and 294=*A.* 133.21gal. ale, and 162.66gal. wine.

19. What is the content of a cask in wine and ale gallons, whose bung diameter is 36 inches, head diameter 30 inches, and length 48 inches? *A.* 153.65+ale gal.; 187.62+wine gal.

20. To find the capacity of a vessel, which is in the form of a lower frustrum of a cone, that is round, and larger at one end than at the other.

RULE.

21. *To the product of the diameters add* $\frac{1}{3}$ *of the square of their difference; which result multiply by the height, and divide as above directed.*

22. What is the capacity both in wine and ale gallons, of a tub 40 inches in diameter at the top, 32 inches at the bottom, and its perpendicular height 48 inches?

A. 174 ale gal. nearly+; 212.46+ wine gal.

TONNAGE OF VESSELS.

CXIV. 1. SHIP CARPENTERS' RULE. *For single-decked vessels, multiply the breadth of the main beam, the depth of the hold, and the length together, and divide the product by* 95; *the quotient will be tons. For double-decked vessels, take one half of the breadth of the main beam for the depth of the hold, with which proceed as before.*

2. A single-decked vessel is 80 feet long, 25 feet broad, and 12 feet deep. Required its tonnage. *A.* 252$\frac{12}{19}$ tons.

3. Required the tonnage of a double-decked vessel, whose length is 80 feet, and breadth 26 feet. *A.* 284$\frac{13}{19}$.

4. GOVERNMENT RULE. "If the vessel be double-decked, take the length thereof from the fore part of the main stern, to the after part

---

Q. What is the rule for ascertaining the capacity of casks? 14, 15, 16.

CXIV. *Q.* What is the rule employed by ship carpenters in estimating the tonnage of vessels? What is the government rule? 4.

of the stern post, above the upper deck; the breadth thereof at the lowest part above the main wales, half of which breadth shall be accounted the depth of such vessel, and then deduct from the length $\frac{3}{5}$ of the breadth; multiply the remainder by the breadth, and the product by the depth, and divide this last product by 95, the quotient whereof shall be deemed the true contents or tonnage of such ship or vessel; and if such ship or vessel be single-decked, take the length and breadth as above directed, deduct, from the said length, $\frac{3}{5}$ of the breadth, and take the depth from the under side of the deck plank to the ceiling in the hold, and then multiply and divide as aforesaid, and the quotient shall be deemed the tonnage."

5. A single-decked vessel is 75 feet long, breadth 25 feet, and depth 12 feet. Required the government tonnage of it.
A. $189\frac{9}{19}$ tons.

6. A double-decked vessel is 97 feet in length, and breadth 30 feet. Required the government tonnage. *A.* $374\frac{4}{19}$.

7. What is the government tonnage of a double-decked vessel, with a keel of 115 feet and 32.6 breadth? *A.* 533.841+ tons.

8. What is the government tonnage of a single-decked vessel having 80 feet keel and 35 feet breadth at the beam, and 14 feet deep in the hold? *A.* $304\frac{6}{19}$.

9. What must have been the government tonnage of Noah's ark, the length of which was 300 cubits, the breadth by the midship beam 50 cubits, and the depth in the hold 30 cubits, allowing the cubit to be 22 inches?

As the ark was differently constructed from modern vessels we must, although it had more than one deck [" with lower, second and third stories shalt thou make it"] calculate its tonnage by the rule for single decked vessels. *A.* $26,269\frac{14}{19}$.

EXCHANGE.*

CXV. 1. EXCHANGE is the method by which we find what sum of money of one country, is equivalent to a given sum of another, according to some settled rate of commutation.

2. The COURSE OF EXCHANGE is the quantity of money of one country, which is given for a fixed sum of another; the former is called the *uncertain* price, the latter the *certain* price.

3. The PAR OF EXCHANGE may be considered either as *intrinsic* or *commercial*.

4. The INTRINSIC PAR is the value of the money of one country compared with that of another, both with respect to its weight and fineness.

CXV. *Q.* What is Exchange? 1. The Course of Exchange? 2. The Par of Exchange? 3. The Intrinsic Par? 4.

* Most of what was said on Notes [LXXXI.] and Banking [LXXXV.] in respect to the obligations of the parties, is applicable to Bills of Exchange. The subject, however, has some peculiarities which it may be proper to notice.

5. The COMMERCIAL PAR is when a comparison is made in respect to the market prices of the metals.

7. AGIO is the difference between bank notes and current coins; it also means, in places where foreign coins are current, the difference between the actual value of such coins, and their current value as fixed by government.

8. USANCE is the usual time at which bills are drawn between certain places, as one, two, or three months after date, or sight.

9. When the course of exchange in any place runs low, it is favorable to that place, that is, to its buyers or remitters, but unfavorable to drawers and sellers.

10. For when the exchange is against a place, it is an object with remitters to pay their foreign debts in specie; the exportation of which is considered a national disadvantage.

11. A BILL OF EXCHANGE, or DRAFT, is a written obligation to pay a certain sum of money at a specified time, and is either *Foreign* or *Inland.*

12. An INLAND BILL or DRAFT is one payable in the same country where it is drawn.

13. FORM OF A DRAFT.

$2,000. HARTFORD, Jan. 15, 1840.

Two months after date, pay to the order of Messrs. Spalding and Storrs, two thousand dollars, value received, with or without further advice from me. D. BURGESS.

Messrs. J. & J. Harper,
Booksellers, New York.

14. A FOREIGN BILL is an order from a person of one country to a person of another country, requesting him to pay to a third person, or to his order, either on demand or at a specified time.

15. FORM OF A FOREIGN BILL.

£500 sterling. BOSTON, June 18, 1840.

At ninety days sight, pay to Rufus Smallet, or order, five hundred pounds sterling, value received, and charge the same to the account of NORMAN WELLS.

To Messrs. Stimpson & Co.
Merchants, London.

16. It seems that Bills of Exchange are the same thing in reality as a common order, the only distinctions being in respect to the places of residence of the parties, and the ceremony of collection.

17. In a Bill of Exchange there are usually concerned four persons, viz.

Q. The Commercial Par? 5. What is Agio? 7. What is Usance? 8. When is the course of exchange favorable to any place? 9. What is the explanation? 10. What is a Bill or Draft? 11. An Inland Bill? 12. What is the form of a Draft? 13. What is a Foreign Bill? 14. Its form? 15. What similarity has a bill to a draft? 16. How many persons are generally concerned in a Bill of Exchange? 17

18. *The Drawer*, to whom the bill is made payable, and who is also called the *Maker* and *Seller* of the bill.

19. *The Drawee*, on whom the bill is drawn, and who is also called the *Acceptor*, if he accepts it, which is done by writing his name either at the bottom or across the back, with the word *accepted* over it, by which act he becomes responsible for its payment.

20. *The Buyer*, who purchases it, and who is also called the *Taker* and *Remitter*.

21. *The Payee*, to whom it is ordered to be paid by indorsement, and who may pass it to any other person in the same manner.

22. An indorsement may be either blank or special.

23. A blank indorsement is merely the name of the person written on the back of the bill, which then becomes transferable, like any article of merchandize.

24. A *special* indorsement has with the name a specification directing to whom the bill shall be paid; consequently, the holder must put his own name on it before he can negotiate the bill.

25. Special indorsements are preferable when bills are to be transmitted a great distance; for, should they fall into improper hands the indorser's name must be forged before they can be negotiated and consequently fraud or imposition is prevented as much as possible

26. All the indorsers are severally responsible to the holder of the bill, and must pay it in case the acceptor fails to do so at the proper time.

27. A Set of Exchange implies that there are two or more bills drawn at the same time, and of the same tenor and date.

28. These bills are drawn for the purpose of being transmitted by different ships, or posts, as a security against accidents and delays, and when one of them is accepted and paid, the others are of no further use.

29. FORM OF A BILL IN A SET OF EXCHANGE.

$4,500$\frac{75}{100}$. London, June 5, 1840

Sixty days after sight of this, my first of Exchange, (second and third unpaid,) pay to the order of James Cornwall, Esq four thousand five hundred dollars and seventy-five cents, value received. John Smith.

To Messrs. Williams & Co.
Merchants, New York.

30. Quotations are the lists of the courses of exchange which are transmitted from one country or place to another, for the brokers and others, and are made in the following manner.

31. Thus, "London on the United States 2 per cent. advance"

Q. What is the Drawer? 18. The Drawee? 19. The Buyer? 20. The Payee? 21. What is an Indorsement? 22. A Blank Indorsement? 23. A Special Indorsement? 24. Why is the latter preferable in any case? 25. What liabilities do indorsers incur? 26. What is a Set of Exchange? 27. What is their use? 28. Give an example. 29. What are Quotations? 30. What is meant by "London on the United States at 2 per cent. advance?" 31 Repeat the table of sterling money.

means, that a bill drawn by a merchant of London on his broker in the United States is worth 2 per cent. more than the par value in the latter place. The par value, we have seen, is 4s. 6d. sterling for every dollar in federal money.

TABLES OF FOREIGN CURRENCIES.

32. These tables show the principal denominations of the currencies of those nations with which the United States has the most inercourse, together with the *nominal* or *par* value of each, expressed in federal money. (LXXVI. 3, 4, 5.)

33. GREAT BRITAIN AND IRELAND.*

4 farthings sterling = 1 penny × 12 = 1 shilling × 20 = £1 = \$4.444+.
4s. 6d. sterling = \$1 = £$\frac{9}{40}$. 21 shillings = 1 guinea = \$4.666$\frac{2}{3}$

34. BRITISH AMERICA.

5 shillings = \$1, or £$\frac{1}{4}$. 20 shillings = £1 - - - - = \$4.00.
In Jamaica 6s. 8d. = \$1 or £$\frac{1}{3}$, and £1 - - - - - = \$3.00.

35. SPAIN AND HER DEPENDENCIES—SPANISH AMERICA.

10 reals vellon = 1 real plate × 10 = 1 dollar plate - - = \$1.00.

36. HOLLAND AND BELGIUM.

100 Flemish cents = 1 florin or guilder - - - - - - = \$.40
2$\frac{1}{2}$ florins = \$1. 20s. Flemish = £1 Flemish - - - = \$2.40

37. PORTUGAL.

1,000 rees = 1 milree - - - - - - - - - - - - = \$1.24

38. FRANCE.

10 centimes = 1 aecime × 10 = 1 franc - - - - - - - = \$.18$\frac{3}{5}$.

39. RUSSIA.

10 copecks = 1 grievener × 10 = 1 rouble - - - - - = \$.75.

40. PRUSSIA.

12 pfennings = 1 good groschen × 24 = 1 rix dollar - - = \$.68$\frac{1}{3}$.

41. SWEDEN AND NORWAY.

12 runstycken = 1 skilling × 48 = 1 rix dollar - - - = \$1.05.

42. DENMARK.

16 skillings = 1 mark × 6 = 1 rigsbank dollar - - - - = \$.53.

43. GENOA AND LEGHORN.

12 denari de pezza = 1 soldo de pezza × 20 = 1 pezza - = \$.90.
12 denari de lira = 1 soldo de lira × 20 = 1 lira - - - = \$.15$\frac{3}{4}$

44. ROME.

10 bajocchi = 1 paolo × 10 = 1 scudo or Roman crown = \$1.00

45. NAPLES.

10 grani = 1 carlino × 10 = 1 ducato - - - - - - - = \$.80.

Q. How many dollars and cents in £1 sterling? 33. In 1 guinea? 33. In 4s.6d.? 33. What is the ratio of the dollar to the pound sterling? 33. How many dollars make £1 in British America? 34. What sum in federal money is equal to 1 dollar plate? 35—to 1 florin or guilder? 36—to 1 milree? 37—to 1 franc? 38—to 1 centime? 38—to 1 rouble? 39—to 1 rix dollar? 40—to 1 Roman crown? 44—to 1 rupee? 50—to 1 tale? 51—to 1 mace? 51—to 1 guilder? 52.

---

* Formerly the Irish pound was about \$4.10$\frac{1}{4}$.

46. MALTA.

20 grani= 1 taro × 12= 1 scudo = $.40. 30 tari= 1 pezza = $1.00.

47. SICILY.

20 grani= 1 taro × 12= 1 scudo or Sicilian crown - - = $.95.
30 tari= 1 oncia - - - - - - - - - - - - - - - = $2.40

48. VENICE.

10 millesimi= 1 centesimo × 10= 1 lira de Austria - - = $.16¼.

49. AUSTRIA.

4 pfennings= 1 crewtzer × 60= 1 florin or guilder - - = $.48
1½ florin or guilder= 1 rix dollar of account - - - - = $.72

50. EAST INDIES.

8 pices= 1 anna × 10= 1 rupee - - - - - - - - - = $.55½

51. CANTON, (CHINA.)

10 cash= 1 candarine × 10= 1 mace × 10= 1 tale - - = $1.48.

52. BATAVIA, (JAVA.)

3 dubbles= 1 skilling × 4= 1 florin or guilder - - - = $.40.

53. SURINAM, BERBICE, DEMARARA, AND ESSEQUIBO.

8 doits= 1 stiver × 20 = 1 guilder - - - - - - - - = $.33⅓.

54. ST. DOMINGO AND HAYTI.

The currency is federal money, as in the United States.

COINS.

CXVI. 1. When the American mint was established, in 1770, pure gold was worth a trifle more than 15 times as much as an equal quantity of pure silver.

2. When gold had become more valuable, being worth about 16 times as much as the same quantity of silver, congress passed the act of 1834, by which it was ordered that our gold coins should contain 15½ grains less of pure gold than formerly, in order that the eagle should not be worth any more than 10 dollars, its original value.

3. Hence we see why old eagles are worth more than new ones; also, why foreign coins pass for more than formerly.

4. The eagle of the old coinage was to be of the value of 10 dollars, and to contain 270 grains of standard gold, or 247¼ grains of pure gold, and 22¾ grains of alloy. The standard was 22 carats or $\frac{11}{12}$ fine gold, and $\frac{1}{12}$ alloy.

5. By the act of 1834, the eagle is to contain 258 grains of standard gold, or 232 grains of pure gold and 26 grains of alloy, and to be of the value of 10 dollars. Old eagles are worth about $10.665.

6. The dollar is to possess the same value as the Spanish milled

---

CXVI. *Q.* What was the comparative value of gold and silver in 1790? 1. What has occasioned a difference in the value of the eagle? 2, 3. What were the component parts of the old eagle? 4. What are the component parts of the new? 5. What of the dollar? 6.

dollar, and to contain $371\frac{1}{4}$ grains of pure silver and $44\frac{3}{4}$ grains of pure copper, forming a standard of 416 grains.

7. The price established at the mint for pure silver is $0.26936 per grain.

8. The cent must be of the value of the one hundredth part of a dollar, and contain 208 grains of copper.

9. TABLE.

Value of the principal Coins according to the Act of Congress, 1834

UNITED STATES.	
Eagle, p.[1] - -	$10.000
Eagle, (old) - -	10.665
Dollar, s.[1] p. - - -	1.000
GREAT BRITAIN.	
Guinea, (21s.) p. c. -	$5.075
7 shilling piece, c. - -	1.698
Sovereign, (20s.) p. c. -	4.846
Crown, (5s.) s. p. - -	1.100
Half Crown, s. - -	.550
Shilling, s. - - -	.220
Sixpence, - -	.110
FRANCE.	
Double Louis, p. c. -	$9.697
Louis[2] p. c. - -	4.846
Double Louis,[3] c. - -	9.153
Louis,[3] c. - - -	4.576
Napoleon, (20 francs) c. -	3.851
Double Napoleon, c. -	7.702
Guinea, p. c. - - -	4.655
5 Franc piece, s. c. -	.930
2 Franc piece, s. - -	.372
Franc, s. - - -	.186
50 centimes, s. - -	.093
25 centimes, s. - -	$.04\frac{13}{20}$
SPANISH POSSESSIONS, MEXICO, COLOMBIA.	
Doubloons, p. c. . -	$16.028
Patriot Doubloons, c. -	15.535
Pistole, c. - -	3.884
Coronilla (dollar) c. -	.983
Dollar, s. p. c. - ·	1.000
PORTUGAL AND BRAZIL.	
Dobraon, p. c. - -	$32.706
Dobra, p. c. - -	17.301
Johannes, c. - -	17.064
Moidore, c. - -	6.557
Piece of 16 Testoons,[5] c.	$2.121
Old Crusado, (400 rees) c.	.588
New Crusado, (480 rees) c.	.635
Milree, (1775) c. - -	.730
HOLLAND AND BELGIUM.	
Gold Lion, or 14 florin piece, - -	$5.046
Ten florin piece,[6] - -	4.020
Florin, s. - - -	.400
RUSSIA.	
Ducat, (1796) - -	$2.297
Ducat, (1763) - -	2.667
Gold Ruble, (1756) -	.967
Gold Ruble, (1799) -	.737
Gold Polten, 1777) -	.355
Imperial, (1801) p. -	7.829
Half Imperial, (1801)	3.918
Half Imperial, (1818) -	3.933
Ruble, s. - - -	.750
PRUSSIA.	
Ducat, (1748) - -	$2.270
Ducat, (1787) - -	2.267
Frederick, double, (1769)	7.975
Frederick, double, (1800)	7.950
Rix Dollar, - - -	.690
SWEDEN AND NORWAY.	
Ducat, - - -	$2.235
Rix dollar, s. - -	1.070
HAMBURG.	
Ducat, - - -	$2.800
Crown dollar, s. - -	1.090

Q. What are the component parts of the cent? 8. What does pure silver bring at the mint? 7. What is the difference in value between the eagles coined before 1834 and those coined since? 9. What is the value of the guinea, sovereign, and crown of England?—of the double Louis, Louis, Napoleon, and franc of France?—of Spanish doubloons and pistoles?

1. The letter s. stands for silver coins, p. for shares in proportion, c. for those rendered current by act of Congress, 1834. Coins whose metal is not indicated by any letter are gold coins.

2. Coined since 1786. 3. Coined since 1786. 4. Coined since 1772. 5. Or 1600 rees 6. Coined since 1820.

DENMARK.

Ducat, current, - -	$1.812
Ducat, specie, - -	2.670
Christian d'or, - -	4.021
Rigsbank dollar, s. -	.530

GENOA AND LEGHORN.

Sequin, - - -	$2.300
Francescone, s. -	- 1.05

NAPLES.

6 Ducat piece (1783) -	$5.30
2 Ducat piece, (Sequin)	- 1.591
3 do. or Oncetta, s. -	2.490
Ducat, s. - - -	.800

SICILY.

Ounce, (1751) - -	$2.500
Double ounce, (1758)	5.044
Ducat, - - -	.800

ROME.

Sequin, (1760) - -	$2.511
Scudo of Republic, -	15.811
Scudo, or Roman crown, s.	.800

VENICE.

Sequin, - - -	$2.310
Ducat, effective, - -	.770

TURKEY.

Sequin, (1818) - -	$1.855
Spanish dollar, - -	1.000

MALTA.

Double Louis, - -	$9.278
Louis, - - - -	4.852
Demi Louis, - -	2.336
Spanish dollars, - -	1.000

AUSTRIAN DOMINIONS.

Souverein, - - -	$3.377
Double ducat, - -	4.589
Hungarian do. - -	2.996
Rix dollar, s. - -	.960

BAVARIA.

Carolin, - - -	$4.957

BERNE.

Ducat, p. - -	$1.986

BRUNSWICK.

Pistole, - - -	$4.548

COLOGNE.

Ducat, - - -	$2.667

EAST INDIES.

Rupee, Bombay, (1818)	$7.096
Rupee, Madras, (1818)	7.110
Pagoda Star, - -	1.798
Sicca Rupee, (Calcutta) s.	.480
Rupee, Bombay, Madras, s.	.450

WEST INDIES.

Spanish gold and silver coins.

FRANKFORT ON THE MAINE.

Ducat, - - -	$2.799

SWITZERLAND.

Pistol Hel'c Repub. (1800)	$4.560

REDUCTION OF FOREIGN CURRENCIES.

10. To change English or Sterling money to American or Federal money, and the reverse :—

RULE.

11. *Reduce the given sum to the decimal of a pound, then multiply it by* 40 *and divide by* 9, *to produce Federal money: and reverse the process to produce sterling money again.*

12. Reduce £370. 6s. 3d sterling to Federal money.

13. Reduce $1645.83333+ to sterling money.

14. Reduce £457. 17s. 6d. sterling to Federal money.

15. Reduce $2,035 to sterling money.

16. The reduction of sterling money is of so frequent occurrence among commercial men, that it is very desirable to have the process rendered as easy and as concise as possible, as is attempted in the following rule.

RULE.

17. *Take half the shillings for the tenths of a pound, and if the shillings be odd, call the hundredths* 5 ; *next reduce the pence and*

Q. What is the value of the Johannes of Portugal?—of the ducat and ruble of Russia?—of the rix dollar of Sweden?

farthings to farthings, for thousandths and hundredths respectively, then the sum of the whole increased by 1 *thousandth when the farthings are above* 12, *and by* 2 *thousandths when they are above* 36, *will express the decimal required.**

18. Reduce £815. 16s. 6¼d. to the decimal of a £. *A.* £815.826.

16s.= . 8 = ½ of 16s.
6¼d.= . 0 2 5 = farthings in 6¼d.
. 0 0 1 for the excess above 12.
£ . 8 2 6

19. Reduce £182. 9s. ¾d. to the decimal of a £. 9s.=£ . 45.
¾= . 0 0 3
A. £182. 543. £ . 4 5 3

20. Reduce £240. 13s. 4½d. to the decimal of a £ *A.* £240.669
21. Reduce £203. 17s. 1¾d. to the decimal of a £. *A.* £203.857
22. Reduce £814. 2s. 11d. to the decimal of a £. *A.* £814.146
23. Reduce £317. 0s. ½d. to the decimal of a £. *A.* £317.002
24. Reduce £215. 1s. 1d. to the decimal of a £. *A.* £215.054
25. Reduce £215. 1s. 1d. to Federal money. *A.* \$955.795⅝
26. Reduce \$62,354.62½ to sterling money. *A.* £14,029. 15s. 9½d.
27. To reduce the denomination of any foreign currency to Federal money:—*Multiply by its value in Federal money found in the table.*
28. Reduce 500 reals plate to Federal money.
29. Reduce 50 dollars Federal money, to reals plate.
30. Reduce 250 guilders to Federal money.
31. Reduce 100 dollars, Federal money, to florins.
32. Reduce 37 francs and 5 decimes to \$. *A.* \$6.975.
33. Reduce 215 roubles and 5 grievener to \$. *A.* \$161.62½.
34. Reduce 417 rix dollars, (Prussia) to Fed. money *A.* \$285.41⅓.
35. Reduce 240 skillings, (Sweden) to Fed. money. *A.* \$5.25
36. Reduce 415 marks, (Denmark) to Fed. money. *A.* 36.65⅚.
37. Reduce \$4.50 to pezzas, (Genoa.) *A.* 5 pezzas.
38. Reduce \$8.50 to paoli, (Rome.) *A.* 85 paoli.
39. Reduce 200 ducati and 6 carlini to \$. *A.* 160.48.
40. Reduce 700 tari, (Malta,) to Federal money. 4. \$23.33⅓.
41. Reduce 1 oncia and 15 tari to \$. *A.* \$3.60.
42. Reduce 1100 rupees 5 annas to Federal money. *A.* \$610.77¾.

Q. What is the rule for changing sterling money to Federal money and the reverse? 11. What is a more concise rule? 17. How is the denomination of any foreign currency reduced to Federal money, and the reverse?

* Since shillings are twentieths of a pound, half their number must be tenths, and if there be a remainder it is of course .1=.10, the half of which is .05 or five hundredths. Again, £1=960qr. or 960ths of a pound, but if 1,000 farthings instead of 960 were equal to 1£, any number of farthings would make the same number of thousandths. But 960 increased by 1-24th part of itself is 1,000; if therefore, we increase any number of farthings by the 24th part of itself, the sum will be an exact decimal; wherefore, if the number of farthings exceed 12, the 1-24th part of them is greater than ½qr. and therefore 1 must be added, and when their number exceeds 36, 1-24th part is greater than 1½qr and therefore 2 must be added. The 1 and 2 are properly 1 and 2 thousandths, because they occupy those places in the decimal.

43. Reduce \$203 to guineas, (England.) See Table. *A.* 40.
44. Reduce 40 guineas to American eagles, (new.) *A.* 20E. \$3
45. Reduce 450$\frac{1}{2}$ sovereigns to Federal money. *A.* \$2,183 123.
46. Reduce 100 shilling pieces, (England,) to dollars. *A.* \$22.
47. Reduce 40 double louis (old) to English crowns. *A.* 352$\frac{34}{55}$.
48. Reduce 80 Napoleons to American eagles. *A.* 30E. \$8.08cts.
49. Reduce 100 Spanish coronilla to sovereigns. *A.* 20S. 5s. 8$\frac{1}{4}$d.
50. Reduce 17,000 johannes, (Port.) to Napoleons, *A.* 75,327.+
51. Reduce 2,523 gold Lions, (Belgium,) to \$. *A.* \$12,731.058.
52. Reduce 1,333$\frac{1}{2}$ ducats, (new, Russia,) to \$. *A.* \$3,063.049$\frac{1}{2}$
53. Reduce 8,042 Chr. d'ors to ducats, (Sicily.) *A.* 40,421$\frac{1}{16}$+.
54. Reduce 10,668 duc. (Cologne,) to pistoles, (Bk.) *A.* 6,255$\frac{318}{379}$
55. Reduce 3,500 silver rupees, (Madras, E. Indies,) to Carolins, (Bavaria.) *A.* 317$\frac{3}{4}$ nearly.

56. How many dollars, in Federal money, will pay a debt of £450 3s. 6d. in Great Britain? *A.* \$2,000.777$\frac{7}{9}$.

57. How many reals plate, in Spain, may be purchased for £234 sterling? *A.* 10,400 reals plate.

58. In cases like the last, either proceed as in the Rule of Three, or first reduce to Federal money.

59. How many rix dollars, (Prussia,) will discharge a debt of 1800 francs, (France?) *A.* 489$\frac{12}{77}$ rix dollars.

60. A bought 400 yards of silk, in Paris, at 3 francs per yard. What was its cost in Federal money? *A.* \$223.20.

61. How many rigsbank dollars, (Denmark,) will purchase 1,575 liras in Genoa? *A.* 468$\frac{9}{212}$.

62. How many American dollars will cancel a debt of 7,400 tales in China? *A.* \$10,952.

63. United States on England, (cxv. 31.) Reduce £840. 14s. 6d sterling, to Federal money, exchange at 3 per cent. advance. *A.* \$3,848.652$\frac{2}{9}$.

64. Great Britain on the United States. What sum, in London will purchase a Bill on Boston for \$8,967.75, exchange at 4 per ct discount? *A.* £1937. 8d.+

65. What sum in American eagles, (old,) will purchase 4,500 guineas, at a premium of 3 per ct.? (cxv. 33.) *A.* 2,027E. \$5.59.+

66. United States on France. Reduce 1,174 francs 60 centimes to Federal money, exchange at 5 francs 40 centimes per dollar. *A.* \$217.519 nearly.

67. France on the United States. Reduce \$15,000,000, the cost of the Louisiana territory, to the currency of France, exchange at 5 francs 39 centimes per dollar. *A.* 80,850,000 francs.

68. United States on Amsterdam. Reduce 1,250 florins to Federal money, exchange 39 cents per florin. *A.* \$487.50.

69. Russia on the United States. Reduce 240 rubles 5 grieveners to Federal money. Exchange at par. *A.* \$180.37$\frac{1}{2}$.

70. United States on Denmark. Reduce 424 rigsbank dollars to Federal money, exchange at 2 per cent. in favor of Denmark.
A. \$229.214$\frac{4}{10}$.

71. A merchant in Boston bought in London 40 pieces of black broadcloth, each containing 29$\frac{3}{4}$ yards, for 13s. 6d. per yard. How many eagles (new) at a premium of 2 per cent., must he remit to settle the bill? *A.* 350 eagles.

72. If you purchase in China, 20,000lb. of tea, for 2 maces per lb. which you sell in Amsterdam for 1 guilder per pound, worth in the United States 2 per cent. advance, and with the proceeds purchase a bill on New York at 5 per cent. discount, what will be your profit in the whole transaction? *A.* \$2,669.473$\frac{13}{19}$.

MISCELLANEOUS EXAMPLES.

CXVII. 1. Suppose you employ a capital of \$3,000, and invest 315 dollars 9 cents in cloths, 176 dollars 6$\frac{1}{4}$ cents in linens, 3$\frac{1}{2}$ times the last amount in silks, 518 dollars 8 dimes 5 mills in various other foreign articles, and the balance in domestic cottons at 12$\frac{1}{2}$ cents per yd.; how many yds. could you purchase? *A.* 10,990yd. 2qr. 1$\frac{1}{25}$na.

2. Suppose a person at Boston, and another at Philadelphia, the distance between each place being about 300 miles, set out to meet each other on the road. Required how far they are distant when each has traveled 79m. 5fur. 200 yd.? *A.* 140m. 4fur. 7rd. 1$\frac{1}{2}$yd.

3. The sum of £1,261 was left a person in the United States, by a relative in England, to be divided between a mother, son, and daughter, so that the son shall have three times as much as his mother, and the mother double that of the daughter; required the share of each.

A. The daughter's, £140$\frac{1}{9}$; the mother's, £280$\frac{2}{9}$; the son's, £840$\frac{2}{3}$.

4. A tract of land measuring 7,495A. 3R. 32rd. is to be divided among a regiment consisting of a colonel, a major, 5 captains, 9 lieutenants, 6 ensigns, 20 sergeants, and 450 privates, so that a sergeant is to have twice as much as a private, an ensign 8 shares, a lieutenant 12, a captain 20, the major 30, and the colonel 50; required the share of each, and the value of the land at 3 dollars per acre?

Answers.—Soldiers, 9A. 12rd.=\$27.22$\frac{1}{2}$; sergeants, 18A. 24rd. =\$54.45; ensigns, 72A. 2R. 16rd.=\$217.80; lieutenants, 108A. 3R. 24rd.=\$326.70; captains, 181A. 2R.=\$544.50; major, 272A. 1R.=\$816.75; colonel, 453A. 3R.=\$1,361.25.

5. Jacob, by contract, was to serve Laban, for his two daughters, 14 years; when he had accomplished 10Y. 10mo. 10wk. 10da. 10h 10m., how many minutes had he then to serve? *A.* 1,416,350.

6 Suppose you were 15 years old on the first day of January, 1810, how many seconds old must you have been on the first day of August, 1840, making due calculations for the leap years and the days in each month from one date to the other? *A.* 1,438,473,600 seconds.

7. How much time has elapsed from Nov. 19th, 1826, to Jan. 15th, 1830; from April 19th, 1815, to Sept. 20th, 1837; and from June 12th, 1836, to March 1st, 1839?

A. 3y. 1m. 26d.; 22y. 5m. 1d.; 2y. 8m. 19d.

8. How many barrels can be filled with 60 hogsheads of molasses, each containing 62 gallons, 3 quarts, 1 pint, 3 gills? *A.* 119$\frac{70}{84}$bl.

9. A vintner bought 138 gallons of wine, at 10s. a gallon, of which he retained 18 gallons for his own use; at what rate must he sell the remainder, that he may have his own for nothing? *A.* \$1.91$\frac{2}{3}$.

10. The national debt of England, some time ago, amounted to 820 millions sterling.

11. Required the number of shillings and guineas (21s.) in that sum. *A.* 16,400,000,000s.; 780,952,381 guineas.+

12. Required the weight of the debt in guineas, (gold,) each weighing 5dwt. 9gr. *A.* 17,490,079lb. 4oz. 7dwt. 21gr.

13. Required the weight of the debt in shilling pieces, (silver,) each weighing 3dwt. 21gr. *A.* 264,791,666$\frac{2}{3}$lb.

14. Required the weight in one pound notes, (£1,) 120 weighing 1 ounce. *A.* 213T. 10cwt. 3qr. 8lb. 5$\frac{1}{3}$oz.

15. Required the time it would take a person, allowing him to count 100 pieces or notes in a minute for 10 hours of the day, (Sundays excepted,)—

To count the debt in guineas. *A.* 41Y. 182da. 8h. 44m. nearly.
To count the debt in shillings. *A.* 873Y. 84da. 3h. 20m.
To count the debt in notes, (£1.) *A.* 43Y. 207da. 6h. 40m.

16. Required how many wagons, loaded with 1,200lb. each, would be sufficient—

To carry the debt in shilling pieces. *A.* 220,660, nearly.
To carry the debt in gold (guineas). *A.* 14,576, nearly.
To carry the debt in pound notes. *A.* 356, nearly.

17. Required the number of miles the wagons would extend, allowing 30 feet to each wagon and horses—

When they are loaded with shillings. *A.* 1,253m. 6fur.
When they are loaded with guineas. *A.* 82m. 6$\frac{1}{2}$fur.
When they are loaded with pound notes. *A.* 2m. 7rd. 1$\frac{1}{2}$yd.

18. Required the yearly interest of the debt. *A.* \$218,666,666.66$\frac{2}{3}$.

19. Required the number of years that would be required to pay off the debt by an annual assessment of 9 pence per pound on the value of the whole property in Britain, which, according to the estimate of Dr. Colquhoun, is £2,736,640,000. *A.* Almost 8 years.

20. Required what per cent. and what sum on the pound the British proprietors must assess themselves, supposing that, by a generous exertion, they agree to pay off the debt at once.

A. 30 per cent. nearly.

21. What is the greatest common divisor of 204, 1,190, 1,445, and 2,006? *A.* 17.

22. Find the greatest common measure of 63 and 168, with which divide their sum, difference, and common multiple. *A.* 21.

23. *Hence, if any number will measure each of two others, it will measure their sum, their difference, and their common multiple.*

24. Since the common measure of two or more numbers becomes, when those numbers are multiplied together, or when each is multiplied into itself, a factor in every product, therefore—

25. *A common measure of two or more numbers will measure the squares, cubes, &c. of those numbers, also the continued product of those same numbers into each other.*

26. What is the least common multiple of 2, 3, 4, 6, 7, 12, and 14? *A.* 84.

27. Reduce $2\frac{5}{7}$, $5\frac{4}{9}$, $12\frac{7}{12}$, and $54\frac{8}{11}$, to their equivalent improper fractions. *A.* $\frac{19}{7}$; $\frac{49}{9}$; $\frac{151}{12}$; $\frac{602}{11}$.

28. Reduce to improper fractions $41\frac{7}{13}$, $123\frac{4}{17}$, $275\frac{14}{15}$, and $374\frac{54}{103}$. *A.* $\frac{540}{13}$; $\frac{2095}{17}$; $\frac{4139}{15}$; $\frac{38576}{103}$.

29. Reduce to mixed numbers $13\frac{5}{9}$ of $7\frac{1}{3}$, $\frac{3}{4}$ of $\frac{4}{9}$ of $12\frac{1}{2}$, and $15\frac{7}{11}$ of $8\frac{4}{5}$ of $13\frac{3}{7}$. *A.* $99\frac{11}{27}$; $4\frac{1}{6}$; $1{,}847\frac{27}{35}$.

30. Reduce $\dfrac{2\frac{1}{5}}{3\frac{2}{9}}$ to a simple fraction. *A.* $\frac{99}{145}$.

31. Reduce to mixed numbers $\frac{8357}{278}$; $\frac{18793}{359}$; $\frac{1}{3}$ of $\frac{4}{7}$ of $6\frac{3}{8}$; $\frac{5}{6}$ of $\dfrac{13\frac{3}{5}}{4\frac{7}{9}}$. *A.* $30\frac{17}{278}$; $52\frac{125}{359}$; $1\frac{3}{14}$; $2\frac{29}{86}$.

32. Reduce to prime terms $\frac{8398}{29393}$; $\frac{11050}{35581}$; $\frac{109375}{10000000}$; $\frac{135795}{222210}$. *A.* $\frac{2}{7}$; $\frac{50}{161}$; $\frac{7}{640}$; $\frac{11}{18}$.

33. Find the least common denominator of $\frac{1}{2}$, $\frac{2}{3}$, $\frac{3}{4}$, $\frac{4}{5}$, and $\frac{5}{6}$. *A.* $\frac{30}{60}$; $\frac{40}{60}$; $\frac{45}{60}$; $\frac{48}{60}$; $\frac{50}{60}$.

34. Find the least common denominator of $2\frac{1}{2}$, $\frac{2}{7}$ of $9\frac{3}{5}$, and $\dfrac{7\frac{1}{2}}{2\frac{5}{6}}$. *A.* $\frac{2975}{1190}$; $\frac{3196}{1190}$; $\frac{3150}{1190}$.

35. What is the value of $1\frac{1}{3}+\frac{8}{3}$ of $\frac{41}{34}+\dfrac{4}{5\frac{1}{10}}$? *A.* $5\frac{1}{3}$.

36. What is the difference between $\frac{3}{5}$ of $\dfrac{4\frac{1}{6}}{5\frac{1}{7}}$ and $\frac{2}{3}$ of $\frac{15}{12}$?—between $\dfrac{3\frac{3}{5}}{4\frac{7}{9}}$ and $\dfrac{6\frac{2}{3}}{12\frac{2}{7}}$? *A.* $\frac{25}{72}$; $\frac{109}{645}$.

37 How many times is the sum of $5\frac{1}{3}$ and $3\frac{1}{5}$ greater than their difference? *A.* 4 times.

38. Multiply together $\dfrac{2\frac{3}{4}}{5\frac{7}{9}}$ of $\frac{1}{3}$, and $\frac{3}{5}$ of $\dfrac{4\frac{3}{8}}{7\frac{1}{2}}$. *A.* $\frac{231}{4160}$.

39. Find the continued product of $\frac{324}{361}$, $\frac{1444}{1296}$, $\frac{441}{529}$, and $\frac{2116}{1764}$. *A.* 1.

40. Find the value of $\frac{3}{8}$ of $\frac{4}{7}-\frac{2}{11}$ of $3\frac{1}{7}+\frac{5}{9}$ of $3\frac{3}{8}$. *A.* $1\frac{29}{56}$.

41. Express $\frac{7}{342}$ of a yard as the fraction of an inch, and $\frac{108}{143}$ of an inch as that of a pole. *A.* $\frac{14}{19}$; $\frac{6}{1573}$.

42. Required the sum and difference of $\frac{2}{3}$ of a pound, and $\frac{4}{9}$ of a guinea, (21s.) *A.* £1. 2s. 8d.; 4s.

43. Reduce $\frac{1}{21}$ of a pound, $\frac{1}{22}$ of a guinea, and $\frac{1}{4}$ of 3s. $9\frac{1}{2}$d. to fractions of the same denomination, and to the same denominator *A.* $\frac{7040}{7392}$; $\frac{7056}{7392}$; $\frac{7007}{7392}$.

44. The aggregate of $\frac{2}{3}$ of $\frac{3}{5}$ of a sum of money is $133; what is that sum? *A.* $$332\frac{1}{2}$.

45. Find the fraction which, when multiplied by $\frac{2}{3}$ of $\frac{4}{5}$ of $3\frac{1}{2}$, gives a result equal to $\frac{7}{9}$. *A.* $\frac{5}{12}$.

46. A person having $\frac{2}{5}$ of a coal mine, sells $\frac{3}{4}$ of his share for $2,000; what is the whole mine worth? *A.* $$6,666.66\frac{2}{3}$.

47. The third part of an army was killed, the fourth part taken prisoners, and 1,000 fled. How many were there in the army?—how many killed?—how many taken prisoners?
A. 2,400; 800 killed; 600 prisoners.

48. A can do a job of work in 5 days, B in 6, and C in 7; how much can they jointly do in 2 days? *A.* $1\frac{2}{105}$ job.

49. The owner of $\frac{4}{17}$ of a ship sold $\frac{3}{11}$ of $\frac{2}{9}$ of his share for $$12\frac{4}{33}$; what would $\dfrac{2\frac{1}{2}}{4\frac{1}{4}}$ of $\frac{2}{5}$ cost at the same rate? *A.* $200.

50. If a cask be emptied by two taps in 4 and 6 hours respectively, in what time will it be emptied by both of them together, the rates of efflux remaining the same throughout? *A.* 2h. 24m.

51. A, B and C can perform a piece of work in 12 hours; also, A and B can do it in 16 hours, and A and C in 18 hours; what part of the work can B and C do in $9\frac{1}{7}$ hours, and in what time would A do the whole? *A.* $\frac{4}{9}$: 28h. 48m.

52. What decimal fractions are equivalent to $\frac{4}{25}$, $\frac{9}{125}$, and $\frac{17}{2560}$?
A. .16; .072; .006640625.

53. From unity take .123456789. *A.* .876,543,211.

54. Find the continued product of .275, 2.75, and 27.5.
A. 20.796875.

55. Find the sum of the quotients of 1.68 by .024, of 971.7 by 123, and of 142.025 by .0437, and prove the results by vulgar fractions. *A.* 3,327.9.

56. One man owns .6 of a bank, another $\frac{1}{8}$, and Mr. Darby the rest; what is Darby's part, and what the value of each part, allowing the capital of the bank to be $100,000?
A. $27,500; $60,000; $12.500.

57. In a certain school, .125 of the pupils study geography, .3 study grammar, $\frac{1}{2}$ arithmetic, and 18 learn to read. What is the number in each branch? *A.* Geography, 30; grammar, 72; arithmetic, 120; reading, 18. Whole number, 240.

58. A man whose annual income is $3,000, spends .12 of it; how many dollars will he have saved at the end of each year?
A. $2,640.

59. If a man in trading adds annually to his capital 20 per cent., how many years will be required for his capital to double.

A. 5 years.

60. What will be the value of 8cwt. 3qr. 14lb. of sugar at \$5$\frac{1}{4}$ per cwt.? *A.* \$46.67$\frac{1}{4}$.

61. What is the amount of \$300 from Jan. 1st, 1830, to Nov. 19th, 1834? *A.* \$387.90.

62. How much is the present worth of \$1,000, discounted at bank, and payable in 4 months (which includes the 3 days grace) less than the present worth of the same sum for the same time, calculated as in LXXXIII.? *A.* 89c. 2$\frac{8}{51}$m.

63. Suppose a minor, now 15 years old, is to receive a legacy, at 21, of \$27,200; what is its present value? *A.* \$20,000.

64. When a merchant wants a loan at bank of \$3,958, for 60 days, what sum must be specified in the note to receive that sum?

A. \$4,000.

65. What sum in cash, reckoning compound interest, is equivalent to \$3,207.13$\frac{6}{10}$, due 20 years hence, without interest? *A.* \$1,000.

66. A merchant bought cotton goods to the amount of \$1,300 for cash; he kept them on hand 1 year and 8 months, then sold them for 15 per cent. advance on their cost, but on 4 months' credit; what was his profit? *A.* \$35.685+.

67. Suppose a bookseller purchases, at "trade sale," books to the amount of \$500, the terms being indisputable paper for 6 months, or 4 per cent. discount for cash, and he chooses the latter, what does he make by advancing the cash? Discount made at the time of the purchase is usually computed like bank discount. *A.* \$5.436.

68. A merchant buys on credit goods amounting to \$2,060, but for cash down gets a deduction of 5 per cent. bank discount; keeps them on hand 60 days, then sells them on one year's credit, at 20 per cent. advance from the purchase price. What is the present worth of his clear profit at the time of sale? *A.* \$355.50.

69. The bill for the union of the Canadas, in 1840, provided that the governor general should receive an annual salary of £7,000. How much does that exceed the salary of the president of the United States, which is \$25,000? *A.* \$6,111.111$\frac{1}{9}$.

70. To find the equated time for the payment of bills due at different times.—*Multiply each sum by the time it has to run, (computing from the date of the first,) and divide the sum of the products by the amount of the debt.*

71. An agent sells goods for his employer, payable at the following dates—\$120 due 7th of January; \$130 due 9th of February; \$200 due 15th of March; \$500 due 20th of May; \$240 due 15th of August. Required the average time of payment. *A.* May 2d.

72. The prices of goods are often named in pounds, shillings, &c., while their amount is carried out in federal money, as follows:—

Messrs. Bates & Allen, Bought of Brown & Dobson,

Wheat,	1,000 bushels, at 7s. 6d	- - -	$
Salt,	1,300 bushels, at 3s.	- - - -	
Rye,	2,500 bushels, at 4s. 6d.	- - -	
Oats,	1,800 bushels, at 2s. 3d.	- - -	
			$4,450.

73. The notes referred to below are to be calculated according to the rules under which they are given as examples. In these operations, the pupil will derive much aid from the Table in Compound Interest.

FOR RULES SEE LXXXI.

74. What is the balance due at compound interest on Note 25? Results—357,304+; 115,730; 56,047; 12,240. *A.* $63.127+.

75. What is the balance due at compound interest on Note 26? Results—596,996+; 22,472; 43,778; 248,676. *A.* $79.822+.

76. What is the balance due at compound interest on Note 33? Results—2,472; 236,856; 210,476. *A.* $1,300.458, nearly.

77. What is the balance due at compound interest on Note 35? Results—64,236; 472,418; 433,641; 337,228. *A.* $197.057+.

78. What is the balance due at compound interest on Note 39? Results—81,265; 710,869; 662,824; 464,715. *A.* $479.423.

79. Find a 4th proportional to $35 : \frac{1}{20} :: 3\frac{3}{4}$; also to 125 : .0145 : : 35. *A.* $\frac{3}{560}$ and .00406.

80. If two men, A and B, together, can finish a piece of work in 10 days, and A, by himself, in 18 days, what time will it take B to do the whole? *A.* $22\frac{1}{2}$ days.

81. Three agents, A, B, and C, can produce a given effect in 12 hours; also, A and B can produce it in 16 hours, and A and C in 18 hours; in what time can each of them produce it separately?

A. A, $28\frac{4}{5}$h.; B, 36h.; C, 48h.

82. Distribute $200 among A, B, C, and D, so that B may receive as much as A; C as much as A and B together, and D as much as A, B, and C, together. *A.* A's, $25; B's, $25; C's, $50; D's, $100.

83. At what time between 2 and 3 o'clock are the hour and minute hands of a clock together? At two o'clock the hour hand is two of the portions called hours of one hand, and five minutes of the other, in advance of the minute hand; and their rates being as 1 : 12, the minute hand gains 55 in 60, or 11 in 12, upon the hour hand; hence, 11 : 2 : : 12 : 2h. $10\frac{10}{11}$m. *A.* $10\frac{10}{11}$m. past 2 o'clock.

84. A person, on looking at his watch, was asked the "time of day." He replied, that it was between 4 and 5, and the minute and hour hand were together; what was the exact time?

A. 21m. $49\frac{1}{11}$sec. past 4.

85. Two clocks point out 12 at the same instant; one of them gains 7sec. and the other loses 8sec. in 12 hours; after what interval will one have gained half an hour of the other, and what o'clock will each then show? *A.* 60 days; 14m. past 12; 16m. of 12.

86. A father divided his estate among his three sons, giving to A \$10 as often as to B \$6, and to C but \$3 as often as to B \$7, and yet C's dividend was \$4,800. What did the whole estate amount to?
A. \$34,666$\frac{2}{3}$.

87. If 1 dollar or 6 shillings in New England be equal in value to 8 shillings in New York; £1. 8s. in New York to £1. 6s. 3d. in New Jersey; £1. 7s. 6d. in New Jersey to £1. 9s. 4d. in North Carolina; £1. 8s. in North Carolina to 17s. 6d. in Canada; and £2. 10s. in Canada to £2. 5s. sterling; how many dollars in New England are equal to £450 sterling? *A.* \$2,000.

88. A person, by disposing of goods for \$182, loses at the rate of 9 per cent.; what ought they to have been sold for to realize a profit of 7 per cent.? *A.* \$214.

89. A stationer sold quills at 11s. a thousand, by which he cleared $\frac{3}{8}$ of the money, and he afterwards raised them to 13s. 6d. a thousand; what did he clear per cent. by the latter price?
A. £96. 7s. 3d. 1$\frac{1}{11}$qr.

90. At what price must a commodity, purchased at the rate of £14. 5s. per cwt., be sold to gain 21 per cent., and how much must be sold to clear £100?
A. £17. 4s. 10$\frac{1}{5}$d. per cwt.; 33cwt. 1qr. 16lb. 11$\frac{1}{197}$oz.

91. Divide \$64 among A, B, and C, so that A may have three times as much as B, and C may have one third of what A and B together have? *A.* A's \$36; B's \$12; C's \$16.

92. A person paid a tax of 10 per cent. on his income; what must his income have been, when, after he had paid the tax, there was \$1,250 remaining? *A.* \$1,388.888$\frac{8}{9}$.

93. A hare starts 40 yards before a greyhound, and is not perceived by him till she has been up 40 seconds; she scuds away at the rate of 10 miles an hour, and the dog pursues her at the rate of 18 miles an hour; how long will the course last, and what distance will the hare have run? *A.* 60$\frac{5}{22}$sec.; 490yd.

94. If 5 men or 7 women can perform a piece of work in 35 days, in what time can 7 men and 5 women do the same? *A.* 16$\frac{41}{74}$ days.

95. If 15 men, 12 women, and 9 boys, can complete a piece of work in 50 days, what time would 9 men, 15 women, and 18 boys, take to do twice as much, the parts done by each in the same time being as the numbers 3, 2, and 1? *A.* 104 days.

96. If A by himself can do a piece of work in 5 days, B twice as much in 7 days, and C four times as much in 11 days, in what time can A, B, and C, together, do 3 times the said work?
A. 3 days 12h. 46$\frac{26}{109}$m.

97. If A can do a piece of work by himself in 1 hour, B in 3 hours, C in 5 hours, and D in 7 hours, in what time can they do three times as much, all working together? *A.* 1h. 47m. 23$\frac{3}{11}$sec.

98. If 27 men can do a piece of work in 14 days, working 10 hours in a day, how many hours a day must 24 boys work, in order to com-

plete the same in 45 days, the work of a boy being half that of a man? *A.* 7 hours.

99. If 10 cannon, which fire 3 rounds in 5 minutes, kill 270 men in $1\frac{1}{2}$ hours, how many cannon, which fire 5 rounds in 6 minutes, will kill 500 men in 1 hour at the same rate? *A.* 20 cannon.

100. A and B can do a piece of work in 10 days, A and C in 12 days, and B and C in 14 days; in what time can they do it jointly and separately? *A.* Together, $7\frac{91}{107}$ days; A in $17\frac{41}{47}$ days; B in $22\frac{26}{37}$ days; and C in $36\frac{12}{23}$ days.

101. If 120 men, in 3 days of 12 hours each, can dig a trench 30 yards long, 2 yds. broad, and 4 feet deep, how many men would be required to dig a trench 50 yards long, 6 feet deep, and $1\frac{1}{2}$ yards broad, in 9 days of 15 hours each? *A.* 60 men.

102. A watch, which is 10 minutes too fast at 12 o'clock on Monday, gains 3m. 10sec. per day; what will be the time by the watch at a quarter past 10 in the morning of the following Saturday?
A. 40m. $36\frac{7}{48}$sec. past 10.

103. Required the cavity of a well, whose surface measures 9 square yards 5 feet, and depth $28\frac{1}{2}$ fathoms. *A.* 544yds. 18ft.

104. How many square yards in the area of Solomon's Temple, whose dimensions are mentioned in 1 *Kings*, chap. VI., reckoning the cubit to be 18 inches? *A.* 300.

105. The length of Noah's Ark was three hundred cubits, the breadth 50, and the height 30; how many cubic yards did it contain, and how many horses might have been lodged in it, allowing 10 yards to each horse? *A.* 5,625 horses.

106. The *hold* of a vessel is 120 feet long, 33 feet broad, and 6 deep; how many bales of goods, each measuring, at an average, 6 feet by 4, and 3 feet deep, may be stowed in her, leaving a gangway 3 feet broad? *A.* 300 bales.

107. Required the square feet in 7 oak planks, each $23\frac{1}{4}$ feet long, and their several breadths as follows—4 of $13\frac{1}{2}$ inches in the middle, 1 of $14\frac{1}{2}$ inches, and the other 2 each 16 inches at the broader end, and $13\frac{1}{4}$ inches at the narrower; and what will be the cost of the whole in federal money, at 2s. $4\frac{1}{2}$d. sterling per foot?
A. 189ft. 4in. $8\frac{1}{4}''$; \$99.96, nearly.

108. How many solid feet in a hewn stick of timber 30 feet long, 1ft. 6in. square at one end, and 2ft square at the other? *A.* $92\frac{1}{2}$ft.

109. How many solid feet in a round log 24 feet long and 68 inches in girt? *A.* $48\frac{1}{6}$ s. ft. = 48 s. ft. 288 s. in.

110. How much in length of a tree 40 inches in girt, will make a solid foot? *A.* $17\frac{7}{25}$ inches.

111. Required the solid content of an irregular beech log, the larger end of which is 6 feet by 50 inches in girt, and smaller end 5 feet by $35\frac{1}{2}$ inches in girt, allowing $\frac{1}{2}$ inch upon the quarter girt for the bark.
A. 8ft. 5in. $2''$.+

112. A maltster has a kiln 18 feet square, which he intends to take

down, and build a new one which shall be 24 feet in breadth, and to dry 3 times as much as the old one, required its length. *A.* $40\frac{1}{2}$ft.

113. Bought 2,688 yards of cambric at 8s. 8d. a yard, and sold $\frac{1}{4}$ at 10s. 2d. per yard; $\frac{1}{3}$ at 10s. $11\frac{1}{2}$., and the remainder at 11s. $4\frac{1}{2}$d. per yard; what is the whole gain, and what the gain per cent.?

A. £304. 14s. 8d.; £26. 2s. $4\frac{1}{2}$d.+

114. At what times between 2 and 3 o'clock are the hour and minute hands together; at right angles, and in opposite directions?

A. Together at $10\frac{10}{11}$m. past 2; at right angles, $27\frac{3}{11}$m. past 2; in opposite directions, $43\frac{7}{11}$m. past 2.

115. Four men bought a grindstone of 60 inches diameter. Now how much of the diameter must be ground off by each man, one grinding his part first, then another, and so on, that each may have an equal share of the stone, no allowance being made for the eye?

A. 1st., 8.04 in.; 2nd, 9.534 in.; 3d, 12.426 in.; 4th, 30in.

116. The wheels of a carriage are $2\frac{1}{2}$ yards asunder, and the inner wheel describes the circumference of a circle, whose radius is 20yd.; find the difference of the paths of the two wheels. The circumference of the inner circle$=3.14159\times40$; the circumference of the outer circle$=3.14159\times45$; whence their difference is evidently$=3.14159\times5=15.70795$ yards nearly. The 45 is double the radius of the outer circle$=20+20+2\frac{1}{2}+2\frac{1}{2}=45$. *A.* $15\frac{7}{10}$yds. nearly.

117. Suppose a gentleman from London contracts to construct a railroad in the United States, to be 415m. 5fur. 30rd. long, at the rate of £2,115. 19s. 6d. sterling per mile, and is able to complete only $\frac{2}{5}$ of that distance, what sum in federal money ought he to receive at the rate he was to be paid for the whole? *A.* \$1,563,823.0791$\frac{2}{3}$.

118. A person being asked the time of the day, replied, that the day is 12 hours long, and the sun rises at 6 o'clock. Now if you add $\frac{1}{2}$ of the hours that have elapsed since the sun rose, to $\frac{3}{4}$ of those which must elapse before the sun sets, you will have the exact time of the day.

1st. Suppose it was 2 o'clock, then 8 hours must have elapsed since sunrise, $\frac{1}{2}$ of which, added to $\frac{3}{4}$ of 4 hours make 7 hours; then $8-7=1$, 1st error. Next suppose that it was 10 o'clock, then 4h. have elapsed, &c. 2nd error, 4. *A.* 12m. past 1 o'clock.

119. If A, B and C, could reap a field in 18 days; B, C and D, in 20 days; C, D and A in 24 days; and D, A and B in 27 days; in what time would it be reaped by them all together, and by each of them separately?

Answers. All together, in $16\frac{56}{199}$ days; A, in $87\frac{21}{37}$ days; B, in $50\frac{5}{8}$ days; C, in $41\frac{1}{79}$ days; and D, in $170\frac{10}{19}$ days

QUESTIONS FOR EXAMINATION.*

ADDRESSED TO THE LEARNER AND CANDIDATES FOR SCHOOL-KEEPING

TO BE ANSWERED WITHOUT THE SLATE.*

CXVIII. Q. 1. What is Quantity? See page 26. How is it estimated?† Illustrate it by an example. How are different quantities expressed?

Q. 2. What is Number? 3. How is it represented? What are Concrete and Abstract Numbers? 26. Simple and Compound Numbers? 27. What is meant by denomination?

Q. 3. What is Arithmetic? 27. What are its fundamental rules? Describe the different methods of representing numbers? 28.

Q. 4. What is Numeration? 32. Notation and its rule? Repeat Numeration Table III. What is the greatest number that can be formed by thirty 5s?

Q. 5. What is addition? 37. Its rule? Why do you carry 1 for every 10? 36. What is the proof and the reason for it? 37.

Q. 6. What is Subtraction? 41. Its terms? Rule? Why do you borrow from one figure and pay to another? 40. What is the proof and the reason for it? 42.

Q. 7. What is Multiplication? 45. Its terms? Which terms are called factors and why? When may Addition be performed by Multiplication? How do you multiply by 12 or less?—by 13 or more?—by a composite number?—by 10, 100, &c.?—3700 by 210?

Q. 8. What is Division? 52. Its terms? What rule is proved, and what one is performed, by Division? What is Short Division? 53. Rule? Long Division? 54. Rule 57. Rule for dividing by 10, 100, &c.? 60. Rule for dividing 888 by 800 in the most concise manner?

Q. 9. How is the subtrahend found when the minuend and remainder are given? How may the multiplier be found with the multiplicand and product? How may the remainder be found with the dividend and quotient?

Q. 10. What is the amount of 35 and 7500 and 30000 and 4200000 added together? When the minuend is 400 and the remainder 40, what is the subtrahend? When the multiplier is 8 and the product 4000, what is the multiplicand?

Q. 11. What are the results of 34398 multiplied by 100 and the same divided by 100? When the quotient is 50 and the dividend 2504, what is the remainder?

---

* For the benefit of the Teacher the answer to each example is given in the key, together with reasons for the more difficult operations.

† All questions without numbers annexed, refer to the page indicated by the last preceding number.

Q. 12. Two men are traveling in the same direction, one at the rate of 36 miles a day, and the other 40 miles; in how many days will the latter be 100 miles forward of the other?

Q. 13. Suppose a fox, which is 120 rods before a greyhound, runs at the rate of 4 rods in 2 seconds, and the dog 13 rods in 6 seconds. How far must the dog run to catch the fox, and how long will he be in doing it?

Q. 14. If the minuend be 600 and the difference between the remainder and subtrahend 100, what are the last two terms?

Q. 15. Two men having met on a journey, found that they had both traveled 1,200 miles; but one had traveled 200 miles more than the other. What was the distance each traveled?

Q. 16. Suppose four boys together weigh 435 pounds, and that it should so happen that three of them go in the same notch, but the other in a notch 15 pounds higher; what would be the weight of each boy?

Q. 17. Divide 600 dollars so that A may have 50 dollars more than B, and C 100 more than B.

Q. 18. Said Harry to Dick, my purse and money are worth 100 dollars; but the money is worth 19 times more than the purse. How much money was in the purse?

Q. 19. If a clerk, whose salary for 4 years amounted to 2,000 dollars, had received 50 dollars advance for each successive year after the first, what was his annual salary?

Q. 20. Suppose that C resides 3 times as far from Boston as A, and D 5 times as far as C, and that to meet in that city they must all travel 380 miles. What distance from Boston does each reside?

Q. 21. A man bought a horse, saddle and bridle for 318 dollars, and paid 20 times as much for the horse as for the saddle, and 5 times less for the bridle than for the saddle. What did the bridle and saddle both cost?

Q. 22. Fifteen years ago I was three times as old as my eldest son, who was then but 15, but am now only twice as old. What are our present ages?

Q. 23. A company at a tavern spent $\$3\frac{6}{10}$, and each of them had as many dimes to pay as there were persons in the company; how many persons were there?

Q. 24. What is the difference between twice twenty-five and twice five and twenty?

Q. 25. A snail in going up a May-pole 22 feet high, ascended $4\frac{1}{4}$ feet every day, and descended every night $2\frac{2}{8}$ feet; how long would it be in getting to the top of the pole?

26. What is Federal Money? 73. What is the rule for Addition of Federal Money? 75. For Subtraction? 76. For Multiplication? 77. For Division? 78.

Q. 27. Suppose you sell 4 bushels of oats at $37\frac{1}{2}$ a bushel, and pay $62\frac{1}{2}$ cents for 5 pounds of cheese, and lay out the balance in pins at $6\frac{1}{4}$ cents a paper; how many papers of pins will you have, and how much will the cheese cost by the pound?

Q. 28. If you purchase 29 yards of ribbon at $6\frac{1}{4}$ cents per yard, and give the shop keeper a five dollar bill, how much change must he give you? What will 2 barrels of pork cost at $12\frac{1}{2}$ cents per pound? —at 10 cents per pound?

Q. 29. What will 6,404 articles of any thing cost at $6\frac{1}{4}$ cents each?—at $12\frac{1}{2}$ cents?—at 1 shilling?—at 25 cents?—at 50 cents?—at 75 cents?

Q. 30. What is Reduction? 74. Reduction descending and its rule? 89. Reduction Ascending and its rule? What is Compound Addition and its rule? 97. Compound Subtraction and its rule? 101. Compound Multiplication and its rule? 104. Compound Division and its rule? 107.

Q. 31. What are Fractions? 110. What is the denominator? Numerator? How is the integer found from having its fraction given? What is the integer of which $\frac{2}{3}$ is 20? What is the value of a fraction? 111.

Q. 32. What part of 48 is 36? What is the rule and reason for it? 112. What are the two ways for multiplying a fraction, and why? 113. What for dividing fractions, and the reason? 114.

Q. 33. What is the product of $\frac{2}{119}$ multiplied by 47?—$\frac{11}{300}$ multiplied by 20?—$\frac{7}{20}$ divided by 16?—$\frac{9}{53}$ divided by 3?

Q. 34. What is meant by a common measure or a common divisor? 116. What by a common multiple? 118. What are the rules for finding both? 117, 120. What is the greatest common divisor of 48 and 24? What is the least common multiple of 5 and 20?

Q. 35. What is a Vulgar Fraction? 121. Describe the different sorts of Vulgar Fractions? How are mixed numbers reduced to improper fractions? 125. Compound and complex fractions to single ones? 123, 127. Rule, and the reason of it for the reduction of fractions to their lowest terms? 122. For finding the least common denominator? 127.

Q. 36. What fraction of a dollar is equal to $\frac{3}{4}$ of a pound? What is the rule for it? 131. What are the rules for Addition of Fractions? 133. For Subtraction? 134. For Multiplication? 136. For Division? 139.

Q. 37. A man being asked how long he had been in business, replied that the time he spent at school was 6 years, and that $\frac{2}{3}$ of that period is just $\frac{3}{8}$ of the time he had been in business; what was a direct answer to the question?

Q. 38. $\frac{5}{7}$ of 21 is $\frac{3}{4}$ of what number? $\frac{3}{4}$ of 22 is $\frac{2}{3}$ of what number?

Q. 39. One drover said to another, I have 20 cows Well, said the other, $\frac{1}{2}$ of $\frac{2}{3}$ of your drove is only $\frac{4}{15}$ of mine. How many cows had both?

Q. 40 $\frac{5}{8}$ of 24 is $\frac{3}{4}$ of how many times 8.

Q. 41. $\frac{5}{6}$ of 24 is $\frac{2}{3}$ of how many eighths of 40?

Q. 42. What number added to $\frac{3}{4}$ of 16, will make $\frac{2}{3}$ of 60?

Q. 43. A man bought $2\frac{1}{2}$ bushels of rye at one time, and $1\frac{3}{4}$ bushels at another. He sold $3\frac{1}{3}$ bushels; how much had he left?

Q. 44. A man bought a sheep for $5\frac{1}{8}$ dollars and a calf for $7\frac{5}{16}$ dollars; what did he give for both, and what was the difference in their cost? What fraction added to $\frac{2}{3}$ and $\frac{1}{4}$ will make a unit?

Q. 45. There is a mast erected so that $\frac{1}{3}$ of it stands in the ground, $\frac{1}{5}$ in the water, and 28 feet out of the water; how long is the mast?

Q. 46. Suppose A having taken a job of work does $\frac{1}{3}$ of it in 1 day, $\frac{1}{5}$ of it the second day, and $\frac{7}{15}$ the third day; how much of it does he do in the three days?

Q. 47. Suppose A can mow a certain field in 2 days, and B in 4 days; what part of the field would each mow in one day? What part would both mow in one day? How long would it take both together to mow the field?

Q. 48. Suppose a cistern has three spouts, and that one will fill it in 2 hours, another in 3 hours, and another in 4 hours; in how many hours would it be filled by them all together?

Q. 49. Suppose a cistern is filled by two spouts in 4 and 12 minutes respectively, and is emptied by a tap in 16 minutes; how many minutes will have elapsed before it is filled, when they are all left open or running, the influx and efflux being uniform?

Q. 50. What is one half the quarter of? What part of three pence is a third part of two pence?

Q. 51. If a herring and a half cost a penny and a half, how many may be had for 11 pence? What number is that of which 9 is $\frac{2}{3}$?

Q. 52. What number is that, the 3d and 4th parts of which, taken together, make $24\frac{1}{2}$? What part of a dollar is a 3d part of a cent?

Q. 53. What would $\frac{5}{7}$ of a hogshead of molasses cost at 50 cents a gallon?—$\frac{3}{4}$ of a cwt. of sugar cost at 10 cents per pound?

Q. 54. What part of an eagle is a fourth part of a dime? How many pecks are $\frac{5}{16}$ of a bushel?

Q. 55. The aggregate of $\frac{2}{3}$ and $\frac{3}{5}$ of a debt is 60 dollars; what is that debt? What fraction multiplied by $\frac{2}{3}$ of $\frac{3}{4}$ of 3 will make $\frac{2}{3}$?

Q. 56. A person having $\frac{2}{5}$ of a coal mine, sells $\frac{3}{4}$ of his share for 600 dollars; what is the whole mine worth?

Q. 57. A can do a piece of work in 5 days, B in 3 days, and C in 10 days; how long would they jointly be in doing it?

Q. 58. If $\frac{1}{2}$ the trees in an orchard bear apples, $\frac{1}{4}$ pears, $\frac{1}{6}$ plums, 40 of them peaches, and 10 cherries, what number of trees does the orchard contain? What is that number of which $\frac{1}{3}+\frac{1}{4}+\frac{1}{6}$ is 45?

Q. 59 When $5\frac{1}{2}$ bushels of oats cost $2\frac{3}{4}$ dollars, what is the price by the bushel? What number is that, to which, if its half be added the sum will be 12?

Q. 60. What number is that, to which, if its half, its third, and its fourth, be added, the sum will be 25? The double and the half of a certain number, increased by 5 more, will make 26; what is that number?

Q. 61. If a horse will eat 8cwt. of hay in a month, a cow 4cwt., and a calf 3cwt., in what time will they all consume a load of hay?

Q. 62. A person being asked how much money he had, replied evasively as follows—If $\frac{1}{2}$ and $\frac{1}{4}$ of it, with 18 dollars more, be added to it, the sum would be 4 times as much as he had. What sum had he?

Q. 63. What number is that, which, if increased by its half, its third, and 2 more, will be doubled?

Q. 64. A man spent one third of his life in England, one fourth in Scotland, and the remainder, which was 20 years, in the United States. To what age did he live?

Q. 65. A person, after paying away one third of his money, together with ten dollars, finds that he has remaining 15 dollars more than its half; what sum of money had he?

Q. 66. A man having purchased a drove of cattle, was driving them to market, when he was met by a gentleman, who inquired of him where he was going with his 100 head of cattle. Sir, said he, I have not 100, but if I had as many more as I now have, $\frac{1}{2}$ as many more, and $7\frac{1}{2}$ head of cattle, I should have 100. How many had he?

Q. 67. What are Decimal Fractions? 145. What decimals are equal to $\frac{1}{2}$?—to $\frac{3}{4}$?—to $\frac{3}{8}$?—to $\frac{4}{5}$?—What is the rule for these reductions? 149.

Q. 68. What decimal of a pound is 2s. 6d.? What is the value of £125? Rules for the last two examples? 152, 153.

Q. 69. What is the general method of proceeding in decimal rules, and why? 153. Rule for Addition of Decimals? 154. Subtraction? 155. Multiplication? 156. Division? 158.

Q. 70. What is the sum of .6 and .03 and .004 and .0005? How much does unity exceed .123456789?

Q. 71. Multiply 1.234 by 10;—by 100;—by 1,000.

Q. 72. Divide 123.4 by 10;—by 100;—by 1,000.

Q. 73. How is Reduction of Currencies performed? 161. What number of shillings make a dollar in the different states? What number of pence is equal to $12\frac{1}{2}$ cents in these same states?

Q. 74. What is Rate per Cent.? 163. Rule for finding the per centage? When 5 yards of broadcloth, that cost 6 dollars per yard, sells for 25 per cent. profit, how many dollars does it sell for?

Q. 75. What price must be put on molasses that cost 30 cents a gallon, to gain 20 per cent. on the sale of it?

Q. 76. What are Stocks? 166. When you buy stocks, the par value of which is $500, for 10 per cent. advance, and sell them for 15 per cent. discount, what is your loss in the transaction?

Q. 77. What is Commission? 167. What is your commission for selling goods amounting to \$1,000, on $\frac{2}{5}$ of which you are to have 2 per cent., and on the balance 3 per cent.?

Q. 78. What is Insurance? 168. Suppose you have \$10,000 insured on your house, and \$2,000 on your furniture; what will your insurance amount to at the rate of 40 cents or [illegible].00 for your house, and $\frac{1}{2}$ per cent. for your furniture?

Q. 79. What is Interest? 171. What are the rules for calculating the interest for years, months, and days? 174. What is the interest of \$500 for 1 year?—for 2 years?—for 1 year 8 months?—for 2 years 6 months?

Q. 80. What is the amount of \$600 for 2 months?—for 15 days?—for 20 days?—for 25 days?—for 27 days?

Q. 81. What are the three rules for casting interest on notes? 185, 186, 188. Wherein does an indorsed note differ from others? 182. What is Compound Interest? 189. Discount, and its rule? 192. How is bank discount reckoned? 196.

Q. 82. When you offer to the bank a note of \$60, payable in sixty days, and it is discounted, what sum do you receive? What is equation of payments? 197.

Q. 83. What is Proportion? 209. Geometrical Proportion? 210. Rule of Three? 204. On what principle is the Rule of Three based? 213. How is the Rule of Three in Fractions performed? 207.

Q. 84. What is Compound Proportion? 217. Rule? Conjoined Proportion? 220. Rule? Fellowship? 222. Rule? Compound Fellowship? 224. Rule? 225.

Q. 85. When brandy is ten cents a gill, what is it a pint? What is it a quart? What is it a gallon? How many gallons of ale at ten cents a gallon are worth two gallons of brandy at ten cents a pint?

Q. 86. If 60 bushels of oats and 5 tons of hay will keep 2 horses 6 months, what quantity of each will keep 4 horses a year?

Q. 87. A merchant compounded with his creditors for 75 cents on the dollar. What will he receive to whom he owed 100 dollars?

Q. 88. If ten dollars worth of bread is sufficient for a family of 8 persons 4 months, what will it cost them a year for bread at that rate?

Q. 89. If a manufacturer pays daily to the men in his employ, one dollar and one quarter apiece, to the women seventy-five cents apiece, and to them all 60 dollars a day, how many men and women does he employ, provided there are as many of the one as of the other?

Q. 90. Two men bought a barrel of flour for ten dollars, the one paying 6 dollars, and the other 4 dollars. They sold the flour at an advance of two dollars. What part of the gain ought each to have?

Q. 91. Three men bought a drove of cattle, for which A paid 400 dollars, B 600 dollars, and D 200 dollars. They lost 300 dollars; now what sum ought each to pay to make good the total loss?

Q. 92. Two men hired a pasture for 48 dollars; A put in 4 horses 3 months, and B 18 calves 6 months, with the understanding that 3 calves should be considered the same as 1 horse. What sum ought each to pay for the use of the pasture?

Q. 93. A man and his wife consumed in 2 months a barrel of flour, but when the husband was gone, a barrel of flour lasted the wife 5 months; what part did both consume in one month? What is the difference between what they both consumed and what the wife consumed? How many pounds, then, did each consume in the two months?

Q. 94. A man having a horse and cow, found that 3 loads of hay would keep them both 6 months, and when he had no cow, the same quantity would last his horse 10 months. How long would it take each to consume 1 ton of hay?

Q. 95. Suppose A and B can mow a certain field in 6 days, and with the assistance of C will mow the same field in 4 days: how much of it could A and B mow in one day? How much could the three do in one day? How long would C be in doing it alone?

Q. 96. If A and B together can build a boat in 4 days, and with the assistance of C can do it in 3 days, in what time would each do it alone?

Q. 97. If a third of six be 3, what will the fourth of 20 be?

Q. 98. If 12 apples be worth as much as 20 pears, and 3 pears cost 1 cent, what is the price of 100 apples?

Q. 99. If a staff 8 feet long casts a shade 10 feet long, what is the height of that pole which casts at the same time a shade 25 feet long?

Q. 100. If a family of 8 persons, in 24 months, spend 240 dollars, how many dollars would three times as many persons spend in one fifth part of the time?

Q. 101. If 5 men build a wall 20 rods long in 8 days, how many men will it take to build a wall 30 rods long in 4 days?

Q. 102. If A can do a piece of work in 1 hour, B in 2 hours, C in 3 hours, and D in 4 hours, in what time will they all do it, if they work together?

Q. 103. What is Practice? 226. What are Duodecimals? 227. How is unity divided in Duodecimals? What is the rule for Cross Multiplication? 229.

Q. 104. What is Involution? 230. What is the square of 8?—cube of 3?—biquadrate of 2?—5th power of 10?—6th power of 10?—second power of $\frac{2}{3}$?—third power of $\frac{3}{5}$?—fourth power of $3\frac{1}{3}$?—fifth power of .2?—fourth power of .02?—third power of .002?—of .0002?

Q. 105. What is Evolution? 233. What is a root? What is meant by the *extraction* of roots? What are rational and surd members? What are the signs of roots?

Q. 106. What is the square root? 235. Rule? 237. Rule for finding the mean proportional between two numbers? 240.

Q. 107. What is the square root of 36?—of 81?—of .25?—of .0025?—of .0144?—of .000081?

Q. 108. What is a cube? 241. Cube root? What is the form of a cube? Why is the cube root of any number equal to the length of either side of a cube? What is the rule? 243

Q. 109. What is the Cube Root of 216?—of 1,728?—of $\frac{8}{27}$?—of $15\frac{5}{8}$?—of 27?—of .027?—of .008?—of .000027?—of 27,000? What is the general rule for extracting the roots of all powers? 246.

Q. 110. What is Alligation? 248. Alligation Medial? Alligation Alternate? Rule? 249. Arithmetical Progression? 251. Terms? Rules? Geometrical Progression? 254.

Q. 111. What is an Annuity? 256. How is the amount found at imple interest? 257. How at compound interest? 258. How is the present worth found? 259. How the annuity which any sum will purchase?

Q. 112. What is Permutation? 260. What is the rule for finding the different changes which may be made with any given number of things? 261.

Q. 113. What is Position? 261. Single Position? Double Position? 262. Rule? 263.

Q. 114. What is Mensuration? 263. What is an angle? Describe the different angles. 264. How is each formed? Describe the different triangles. To what is the square of the longest side of a right-angled triangle equal?

Q. 115. Describe the different quadrilateral figures. 266. How are they formed? What is a circle? How are its area, circumference, diameter and radius found? 268.

Q. 116. What is a solid? 271. Describe the different solids. Rule for finding the solidity of each? What is Gauging? 277. Rule for finding what weight of water may be put into a given vessel? Rule for gauging a cask? What are the two rules for finding the tonnage of vessels? 278.

Q. 117. What is Exchange? 279. Bills? Set of Exchange? What has occasioned a difference in the value of our gold coins? 283. What is the value of our eagle? What are the two methods for reducing a sum in sterling money to federal money? 285.

Q. 118. How many pounds sterling must a merchant in London remit to New-York to pay a debt of $800?

A PRACTICAL SYSTEM

OF

BOOK-KEEPING;

FOR

FARMERS AND MECHANICS.

Almost all persons, in the ordinary avocations of life, unless they adopt some method of keeping their accounts in a regular manner, will be subjected to continual losses and inconveniences; to prevent which, the following plan of cutline is composed, embracing the principles of Book-Keeping in the most simple form. Before the pupil commences this study, it will not be necessary for him to have attended to all the rules in the Arithmetic; but he should make himself acquainted with the subject of Book-Keeping before he is suffered to leave school A few examples only are given, barely sufficient to give the learner a view of the manner of keeping books; it being intended that the pupil should be required to compose similar ones, and insert them in a book adapted to this purpose.

Book-Keeping is the method of recording business transactions. It is of two kinds—single and double entry; but we shall only notice the former.

Single Entry is the simplest form of Book-Keeping, and is employed by retailers, mechanics, farmers, &c. It requires a Day-Book, Leger, and, where money is frequently received and paid out, a Cash-Book.

DAY-BOOK.

This book should be a minute history of business transactions, in the order of time in which they occur; it should be ruled with head lines, with one column on the left hand for post-marks and references, and two columns on the right for dollars and cents. The owner's name, the town or city, and the date of the first transaction, should stand at the head of the first page. It is the custom of many to continue inserting the name of the town on every page. This, however, is unnecessary. It is sufficient to write only the month, day, and year, at the head of each page after the first. This should be written in a larger hand than the entries.

On commencing an account with any individual, his place of residence should be noted, provided it is not the same as that where the book is kept. If it be the same, this is unnecessary. As it often happens that different persons bear the same name, it is well, in such cases, to designate the individual with whom the account is opened, by stating his occupation, or particular place of residence.

When the conditions of sale or purchase vary from the ordinary customs of the place, it should be stated. Every month, or oftener, the Day-Book should be copied or posted into the Leger, as hereafter directed. The crosses, on the left hand column, show that the charge or credit, against which they stand, is posted, and the figures show the page of the Leger where the account is posted. Some use the figures only as post marks.

Every article sold on credit, except when a note is taken, should be immediately charged, as it is always unsafe to trust to memory. Also, all labor performed, or any transaction whereby another is made indebted to us, should be immediately entered on the Day-Book. If farmers and mechanics would strictly observe this rule, they would not only save many quarrels, but much money. In this respect, at least, follow the example of Dr. Franklin, who never omitted to make a charge as soon as it could be done. Never defer a charge till to-morrow, when it can be made to-day.

FORM OF A DAY-BOOK.

Edward L. Peckham. *Boston, Jan.* 1, 1829.

			$	c
1	×	*James Murray, Jr.* . . . . *Dr.*		
		To 1 gall. Lisbon Wine $1,92		
		" 6 yds. Calico, *a* 37½ cts. 2,25		
		" 2 yds. Broadcloth, *a* $4,50 9,00	13	17
1	×	*Robert Hawkins*, Blacksmith, ... *Dr.*		
		To 217 lbs. Iron, *a* 8 cts.	17	36
1	×	*Thomas Yeomans*, ... *Cr.*		
		By Cash	75	75
		3		
1	×	*Archibald Tracy*, Salem, ... *Dr.*		
		To 1 piece Broadcloth, containing 29 yds., *a* $3 per yd., 90 days' credit	87	00
2	×	*James Warren*, Wartland, ... *Dr.*		
		To 1 cask Nails, 225 lbs., *a* 8 cts.	18	00
		Cr.		
		By 37 lbs. Cheese, *a* 10 cts. $3,70		
		" 41 lbs. Feathers, *a* 70 cts. 28,70		
		Balance to be paid in Corn, at market price.	32	40
2	×	*Isaac Thomas*, Brattle Square, ... *Dr.*		
		To 32 galls. Molasses, *a* 50 cts.	16	00
		4		
2	×	*William Angell*, Roxbury, ... *Dr.*		
		To 300 lbs. Pork, *a* 7 cts. $21,00		
		" 30 bu. Corn, *a* 45 cts. 13,50	34	50
2	×	*Samuel Stone*, ... *Dr.*		
		To 50 lbs. Harness Leather, *a* 30 cts.. . . . . . $15,00		
		" 7 tons Hay, *a* $10 70,00	85	00
		5		
2	×	*George Carpenter*, ... *Dr.*		
		To 17 Brooms, *a* 12 cts. $2,04		
		" 7 lbs. Butter, *a* 20 cts. 1,40		
		" 4 lbs. Cheese, *a* 10 cts. . . . . ,40	3	84
2	×	*Jesse B. Sweet*, Mendon, ... *Dr.*		
		To 1 hhd. Molasses, 98 — 6 = 92 galls.. *a* 30 cts.	27	60
		Cr.		
		By Cash, ...	15	00
2	×	*Jesse Metcalf*, Tanner, ... *Dr.*		
		To 20 Calf Skins, *a* $5 . . . . $100,00		
		" 50 Dried Hides, *a* $4 200,00		
		60 days' credit.	300	00

Jan. 4, 1829.

				$	c
1	×	*James Murray, Jr.*	*Cr.*		
		By 29 bu. Corn, *a* 60 cts.	$12,00		
		" 4 bu. Oats, *a* 40 cts.	1,60	13	60
		6			
2	×	*James Warren,*	*Dr.*		
		To 24 bu. Corn, *a* 60 cts.		14	40
1	×	*Archibald Tracy,*	*Dr.*		
		To 1 cord Wood	$6,00		
		" 30 lbs. Feathers, *a* 70 cts.	21,00	27	00
1	+	*Robert Hawkins,*	*Cr.*		
		By shoeing my Horse	$2,00		
		" " " Oxen	3,00	5	00
		7			
2	×	*Samuel Stone,*	*Dr.*		
		To 2 yds. Broadcloth, *a* $4	$8,00		
		" 4 pr. Shoes, *a* $1	4,00	12	00
1	×	*Thomas Yeomans,*	*Dr.*		
		To 200 bu. Corn, *a* 70 cts.		140	00
		9			
2	×	*Jesse B. Sweet,*	*Dr.*		
		To 30 quintals Fish, *a* $3,75		112	50
2	×	*George Carpenter,*	*Dr.*		
		To 200 lbs. Cheese, *a* 8 cts.	$16,00		
		" 1 firkin Butter, 76 lbs., weight of tub 10 lbs.=66, *a* 20 cts.	13,20	29	20
1	×	*Archibald Tracy,*	*Dr.*		
		To 2 bbls. Flour, *a* $10	20,00		
		" 25 lbs. Lard, *a* 10 cts.	2,50		
		" 3 bu. Salt, *a* 66 cts.	1,98	24	48
2	×	*Isaac Thomas,*	*Dr.*		
		To 50 yds. Calico, *a* 22 cts.	$11,00		
		" 75 yds. brown Sheeting, *a* 14 cts.	10,50	21	50
			Cr.		
		By Order on Goodrich & Lord, for $12,80, ...		12	80
2	×	*Jesse Metcalf,*	*Dr.*		
		To 500 pr. Men's Shoes, *a* 95 cts.		475	00
		10			
1	×	*Thomas Yeomans,*	*Dr.*		
		To 3 bbls. Flour, *a* $9.50		28	50
1	×	*Robert Hawkins,*	*Dr.*		
		To 120 lbs. Blistered Steel, *a* 8 cts.	$9,60		
		" 100 lbs. Russia Iron, *a* 5 cts	5,00	14	60

Jan. 10, 1829.

			$	c
1 ×	*James Murray, Jr.*	*Dr.*		
	To 10 lbs. Sugar, *a* 11 cts.	$1,10		
	" 20 lbs. Coffee, *a* 15 cts.	3,00		
	" 6 galls. Molasses, *a* 37 cts.	2,22		
			6	32
2 ×	*William Angell,*	*Cr.*		
	By 200 lbs. Lard, *a* 6 cts.	$12,00		
	" 350 lbs. Bacon, *a* 12 cts.	42,00		
			54	00
	12			
2 +	*James Hammond,*	*Dr.*		
	To 1 bbl. Flour, ...	$10,00		
	" 3 bu. Corn, *a* 65 cts.	1,95		
	" 6 galls. Wine, *a* 1,25	7,50		
	" 3 lbs. Coffee, *a* 16 cts.	,48		
	" 4 bu. Salt, *a* 70 cts.	2,80		
	" 1 lb. Y. H. Tea, ...	1,25		
	" 14 lbs. Sugar, *a* 12 cts.	1,68		
	" 3 yds. Broadcloth, *a* $2,50	7,50		
	" 12 yds. Shirting, *a* 19 cts.	2,28		
			35	44
	13			
1 ×	*James Murray, Jr.*	*Dr.*		
	To 6 lbs. Raisins, *a* 20 cts.	$1,20		
	" 5 galls. Currant Wine, *a* 75 cts.	3,75		
			4	95

LEGER.

This book is used to collect the scattered accounts of the Day-Book, and to arrange all that relates to each individual into one separate statement. The business of collecting these accounts from the Day-Book, and writing them in the Leger, is called *posting*. This should be done once a month, or oftener. Debts due from others, and entered upon the Day-Book, are placed on the side of *Dr.*; whatever is on the Day-Book as due to another is placed on the side of *Cr.*

When an account is posted, the page of the Leger, in which this account is kept, is written in the left-hand column of the Day-Book.

Every Leger should have an alphabetical Index, where the names of the several persons, whose accounts are kept in the Leger, should be written, and the page noted down.

When one Leger is full, and a new one is opened, the accounts in the former should be all balanced, and the balances transferred to the new Leger.

EXPLANATION OF THE LEGER, AND THE MANNER OF POSTING.

It will be seen that the name of *James Murray, Jr.* stands first on the Day Book; of course, we shall post his account first. We enter his name on the first page of the Leger, in a large, fair hand, writing *Dr.* on the left, and *Cr.* on the right. At the top of the left-hand column we enter the year, under which we write the month and day when the first charge was made in the Day-Book, and in the next column the page of the Day-Book where the charge stands. Then, as there are several articles in the first charge, instead of specifying each article, as in the Day-Book, we merely say, *To Sundries*, and enter the amount in the proper columns. This charge being thus posted, we write the page of the Leger, viz. 1, in the left-hand column of the Day-Book, and opposite to it a ×, to show more distinctly that the charge is posted. We then pass a finger carefully over the names, till we again come to the name of *James Murray, Jr.*, which we find on the second page; but, as this is credit, we enter it on the credit side, with the date and page in their proper columns. We then enter the Leger-page and cross, as before, and then proceed again in search of the same name, until every charge and credit is transferred into the Leger. The next name is to be taken and proceeded with in the same way as the first; and so continue till all the accounts are posted.

As it is uncertain how extensive an account may be when once opened, it is better to take a new page for every name, until all the Leger pages are occupied. By this time, it is probable, several accounts will have been settled; we may then enter a second name on the same pages, and so continue till all the pages are full.

Whenever any account is settled, the amount of the balance is ascertained, and the settlement entered in the Leger. The settlement may also be entered in the Day-Book; and many practise this, although it is not essentially necessary. But it is essentially necessary that one, if not both, the books, should show how every account is settled, whether by cash, note, order, goods, or whatever way the amount or balance is liquidated.

N. B.—In making out bills, the Leger is used as a reference to the charges in the Day-Book, which must be exactly copied.

FORM OF A LEGER.

1]

Dr.			*James Murray, Jr.*					*Cr.*	
1829.			$	c.	1829.			$	c.
Jan. 1.	1	To Sundries,	13	17	Jan. 5.	2	By Corn and Oats,	13	60
" 10.	3	" do.	6	32	" 15.		" Cash, to bal.	10	84
" 13.	3	" do.	4	95					
			$24	44				$24	44

Dr.			*Robert Hawkins,*					*Cr.*	
1829.			$	c.	1929.			$	c.
Jan. 1.	1	To Iron,	17	36	Jan. 6.	2	By Work,	5	00
" 10.	2	" Sundries,	14	60	" 12.		" Note, *a* 60 days,	26	96
			$31	96				$31	96

Dr.			*Thomas Yeomans,*					*Cr.*	
1829.			$	c	1829.			$	c.
Jan. 7.	2	To Corn,	140	00	Jan. 1.	1	By Cash,	75	75
" 10.	2	" Flour,	28	50	" 11.		" Check, for bal.	92	75
			$168	50				$168	50

Dr.			*Archibald Tracy,*					*Cr.*	
1829.			$	c.	1829.			$	c.
Jan. 3.	1	To Broadcloth,	87	00	Apr. 2.		By Cash,	138	48
" 6.	2	" Sundries,	27	00					
" 9.	2	" do.	24	48					
			$138	48					

Dr.			James Warren,					Cr.	
1829.			$	c.	1829.			$	c.
Jan. 3.	1	To Nails,	18	00	Jan. 3.	1	By Sundries,	32	40
" 6.	2	" Corn,	14	40					
			$32	40					

Dr.			Isaac Thomas,					Cr.	
1829.			$	c.	1829.			$	c
Jan. 3.	1	To Molasses,	16	00	Jan. 9.	2	By Order,	12	80
" 9.	2	" Sundries,	21	50	" 20.		" Note *a* 90 days,	24	70
			$37	50				$37	50

Dr.			William Angell,					Cr.	
1829.			$	c.	1829.			$	c
Jan. 4.	1	To Sundries,	34	50	Jan. 10.	3	By Sundries,	54	00
" 16.		" Cash,	19	50					
			$54	00					

Dr.			Samuel Stone,					Cr.	
1829.			$	c.	1829.			$	c.
Jan. 4.	1	To Sundries,	85	00	Jan. 30.		By Cash,	97	00
" 7.	2	" do.	12	00					
			$97	00					

Dr.			George Carpenter,					Cr.	
1829.			$	c.	1829.			$	c.
Jan. 5.	1	To Sundries,	3	84	Jan. 15.		By Note, *a* 60 days,	33	04
" 9.	2	do.	29	20					
			$33	04					

Dr.			Jesse B. Sweet,					Cr.	
1829.			$	c.	1829.			$	c.
Jan. 5.	1	To Molasses,	27	60	Jan. 5.	1	By Cash,	15	00
" 9.	2	" Fish,	112	50	" 20.		" Cash to bal.	125	10
			$140	10				$140	10

Dr.			Jesse Metcalf,					Cr.	
1829.			$	c.	1829.			$	c.
Jan. 5.	1	To Sundries,	300	00	Apr. 7.		By his Check,	775	00
" 9.	2	" Shoes,	475	00					
			$775	00					

Dr.			James Hammond,					Cr	
1829.			$	c.	1829.			$	c.
Jan. 12.	3	To Sundries,	35	44	Jan. 30.		By Order on Brown & Ives,	35	44

INDEX TO THE LEGER.

CASH-BOOK.

This book records the payments and receipts of cash.

It is kept by making cash *Dr.* to cash on hand and what is received, and *Cr.* by whatever is paid out.

At the end of every day or week, as may best suit the nature of the business, the cash on hand is counted, and entered on the *Cr.* side.

If there is no error, this will make the sum of the *Dr.* equal to that of the *Cr.* A balance is then struck, and the cash on hand carried again upon the *Dr.* side.

FORM OF A CASH-BOOK.

Dr. CASH, *Cr.*

1827.		$	c.	1827.		$	c.
Jan. 1.	To Cash on hand,	637	50	Jan. 2.	By rent of Store for one quarter, paid Thomas Taylor,	62	50
2.	" J. Thompson,	37	94	4.	" Paid note to R. Thachor,	127	83
2.	" J. Hart, paid acc't,	65	43	5.	" Family expenses,	27	61
3.	" H. Palmer, on note,	127	28	6.	" Mdse. bought of T. Thamor,	614	27
4.	" S. Snowdon,	84	73		Cash on hand,	550	65
5.	" J. Mervin on acc't.	17	90				
6.	" S. Crane,	100	90				
6.	Sales of Mdse.	311	18				
		1382	86			1382	86
8	Cash on hand,	550	65				

Form of a Bill from the preceding Work.

Mr. Jesse Metcalf, To E. L. Peckham, Dr.

1829.		$	c.
Jan. 5.	To 20 Calf-Skins, *a* $5	100	00
" "	" 50 Dried Hides, *a* $4	200	00
" 9	" 500 pair Men's Shoes, *a* 95 cts.	475	00
	Received Payment, by his check on N. E. Bank.	$775	00

Boston, *April* 7, 1829. EDWARD L. PECKHAM.

☞ *For Forms of Notes, Drafts, &c., see LXXXI., p.* 182, *and CXV. p.* 279

CONTENTS.

PART FIRST.

PART SECOND.

APPENDIX.

PART THIRD.

www.ingramcontent.com/pod-product-compliance
Lightning Source LLC
LaVergne TN
LVHW020231110826
845151LV00003B/896

* 9 7 8 1 4 2 5 5 3 0 6 2 4 *